Springer
*Berlin*
*Heidelberg*
*New York*
*Barcelona*
*Budapest*
*Hong Kong*
*London*
*Milan*
*Paris*
*Santa Clara*
*Singapore*
*Tokyo*

E. M. Chirka P. Dolbeault
G. M. Khenkin A. G. Vitushkin

# Introduction to Complex Analysis

Springer

Consulting Editors of the Series:
N. M. Ostianu, L. S. Pontryagin
Scientific Editors of the Series:
A. A. Agrachev, Z. A. Izmailova, V. V. Nikulin, V. P. Sakharova
Scientific Adviser:
M. I. Levshtein

Title of the Russian edition:
Itogi nauki i tekhniki, Sovremennye problemy matematiki,
Fundamental'nye napravleniya, Vol. 7,
Kompleksnyĭ analiz – mnogie peremennye 1
Publisher VINITI, Moscow 1985

Second Printing 1997 of the First Edition 1990, which was originally published as Several Complex Variables I, Volume 7 of the Encyclopaedia of Mathematical Sciences.

Die Deutsche Bibliothek – CIP-Einheitsaufnahme

**Introduction to complex analysis** / E. M. Chirka ... Ed.: A. G. Vitushkin. – 2. printing. – Berlin; Heidelberg; New York; Barcelona; Budapest; Hongkong; London; Mailand; Paris; Santa Clara; Singapur; Tokio: Springer, 1997
ISBN 3-540-63005-8

Mathematics Subject Classification (1991):
32-02, 32A25, 32A27, 32B15

ISBN 3-540-63005-8 Springer-Verlag Berlin Heidelberg New York



Typesetting: Macmillan India Limited, Bangalore

SPIN: 10629686
41/3143-5 4 3 2 1 0 – Printed on acid-free paper.

# List of Editors, Contributors and Translators

*Editor-in-Chief*

R.V. Gamkrelidze, Russian Academy of Sciences, Steklov Mathematical Institute, ul. Gubkina 8, 117966 Moscow, Institute for Scientific Information (VINITI), ul. Usievicha 20 a, 125219 Moscow, Russia; e-mail: gam@ipsun.ras.ru

*Consulting Editor*

A.G. Vitushkin, Steklov Mathematical Institute, ul. Gubkina 8, 117966 Moscow, Russia

*Contributors*

E. M. Chirka, Steklov Mathematical Institute, ul. Gubkina 8, 117966 Moscow, Russia

P. Dolbeault, Université Paris VI, Analyse Complexe et Géométrie, 4, Place Jussieu, F-75252 Paris Cedex 05, France

G. M. Khenkin, Université de Paris VI, Pierre et Marie Curie, Mathématiques Tour 45–46, 4, Place Jussieu, F-75252 Paris Cedex 05, France

A. G. Vitushkin, Steklov Mathematical Institute, ul. Gubkina 8, 117966 Moscow, Russia

*Translator*

P. M. Gauthier, Université de Montréal, Département de mathématiques et de statistique, C. P. 6128, Succursale A, Montréal, Québec H3C 3J7, Canada

# Contents

# I. Remarkable Facts of Complex Analysis

A.G. Vitushkin

Translated from the Russian
by P.M. Gauthier

## Contents



## Introduction

The present article gives a short survey of results in contemporary complex analysis and its applications. The material presented is concentrated around several pivotal facts whose understanding enables one to have a general view of this area of analysis.

## §1. The Continuation Phenomenon

The most impressive fact from complex analysis is the phenomenon of the continuation of functions (Hartogs, 1906; Poincaré, 1907). We elucidate its significance by an example. If a function $f$ is defined and holomorphic on the boundary of a ball $B$ in $n$-dimensional complex space $\mathbb{C}^n$ ($n \geq 2$), then it turns out that $f$ may be continued to a function holomorphic on the whole ball $B$. Analogously, for an arbitrary bounded domain whose complement is connected, any function holomorphic on the boundary of such a domain admits a holomorphic continuation to the domain itself. Let us emphasize that this holds only for $n \geq 2$. In the one dimensional case, this phenomenon clearly does not occur. Indeed, for each set $E \subset \mathbb{C}^1$ and each point $z_0 \in \mathbb{C} \backslash E$, the function $1/(z - z_0)$ is holomorphic on $E$ but cannot be holomorphically continued to the point $z_0$.

This discovery marked the beginning of the systematic study of functions of several complex variables. Two fundamental notions, originating in connection with this property of holomorphic functions, are "envelope of holomorphy" and "domain of holomorphy". Let $D$ be a domain or a compact set in $\mathbb{C}^n$. The *envelope of holomorphy* $\tilde{D}$ of the set $D$ is the largest set to which all functions holomorphic on $D$ extend holomorphically. The envelope of holomorphy of a domain in $\mathbb{C}^n$ is a domain which in general "cannot fit" into $\mathbb{C}^n$, but rather is a multi-sheeted domain over $\mathbb{C}^n$ (Thullen, 1932). A domain $D \subset \mathbb{C}^n$ is called a *domain of holomorphy* if $\tilde{D} = D$, i.e. if there exists a holomorphic function on $D$ which cannot be continued to any larger domain. Domains of holomorphy are also sometimes called *holomorphically convex domains.*

The theorem on discs (Hartogs, 1909) gives an idea helpful in constructing the envelope of holomorphy of a domain: if a sequence of analytic discs, lying in the domain $D$, converges towards a disc whose boundary lies in $D$, then this entire limit disc lies in the envelope of holomorphy of $D$. An *analytic disc* is the biholomorphic image of a closed disc. The technique of construction of the envelope of holomorphy of a compact set and in particular of a surface relies on a conglomeration of "attached" discs whose boundaries lie on the given surface (Bishop, 1965).

Closely related to the notion of envelope of holomorphy is the notion of hull with respect to some class or other of functions, for example, the polynomial hull, the rational hull, etc. The *polynomial hull* of a set $D \subset \mathbb{C}^n$ is the set of all $z \in \mathbb{C}^n$ for which the following condition holds: for each polynomial $P(\zeta)$,

$$|P(z)| \leq \sup_{\zeta \in D} |P(\zeta)|.$$

Every smooth curve is *holomorphically convex,* i.e. its envelope of holomorphy coincides with the curve itself. The polynomial hull of a curve is in general non-trivial. For example, if a smooth curve is closed and without self-intersections, then its polynomial hull is either trivial or it is a one-dimensional complex analytic set whose boundary coincides with the given curve (Wermer, 1958; Bishop, 1962). We recall that a *set* in $\mathbb{C}^n$ is called *analytic* provided that in the vicinity of each of its points it is defined by a finite system of equations $\{f_j(\zeta) = 0\}$, where $\{f_j\}$ are holomorphic functions.

*An unsolved problem.* Is a set in $\mathbb{C}^n$ consisting of a finite number of pairwise disjoint balls polynomially convex? If the number of balls is at most 3, then the answer is positive; their union is polynomially convex (Kallin, 1964).

Another variant of the continuation phenomenon is the theorem of Bogolyubov, nicknamed the edge-of-the-wedge theorem (S.N. Bernstein, 1912; N.N. Bogolyubov, 1956; ..., V.V. Zharinov, 1980). Let $C^+$ be an acute convex cone in $\mathbb{R}^n$ consisting of rays emanating from the origin. Let $C^-$ be the cone symmetric to $C^+$ with respect to the origin. Let $\Omega$ be a domain in $\mathbb{R}^n$, and $D^+$ and $D^-$ two wedges, i.e. domains in $\mathbb{C}^n$ of the type

$$D^+ = \{z \in \mathbb{C}^n\colon \operatorname{Re} z \in \Omega, \operatorname{Im} z \in C^+\}$$

and

$$D^- = \{z \in \mathbb{C}^n\colon \operatorname{Re} z \in \Omega, \operatorname{Im} z \in C^-\}.$$

Suppose $f$ is a function holomorphic on $D^+ \cup D^-$ and suppose the functions $f|_{D^+}$ and $f|_{D^-}$ have boundary values which agree in the sense of distributions along the edge of these cones, i.e. on the set $D^0 = \{z \in \mathbb{C}^n\colon \operatorname{Re} z \in \Omega, \operatorname{Im} z = 0\}$. Then, $f$ has a holomorphic extension to some neighbourhood of the set $D^0$. The theorem on $C$-convex hull (V.S. Vladimirov, 1961) gives an estimate on the size of this neighbourhood. For example, if $\Omega = \mathbb{R}^n$, then $(D^+ \cup D^0 \cup D^-)^\sim = \mathbb{C}^n$ (Bochner, 1937).

The theorem of Bogolyubov has been used to establish several relations in axiomatic quantum field theory. This theorem also laid the foundations of the theory of hyperfunctions (Sato, 1959; Martineau, 1964; ..., V.V. Napalkov, 1974). For more details, see articles II, III and volume 8, article IV.

## §2. Domains of Holomorphy

Domains of holomorphy are of interest because in such domains one can solve traditional problems of analysis. In certain of these domains holomorphic functions have integral representations and admit approximation by polynomials. In domains of holomorphy the Cauchy–Riemann equations are solvable; it turns out to be possible to interpolate functions; the problem of division is solvable; etc.

Two of the simplest types of domains of holomorphy are polynomial polyhedra and strictly pseudoconvex domains. A *polynomial polyhedron* is a domain given by a system of the type $|P_j(z)| < 1$, $j = 1, 2, \ldots, k$, where each $P_j(z)$ is a polynomial in $z$. Polynomial polyhedra were introduced by Weil (1932) and are also called *Weil polyhedra*. A domain is called *strictly pseudoconvex* if in the neighbourhood of each of its boundary points the domain is strictly convex for a suitable choice of coordinates. Suppose the hypersurface bounding a domain is given by an equation $\rho(z, \bar{z}) = 0$. If in each point of the hypersurface the Levi form of the hypersurface is positive definite, then the domain in question is strictly pseudoconvex (E. Levi, 1910). The *Levi form* is the form $\sum_{i,k} \frac{\partial^2 \rho}{\partial z_i \partial \bar{z}_k} dz_i \cdot d\bar{z}_k$ restricted to the complex tangent space to the hypersurface at the point $z_0$.

The solution of various forms of the problem of Levi concerning the holomorphic convexity of strictly pseudoconvex domains remained the central problem of complex analysis for several decades. Oka (1942) showed that each strictly pseudoconvex domain is holomorphically convex and conversely each domain of holomorphy can be exhausted from the interior by domains of this type. Polynomial polyhedra are easily seen to be polynomially convex and consequently holomorphically convex.

Boundary points of a domain of holomorphy are not equivalent. A particularly important role is played by that part of the boundary which is called the distinguished boundary or the Shilov boundary. The *Shilov boundary* of a bounded domain is the smallest closed subset $S(D)$ of the boundary of $D$ such that, for each function $f$ continuous on the closure of $D$ and holomorphic in $D$ and for each point $z \in D$ the inequality $|f(z)| \leq \max_{\zeta \in S(D)} |f(\zeta)|$ holds. For a ball the Shilov boundary coincides with its topological boundary. The Shilov boundary of the polydisc $|z_j| < 1$, $j = 1, 2, \ldots, n$, is the $n$-dimensional torus $|z_j| = 1$, $j = 1, 2, \ldots, n$. For domains whose boundary is $C^2$, the Shilov boundary is the closure of the set of strictly pseudoconvex points (Basener, 1973).

For domains of holomorphy, a strong maximum principle holds. If $D$ is a domain of holomorphy and $f$ is non-constant, continuous on the closure of $D$, holomorphic in $D$, and attains a local maximum at some point, then that point lies in $S(D)$ (Rossi, 1961). In simple cases the non-Shilov part of the boundary has analytic structure, i.e. folliates into analytic sets. This was shown, for example, for domains in $\mathbb{C}^2$ having $C^1$ boundary (N.V. Shcherbina, 1982).

Concerning the topology of domains of holomorphy, it is known that the homology groups $H_k$ of order $k$ are trivial for all $k > n$. For polynomially convex domains, the $n$-th homology group is also trivial (Serre, 1953; Andreotti and Narasimhan, 1962).

Several classical problems of analysis are solvable only for domains of holomorphy. For example a domain is a domain of holomorphy if and only if each function holomorphic on a complex submanifold of the domain is the restriction of some function holomorphic on the whole domain (Oka, H. Cartan, 1950). Analogously, a domain is a domain of holomorphy if and only if the problem of division is solvable (Oka, H. Cartan, 1950). The *problem of division* is said to be solvable in the domain $D$ if for any functions $f_1, \ldots, f_k$ holomorphic in $D$, and any holomorphic function $f$ in $D$ whose zero set contains (taking into account multiplicities) the set of common zeros of the functions $f_1, \ldots, f_k$, there exist functions $g_1, \ldots, g_k$, holomorphic in $D$, such that $\sum_j f_j g_j = f$. We recall that on account of the Weierstrass preparation Theorem (1885), the local problem of division is always solvable.

One can define the notion of holomorphic convexity in terms of plurisubharmonic functions. A function is called *plurisubharmonic* if its restriction to each complex line is a subharmonic function. A domain $D$ is a domain of holomorphy if and only if the function $-\ln \rho(z)$ is plurisubharmonic on $D$, where $\rho(z)$ is the distance from the point $z$ to the boundary of $D$ (Lelong, 1945).

For further details see article II and Volume 8, article II.

## §3. Holomorphic Mappings. Classification Problems

By the Riemann Mapping Theorem, in $\mathbb{C}^1$ any two proper simply-connected domains are holomorphically equivalent. In the multidimensional case, the situation is substantially different. For example, a ball and a polydisc are not equivalent (Reinhardt, 1921). Moreover, almost any two randomly chosen domains turn out to be non-equivalent (Burns, Shnider, Wells, 1978).

Let us consider the class of strictly pseudoconvex domains having analytic boundary. In this situation any biholomorphic mapping from one domain onto another extends to a biholomorphic correspondence between the boundaries (Fefferman, 1974; S.I. Pinchuk, 1975), and by the same token, the classification problem for such domains reduces to that of classifying hypersurfaces. There are two approaches to this problem. The first is geometric; the hypersurface is characterized by a system of differential-geometric invariants (E. Cartan, 1934; Tanaka, 1967; Chern, 1974). In the second approach, the characterization is by a special equation, the so-called normal form (Moser, 1974). Both of these constructions enable one to distinguish the infinite-dimensional space of pairwise non-equivalent analytic hypersurfaces.

In connection with the classification problem, a description of mappings realizing the equivalence between two surfaces has been obtained. The results for mappings are described as for the case of functions by properties of continuation. In the case of mappings a new variant of this phenomenon appears. For example, it turns out that a holomorphic mapping of a sphere to itself given in a small neighbourhood of some point of the sphere can be holomorphically extended to the entire sphere and moreover, is in fact a fractional linear transformation (Poincaré, 1907; Alexander, 1974). If the surface is *not spherical*, i.e. cannot, by a local change of coordinates, be transformed into the equation of a sphere, then the germ of such a mapping of the surface into itself can be continued, not only along the surface, but also, in a direction normal to the surface. Namely, if a strictly pseudoconvex analytic hypersurface is not spherical, then the germ of any holomorphic mapping of this surface into itself has a holomorphic continuation (with an estimate on the norm) to a "large" neighbourhood of the center of the germ. Moreover, a guaranteed size, for both the neighbourhood as well as for the constant estimating the norm, is determined by the two characteristics of the surface, namely, the parameters of analyticity of the surface and its constant of non-sphericity (A.G. Vitushkin, 1985). In particular, a surface of the indicated type has a neighbourhood to which all automorphisms of the surface extend. It is worth emphasizing that in both examples we have presented, the mappings, in contrast to functions, extend not only to the envelope of holomorphy of the domain on which they are defined, but also to some domain lying outside the domain of holomorphy. The theorem on germs of mappings concludes a lengthy chain of works on holomorphic mappings of surfaces (Alexander, 1974; Burns and Shnider, 1976; S.I. Pinchuk, 1978; V.K. Beloshapka and A.V. Loboda, 1980; V.V. Ezhov and N.G. Kruzhilin, 1982).

From the Theorem on Germs, it follows that a stability group of a surface (group of its automorphisms which leave a certain point fixed) is compact. Hence, by Bochner's theorem on the linearization of a compact group of automorphisms (1945), one obtains that a stability group of a non-spherical surface can be linearized, i.e. by choosing appropriate coordinates, every automorphism can be written as a linear transformation (N.G. Kruzhilin and A.V. Loboda, 1983). Together with the theorem of Poincaré, this means that for each pair of locally given strictly pseudoconvex analytic hypersurfaces, every mapping sending one hypersurface into the other can be written as a fractional-linear transformation by an appropriate choice of coordinates in the image and preimage. The problem on the linearization of mappings of surfaces having a non-positive Levi form remains open. For further details see article IV and Volume 9, articles V and VI.

We have considered here only one aspect of the problem of classification. Large sections of complex analysis are concerned with the study of invariant metrics (Kähler, 1933; Carathéodory, 1927; Bergman, 1933; Kobayashi, 1967; Fefferman, 1974, ... ); classification of manifolds (Hodge, Kodaira, 1953; Yau, Siu, 1980; ... ); description of singularities of complex surfaces (Milnor, 1968; Brieskorn, 1966; Malgrange, 1974; A.N. Varchenko, 1981; ... ).

## §4. Integral Representations of Functions

A smooth function in a closed domain $\bar{D} \subset \mathbb{C}$ can be expressed using the Cauchy-Green formula

$$f(z) = \frac{1}{2\pi i} \int_{\partial D} \frac{f(\zeta)}{\zeta - z} d\zeta + \frac{1}{2\pi i} \int_{D} \frac{\partial f}{\partial \bar{\zeta}}(\zeta) \frac{1}{\zeta - z} d\bar{\zeta} \wedge d\zeta.$$

The first term on the right side is the formula which reproduces a holomorphic function in a domain in terms of its boundary values. The second term isolates the non-holomorphic part of $f$ and yields a solution to the $\bar{\partial}$-equation $\frac{\partial f}{\partial \bar{z}} = g$. For functions of several variables, there does not exist such a simple and universal formula, and hence it is suitable to consider the problem of integral formulas for holomorphic functions and the solvability of the $\bar{\partial}$-equations separately.

For some classes of domains in $\mathbb{C}^n$, there are explicit formulas which reproduce a holomorphic function in terms of its boundary values. For polynomial polyhedra such a formula was obtained by A. Weil (1932); for strictly pseudoconvex domains, by G.M. Khenkin (1968). Such a formula was given for the polydisc by Cauchy (1841) and for the ball, by Bochner (1943). There is a formula of Bochner–Martinelli (1943) for smooth functions on arbitrary domains having smooth boundary. In this formula, in contrast to the previous ones, the kernel is not holomorphic, and this often makes it difficult to apply. For polynomial polyhedra there is still another formula which distinguishes itself from the Weil formula and other formulas in that its kernel is not only holomorphic but also integrable (A.G. Vitushkin, 1968).

Let us introduce the formulas for the polydisc and the ball. If $f$ is holomorphic on the closure of the polydisc $D^n$, then

$$f(z) = \left(\frac{1}{2\pi i}\right)^n \int_{(\partial D)^n} \frac{f(\zeta)}{(\zeta_1 - z_1) \dots (\zeta_n - z_n)} \, d\zeta_1 \wedge \dots \wedge d\zeta_n.$$

If $f$ is holomorphic on the closed ball $\bar{B}: |z| \le 1$, then inside the ball,

$$f(z) = \frac{1}{V} \int_{\partial B} \frac{f(\zeta)}{(1 - \bar{\zeta}_1 z_1 - \dots - \bar{\zeta}_n z_n)^n} \, dV,$$

where $V$ is the $(2n-1)$-dimensional volume of the sphere $\partial B$ and $dV$ is its element of volume.

All of the formulas which we have mentioned above differ from one another in appearance. The appearance of the formula depends on the type of domain. There is a formula due to Fantappiè–Leray (1956) which gives a general scheme for writing such formulas. Let $D$ be a domain in $\mathbb{C}^n_z$, where $z = (z_1, \dots, z_n)$ is a set of coordinate functions, and let $f$ be holomorphic on the closure of $D$. Then

$$f(z) = \frac{(n-1)!}{(2\pi i)^n} \int_\gamma \frac{f(\zeta)}{[\eta_1(\zeta_1 - z_1) + \dots + \eta_n(\zeta_n - z_n)]^n}$$
$$\cdot \sum_{k=1}^{n} (-1)^k \eta_k \wedge d\eta_1 \wedge \dots \wedge d\eta_{k-1} \wedge d\eta_{k+1} \wedge \dots \wedge d\eta_n \wedge d\zeta_1 \wedge \dots \wedge d\zeta_n,$$

where $\gamma$ is a $(2n-1)$-dimensional cycle in the space $\mathbb{C}^n_\zeta \times \mathbb{C}^n_\eta$ lying over the boundary of the domain $D \subset \mathbb{C}^n_\zeta$ and covering it once. By choosing suitably the form of the cycle $\gamma$, having chosen $\eta$ as a function of $\zeta$, one can obtain any of the preceding integral formulas.

One of the applications of integral formulas is in solving the problem of interpolation with estimates. If a complex submanifold $M$ of the ball $B$ crosses the boundary of the ball transversally, then every function holomorphic and bounded on $M$ can be continued to a function holomorphic and bounded in the entire ball (G.M. Khenkin, 1971). The extension is constructed as follows. The function $f(z)$ for $z \in M$ can be written as an integral $I(z)$ of $f$ on the boundary of $M$. Moreover, it turns out that the function $I(z)$ is defined for all $z \in B$, and from the explicit formula for $I(z)$, one obtains that the extended function $f(z) = I(z)$ is holomorphic and bounded on $B$.

The problem on the possibility of division with uniform estimates remains open. Namely, it is not known whether for each set of functions $f_1, \ldots, f_k$, holomorphic and bounded in the ball $B \subset \mathbb{C}^n$ and such that $\inf_{\zeta \in B} \sum_{j=1}^{k} |f_j(\zeta)| \neq 0$, there exist functions $g_1, \ldots, g_k$ bounded and holomorphic on $B$ such that $\sum_{j=1}^{k} f_j g_j \equiv 1$. This is a modified formulation of the famous "corona" problem. In the one dimensional case, this problem was solved by Carleson (1962). The answer is positive: in the maximal ideal space for the algebra of bounded holomorphic functions in the one-dimensional disc, the set of ideals, corresponding to points of the disc, is everywhere dense.

The above enumerated formulas are for bounded domains. In the present time analysis on unbounded domains is also flourishing. In particular, integral formulas have been constructed for such domains. There are explicit formulas for tubular domains over a cone (Bochner, 1944), on Dyson domains (Jost, Lehmann, Dyson, 1958; V.S. Vladimirov) and Siegel domains (S.G. Gindikin, 1964). Weighted integral representations for entire functions have also been constructed (Berndtsson, 1983). For further results see article II and Volume 8, articles I, II and IV.

## §5. Approximation of Functions

Let us denote by $CH(E)$ the set of all continuous functions on the compact set $E \subset \mathbb{C}^n$ which are holomorphic at interior points of $E$. It is clear that functions which can be uniformly approximated on $E$ with arbitrary accuracy by complex polynomials or by functions holomorphic on $E$ belong to the class $CH(E)$. When we speak of the possibility of approximating functions on the compact set $E$, we shall mean the following: each function in $CH(E)$ can be approximated uniformly with arbitrary precision by functions holomorphic on $E$.

If a compact set $E$ in $C^1$ has a connected complement, then each function holomorphic on $E$ can be approximated by polynomials (Runge, 1885). This is equivalent to a theorem of Hilbert (1897): on each polynomial polyhedron in $\mathbb{C}^1$, any holomorphic function can be represented as the sum of a series of polynomials. Runge's Theorem reduces the question of the possibility of approximating functions by polynomials to that of constructing holomorphic approximations of functions. The criterion for the possibility of approximation by holomorphic functions (A.G. Vitušhkin, 1966) is formulated as follows. The assertion that each function in $CH(E)$, where $E \subset \mathbb{C}^1$, can be uniformly approximated with arbitrary accuracy by functions holomorphic on $E$ is equivalent to the following condition on the compact set $E$: for each disc $K$, $\alpha(K \backslash E) = \alpha(K \backslash \mathring{E})$, where $\mathring{E}$ denotes the interior of $E$, and $\alpha(M)$ is the *continuous analytic capacity* of a set $M$. By definition

$$\alpha(M) = \sup_{M^*; f} \left| \lim_{z \to \infty} z f(z) \right|.$$

The supremum is taken over all compact sets $M^* \subset M$ and all functions $f$ which are everywhere continuous on $\mathbb{C}^1$, bounded in modulus by 1 and holomorphic outside of $M^*$. In particular, approximation is possible if the inner boundary of $E$ is empty, i.e. each boundary point of $E$ belongs to the boundary of some complementary component of $E$. For example, all compact sets with connected complement belong to this class. The above criterion emerged as a result of a long series of works on approximation (Walsh, 1926; Hartogs and Rosenthal, 1931; M.A. Lavrentiev, 1934; M.V. Keldysh, 1945; S.N. Mergelyan, 1951 and others).

The notion of analytic capacity is useful not only in approximation. It appears along with its analogues in integral estimates (M.S. Mel'nikov, 1967). Such capacities are used for describing the set of removeable singularities of a function (Ahlfors, 1947; . . . E.P. Dolzhenko, 1962; . . . Mattila, 1985). Among the unsolved problems, we draw attention to the problem of the subadditivity of analytic capacity: is it true that for any two compact sets, the capacity of their union is no greater than the sum of their capacities?

The integral formula of Weil is a generalization of Hilbert's construction. Using this formula, A. Weil (1932) showed that on any polynomially convex compact set in $\mathbb{C}^n$, each holomorphic function can be approximated by polynomials. Thus in $\mathbb{C}^n$ as in $\mathbb{C}^1$, polynomial approximation reduces to holomorphic approximation. The integral formula of G.M. Khenkin emerged as a result of attempting to construct holomorphic approximations on arcs. While developing such approximations, the technique of integral formulas found various applications. Nevertheless, the initial question on the possibility of approximating continuous functions on polynomially convex arcs by polynomials remains open.

The possibility of holomorphic approximation has been established for the following cases: arcs having nowhere dense projection on the coordinate planes (E.M. Chirka, 1965); strictly pseudoconvex domains (G.M. Khenkin, 1968); non degenerate Weil polyhedra (A.I. Petrosyan, 1970); and C.R.-manifolds (Baouendi and Trèves, 1981). There are several examples of compact sets on which approximation is not possible. Diederich and Fornaess (1975) constructed a domain of holomorphy in $\mathbb{C}^2$, with $C^\infty$-boundary, whose closure is not a compact set of holomorphy, i.e. it cannot be represented as the intersection of a decreasing sequence of domains of holomorphy. Moreover, on this domain one can define a holomorphic function, infinitely differentiable up to the boundary of the domain, which cannot be approximated by functions holomorphic on the closure of the domain.

For related results, see papers II and III.

Above we discussed only the possibility of approximation. There is a lengthy series of works devoted to the explicit construction of approximating functions. In recent years in connection with applications, there has been a renewed interest in classical rational approximation (continuous fractions, Padé approximation, etc.). We mention one example concerning rational approximation in connection with the holomorphic continuation of functions. Let $f$ be holomorphic on the ball $B \subset \mathbb{C}^n$, and set $r_k(f) = \inf_{\varphi} \sup_{z \in B} |f(z) - \varphi(z)|$, where the infimum is taken over all rational functions $\varphi$ of degree $k$. Then, if for each $q > 0$, $\lim_{k \to \infty} r_k(f) q^{-k} = 0$, then the global analytic function, generated by the element $f$, turns out to be single-valued, i.e., its domain of existence is single-sheeted over $\mathbb{C}^n$ (A.A. Gonchar, 1974). See Vol. 8, paper II.

## §6. Isolating the Non-Holomorphic Part of a Function

Sometimes in order to construct a holomorphic function with given properties, one proceeds as follows. One constructs some smooth function $\varphi$ with the desired properties and then one breaks up $\varphi$ as the sum of two functions the first of which is holomorphic while the second is in some sense small. In this situation, the first function may turn out to be the function we require. The second term is sought in the form of a solution to the equation $\bar{\partial} f = g$, where $\bar{\partial} = \frac{\partial}{\partial \bar{z}_1} d\bar{z}_1 + \ldots + \frac{\partial}{\partial \bar{z}_n} d\bar{z}_n$, and $g = \bar{\partial}\varphi$. This scheme is used for constructing functions with prescribed zeros, in approximation, etc. Equations of the type $\bar{\partial} f = g$ are called the Cauchy–Riemann equations or $\bar{\partial}$-equations.

Let us consider a more general case of the equation $\bar{\partial} f = g$, namely, we shall take for $g$ a differential $(p, q)$-form, i.e., a form having degree $p \geq 0$ in $dz$ and degree $q \geq 1$ in $d\bar{z}$. A necessary condition for the solvability of this equation is that the form $g$ be $\bar{\partial}$-closed, i.e. $\bar{\partial} g = 0$. This is a necessary compatibility condition and so it is always assumed to be satisfied. The Cauchy–Riemann equations are solvable on each domain of holomorphy (Grothendieck,

Dolbeault, 1953). If the domain is bounded and $g \in L_2$, then there exists a solution to the C.-R. equations which lies in $L_2$ and is orthogonal to the subspace of $\bar{\partial}$-closed $(p, q-1)$-forms (Morrey, Kohn, Hörmander, 1965). For strictly pseudoconvex domains there are explicit formulas for the solution of these equations and estimates on the solution in the uniform norm and in several other metrics (G.M. Khenkin, Grauert, Lieb, 1969).

For some simple domains, the question of the possibility of solving the $\bar{\partial}$-equations with uniform estimates remains open. For example there are no such estimates on a Siegel domain, also called a generalized unit disc. This is the domain, in the $n^2$-dimensional space, of square matrices $Z$ determined by the condition $E - Z \cdot Z^* \gg 0$, i.e., consisting of matrices $Z$, for which the indicated expression is a positive definite matrix.

To every complex manifold is associated a system of cohomology groups called the *Dolbeault cohomology* (1953). The Dolbeault group of type $(p, q)$ is the quotient of the group of $\bar{\partial}$-closed $(p, q)$-forms by the group of $\bar{\partial}$-exact $(p, q)$-forms. In many cases (for example, for compact Kähler manifolds), these groups can be calculated using de Rham cohomology. However, on domains of holomorphy, the Dolbeault cohomology is trivial while the de Rham cohomology may be non-trivial.

Interest in the $\bar{\partial}$-equations is also connected to the phenomenon that there is a wide class of differential equations which by a change of variables are transformed to the $\bar{\partial}$-equations, and in many cases this yields the possibility of characterizing the solutions of the initial equations in one form or the other. In the general situation, this change of variables leads to the $\bar{\partial}$-equations on a surface (the tangential Cauchy–Riemann equations). In these situations the $\bar{\partial}$-equations are to be understood as follows: $f$ is called a solution to the equation $\bar{\partial} f = g$ on the surface $M$ if this equation is fulfilled for all vectors lying in the complex tangent space to $M$. Each system of linear differential equations in general position, with analytic coefficients, and one unknown function, can be transformed by an analytic change of coordinates to the $\bar{\partial}$-equations (of type $(0, 1)$) on an analytic surface (Rossi, Andreotti, Hill, 1970). Such equations satisfying the natural compatibility conditions, are locally solvable (Spencer, V.P. Palamodov, 1968). If the right-hand side is not analytic, then, such equations are, generally speaking, not solvable. For example, on the sphere in $\mathbb{C}^2$, one can give an infinitely differentiable $(0, 1)$-form such that the equation $\bar{\partial} f = g$ turns out to be not locally solvable (H. Lewy, 1957). Explicit integral formulae for solutions to the $\bar{\partial}$-equations yield criteria for solvability (G.M. Khenkin 1980). Systems of equations with smooth coefficients are, generally speaking, not reduceable to $\bar{\partial}$-equations (Nirenberg, 1971). If we extend the class of transformations acting on these equations, namely, by adding homogeneous simpletic transformations in the cotangent bundle, then almost all linear systems of equations with analytic coefficients can be reduced locally to $\bar{\partial}$-equations on standard surfaces (Sato, Kawai, Kashiwara, 1973). For further results see paper II.

## §7. Construction of Functions with Given Zeros

Let us consider several examples from which it will be clear how the problems on the zeros of functions arise. The first example is the use of the Weierstrass Theorem on the representation of an entire function of one variable in the form of an infinite product. This theorem has applications in information theory. The formula, regenerating an entire function from its zeros, is used in the problem of encoding signals having finite spectrum. A signal with *finite spectrum* is a function of time, whose Fourier transform is a function of compact support, i.e. an entire function of time of finite type. The most economic code for such signals is constructed in the following form. It is necessary to complexify time and to calculate the zeros of the function. As a code, the function takes the coordinates of its zeros. Using the zeros we write the infinite product which gives the function and this is the formula regenerating the original function. It has been shown that for such coding systems which don't increase the density of the code, it is possible to broaden without limit the dynamic range of the connecting channel or of the reproducing system (V.I. Buslaev, A.G. Vitushkin, 1974). The *dynamic range* is the ratio between the maximal and minimal signals which are reproduced with a specified accuracy. Codes which specify a wide dynamic range for the amplitude of a signal are required for example in sound recordings.

It is not known whether there exists an analogous coding system for entire functions of two variables. In this vein we propose a problem. Let us denote by $K_n$ the collection of all sets in $\mathbb{C}^2$ each of which is the intersection of the zero set of some polynomial of degree at most $n$ with the ball $|z| \leq 1$. As a metric on $K_n$ we take the Hausdorff distance between sets. The problem is to calculate the *entropy* $H_\varepsilon(K_n)$. By definition, $H_\varepsilon(K_n) = \log_2 N_\varepsilon(K_n)$, where $N_\varepsilon$ is the number of elements for a minimal $\varepsilon$-net of the compact space $K_n$. The conjecture is that for small $\varepsilon$ and large $n$, $H_\varepsilon(K_n) \approx \frac{1}{2} n^2 \log_2 \frac{1}{\varepsilon}$.

The next example is related to differential equations. Suppose we are given a system of linear differential equations, with constant coefficients and smooth right hand side, defined on a convex domain. Then, provided certain necessary conditions (of compatibility type) are satisfied by the right-hand side, the system admits a solution on this domain (Ehrenpreis, V.P. Palamodov, Malgrange, 1963). If, for example, the right-hand side is of compact support then the Fourier transform carries this system to a system of the type $CX = F$, where $F$ is a vector of entire functions. Here, $C$ is a matrix of polynomials and $X$ is an unknown vector function. By solving this problem and using the inverse Fourier transform, we obtain from $X$ the solution to the original system. If the system has a solution of compact support, then $F$ must be divisible by $C$ and this gives the form of the compatibility conditions for the right-hand side. In order that the inverse Fourier transform be defined, one must solve the division problem with estimates on the growth of the solution at infinity.

Each meromorphic function on $\mathbb{C}^n$ can be represented as the quotient of two entire functions. This assertion was proved by Weierstrass (1874) for $\mathbb{C}^1$, by Poincaré (1883) for $\mathbb{C}^2$, and by Cousin (1895) in general. This was essentially the first series of works on several variables and it layed the foundations of several directions in complex analysis. The modern theory of cohomology comes from the work of Cousin, while potential theory and the theory of currents stems from the work of Poincaré.

The statements of Cousin concerning the solvability of certain problems have come to be called the first and second Cousin problems. The *first Cousin problem* is said to be solvable on some domain or other if it is possible on this domain to construct a meromorphic function with given poles. The *second Cousin problem* is said to be solvable if it is possible to construct, on this domain, a holomorphic function with given zeros. The first Cousin problem is solvable on each domain of holomorphy (Oka, 1937), and if the group of second cohomology with integer coefficients for the domain is trivial, then the second problem is also solvable (Oka, 1939; Serre, 1953). In $\mathbb{C}^n$ the second cohomology is trivial, and so, the second problem of Cousin is solvable. Consequently, any meromorphic function in $\mathbb{C}^n$ can be represented as the quotient of entire functions.

If $M$ is the set of zeros of a function $f$, then $f$ satisfies the *equation of Poincaré–Lelong*: $\frac{i}{\pi}\partial\bar{\partial}\ln|f| = [M]$, where $[M]$ is a certain closed (1, 1)-current of integration on $M$, taking into account the multiplicity of the *divisor* $M$. A *current* is a generalized differential form, i.e. a linear functional on forms of compact support of complementary degree. A current is called *closed* if it is zero on exact forms of compact support. The current $[M]$ is integration on $M$ of the product of a test function and the multiplicity of the divisor $M$. The formula of Lelong, giving a solution of this equation in $\mathbb{C}^n$, yields a completely accurate estimate on the speed of growth of an entire function depending on the density of its set of zeros (1953). For example, an $(n-1)$-dimensional closed analytic subset of $\mathbb{C}^n$ is algebraic (i.e. is the zero set of a polynomial) if and only if the $(2n-2)$-dimensional measure of the intersection of this set with an arbitrary ball of radius $r$ can be estimated from above by the quantity $C \cdot r^{2n-2}$, where $C$ is independent of $r$ (Rutishäuser, Lelong, Stoll, 1953).

Currents were introduced by de Rham. The Lelong theory of closed currents was one of the fundamental tools in the research on analytic sets (Griffiths, 1973; . . . , E.M. Chirka, 1982) and plurisubharmonic functions (Josefson, 1978; Bedford, 1979; A. Sadullaev, 1981). Currents are practical in that they allow one to carry delicate problems on analytic sets over to standard estimates on integrals. Using this technique, Harvey and Lawson (1974) showed that if a $(2k+1)$-dimensional smooth submanifold $M \subset \mathbb{C}^n$ has at each point a complex tangent space of maximal possible dimension, i.e. dimension $2k$, and if $M$ is pseudo-convex, then $M$ is the boundary of a $(2k+2)$-dimensional analytic subset which (together with $M$) is the envelope of holomorphy of $M$.

For related results, see articles II, III; Vol. 8, article II; and Vol. 9, articles I, II.

## §8. Stein Manifolds

A great deal of what we have discussed on domains of holomorphy carries over to manifolds which are called Stein manifolds (1951). A complex manifold $M$ is called a *Stein manifold* if, first of all, it is holomorphically convex, i.e. if the holomorphically convex hull of each compact set in $M$ is compact in $M$, and secondly, if there exists on $M$ a finite set of holomorphic functions such that each point of $M$ has a neighbourhood in which these functions separate points. Each domain of holomorphy is clearly a Stein manifold. Closed complex submanifolds of $\mathbb{C}^n$ are also Stein manifolds. Conversely, each Stein manifold $M$ can be realized as such a submanifold, i.e., $M$ can be imbedded into $\mathbb{C}^n$ by a proper holomorphic mapping (Remmert, 1957).

It is not hard to see that a bounded Weil polyhedron can be imbedded in the polydisc in such a way that its boundary lies on the boundary of the polydisc. It turns out that the ball can also be realized as a closed complex submanifold of a polydisc. (A.B. Aleksandrov, 1984). However, there exists a bounded domain with smooth boundary not admitting such a realization (Sibony, 1985).

A great achievement in complex analysis was the solution of *Whitney's problem*. It has been shown that any real analytic manifold can be analytically imbedded in a real Euclidean space of sufficiently high dimension (Morrey, Grauert, 1958). On a given analytic manifold $S$ we fix an atlas. After complexifying the charts of this atlas, that is, allowing the coordinates to take not only real but also complex values with small imaginary part, we may consider $S$ as a submanifold of some complex manifold $M$. For an appropriate choice of metric on $M$, it turns out that $\varepsilon$-neighbourhoods $M_\varepsilon$ of $S$ are strictly pseudoconvex domains for small $\varepsilon$. The crucial moment in the construction is the generalization of Oka's theorem. Namely it is shown that on a complex manifold, each strictly pseudoconvex domain is holomorphically convex. From the holomorphic convexity of $M_\varepsilon$, it follows easily that for small $\varepsilon$, $M_\varepsilon$ turns out to be a Stein manifold. By Remmert's theorem, $M_\varepsilon$ can be imbedded in $\mathbb{C}^n$, and by the same token, $S$ turns out to be imbedded in $\mathbb{R}^{2n}$.

On Stein manifolds just as on domains of holomorphy the problems of interpolation and division are solvable, the $\bar{\partial}$-problem is solvable for arbitrary type $(p, q)$; the first Cousin problem is solvable and the second Cousin problem is also solvable provided the second integer cohomology group is trivial. All of these problems are solved by one and the same scheme. Let us look at this scheme for the case of solving the $\bar{\partial}$-problem of type (0, 1). First of all one solves the problem locally, i.e. we fix a covering of the manifold such that the equations are solvable on each set of the covering. If an element of the covering is, for example, a ball or a polydisc, then one can give an explicit formula for the solution. On the intersection of two elements of the cover, the local solutions may not agree, i.e., their difference may not be zero. The next stage in constructing a solution consists in determining "correcting factors", in this particular situation, holomorphic functions, defined on elements of the cover and whose difference on any intersection of elements of the cover is the same as for the local solutions constructed above. If such "correcting factors" exist, then subtracting these correcting factors from the corresponding local solutions we obtain new local solutions which agree on intersecting elements of the cover and hence yield a global solution to the equation. The difference of two local solutions on the intersection of two elements of the cover is a one dimensional closed cocycle of holomorphic functions. The existence of the desired correcting factors amounts to the exactness of this cocycle. Thus, on a given manifold, the $\bar{\partial}$-equation is solvable for any choice of the right hand side if and only if the one-dimensional cohomology with holomorphic coefficients (or, as they say, with coefficients in the sheaf of germs of holomorphic functions) is trivial.

The triviality of this group as well as that of other one-dimensional cohomology groups, corresponding to the above enumerated problems, follows from a theorem of H. Cartan (1953): a complex manifold is a Stein manifold if and only if its one dimensional cohomology group with coefficients in an arbitrary coherent analytic sheaf is always trivial. Locally a coherent analytic sheaf is a special type of subspace of the space of germs of holomorphic vector-valued functions on the given manifold or on some submanifold thereof. It can be, for example, the space of germs of vector-valued holomorphic functions itself, the subspace of germs having a given set of zeros, the quotient space of the first sheaf by the second, etc. The theorem of Cartan systematizes the material on domains of holomorphy accumulated till the early 50's. Cartan's theorem successfully combines the results and techniques of Oka with Leray's theory of analytic sheaves (1945). The next step in the development of cohomology theory was the theorem of Grauert (1958) which has come to be called the Oka–Grauert principle.

Let $M$ be a complex manifold. Let $\omega$ be a collection of domains in $M$ forming a cover of $M$. For each two intersecting elements of this cover $\alpha, \beta \in \omega$, let $C_{\alpha\beta}$ be a non-singular square matrix of degree $n$ consisting of functions defined and holomorphic on the intersection $\alpha \cap \beta$. We suppose that the collection $\{C_{\alpha\beta}\}$ forms a cocycle; in other words, they are compatible on a triple intersection, i.e., $C_{\alpha\beta} C_{\beta\gamma} C_{\gamma\alpha} = I$ on $\alpha \cap \beta \cap \gamma$. Such a collection of matrices of functions is called initial data for the Cousin problem for matrices. We say that the second Cousin problem with initial data $\{C_{\alpha\beta}\}$ is solvable if one can find a collection $\{C_\alpha\}$ of matrices of functions defined on $\omega$ and such that on each non-empty intersection $\alpha \cap \beta$, we have $C_\alpha \cdot C_\beta^{-1} = C_{\alpha\beta}$. The Oka–Grauert principle states that on a Stein manifold, the second Cousin problem with given initial data has a solution $\{C_\alpha\}$, where the $C_\alpha$ are matrices of holomorphic functions, provided it has a solution where the $C_\alpha$ are matrices of continuous functions. This was proved by Oka (1939) in the case where $n = 1$ and $M$ is a domain of holomorphy. He made use of this construction in order to find functions with given zeros.

Grauert's theorem has various applications. For example, Griffiths (1975) while working on a problem of Hodge obtained from this theorem that on a Stein manifold every class of even-dimensional cohomology with rational coefficients can be realized as a closed complex submanifold. For related results, see Vol. 10, articles I and II.

## §9. Deformations of Complex Structure

To specify a *complex structure* on a manifold means to specify an atlas with holomorphic transition functions. A deformation of complex structure is a new complex structure obtained from the given one by modifying the transition functions. For example, the extended complex plane can be considered as the two-dimensional sphere with a complex structure. This complex structure does not admit deformation, i.e. on $S_2$ the complex structure is unique. On the $2n$-dimensional sphere $S_{2n}$, for $n \neq 1, 3$, it is in general not possible to introduce a complex structure (Borel, Serre, 1951). It remains unknown, whether one can introduce a complex structure on the six-dimensional sphere.

The various structures of a compact complex one-dimensional manifold of genus $g (g > 1)$ form a manifold of real dimension $6g - 6$ (Riemann, 1857). This manifold of structures has itself a complex structure which can be obtained by factoring by a discrete group on a bounded domain of holomorphy in $\mathbb{C}^{3g-3}$ (Ahlfors, 1953).

Small deformations of the structure of a compact complex manifold of arbitrary dimension can also be parametrized as the points of a complex space which can be realized as an analytic subset of $\mathbb{C}^n$ (Kuranishi, 1964). Compact manifolds, obtained by factoring some simple domain (for example, a ball) by a discrete subgroup, have a rigid structure, i.e. do not admit small deformations of structure. Moreover, if two such spaces are topologically equivalent, then they are also holomorphically equivalent (Mostov, 1973). For non-compact manifolds, the space of structures is as a rule infinite dimensional.

In order to discuss deformations of bundles, we recall that an $n$-dimensional vector bundle is a bundle for which the fibre is the space $\mathbb{C}^n$. The structure of such a bundle is given by the cocycle of transition matrices $\{C_{\alpha\beta}\}$ (of dimension $n$), defined on the intersections $\{\alpha \cap \beta\}$ of elements of a cover of the base. If the base is a complex manifold and the transition matrices $\{C_{\alpha\beta}\}$ are holomorphic, then the *bundle* is called *holomorphic*. For example, the tangent bundle of a complex manifold $M$ is a holomorphic vector bundle of rank $n$. The space of tuples $z_0, \ldots, z_n$ on $n$-dimensional projective space $\mathbb{CP}^n$ can be considered as a one-dimensional holomorphic bundle over $\mathbb{CP}^n$. In this case a fibre is the collection of all tuples which can be obtained from each other by multiplication by a complex number. Two bundles $X$ and $X^*$ on one and the same base $M$ are called *equivalent* if one can find a homeomorphic mapping of $X$ into $X^*$, carrying fibres to fibres, acting linearly on each fibre, and fixing each point of $M$. If there exists a holomorphic mapping having the above properties, then the bundles are called holomorphically equivalent. A deformation of structure of a bundle is a new bundle not holomorphically equivalent to the given one but obtained from it by varying the transition matrices. From the Oka–Grauert principle it follows that a bundle whose base is Stein has a rigid structure and moreover from the topological equivalence of such bundles follows the holomorphic equivalence (Grauert, 1959). On Stein manifolds there are, so to speak, as many holomorphic bundles as continuous ones. If the base is not Stein, then it is usually not so; and this is good. Sometimes a complicated manifold can be interpreted as a space of deformations of a bundle thus yielding significant information concerning the initial object.

The twistor theory of Penrose (1967) is founded on such reductions. The idea of the general plan of Penrose can be taken as follows. Several notions of mathematical physics can be interpreted in terms of complex structure. For example, the metric on Minkowski space satisfying the Einstein equations, i.e. the gravitational field can be interpreted as a holomorphic structure on a domain in $\mathbb{CP}^3$. More precisely, there is a one-to-one correspondence between conformal classes of autodual solutions of the Einstein equations and deformations of structure of domains in $\mathbb{CP}^3$ (Penrose, 1976). A structure of a domain is the same as a choice of functions of three variables. The Penrose transformation, associating to each choice of functions a solution of the Einstein equations, has a sufficiently simple form and hence allows one to write down many solutions to these equations.

There is an analogous correspondence between autodual solutions of the Yang–Mills equations and bundles of rank two over domains in $\mathbb{CP}^3$ (Ward, 1977). In this direction the class of so-called instanton and monopole solutions to the Yang–Mills equations have been obtained which present interest for theoretical physics (Ward, Hitchin, Atiyah, . . . , Ju. I. Manin, 1978). Because of this series of works, this area of mathematics, which Penrose calls "the complex geometry of the real world"; has become very popular. For related results see Vol. 9, article VII and Vol. 10, articles II and III.

Within the limits of this article, we have restricted ourselves to discussing only a few of the outstanding facts from complex analysis. Unfortunately, we have not touched upon several major areas: the theory of residues (see article V and Vol. 8, article 1), the theory of singularities (see Vol. 10, article III), and value distribution (see Vol. 9, articles II–IV). A broad overview with the corresponding bibliographies for the various areas of multidimensional complex analysis is given in the series of articles in volumes 7–10 of this series.

# II. The Method of Integral Representations in Complex Analysis

G.M. Khenkin

Translated from the Russian
by P.M. Gauthier

## Contents

## §0. Introduction

**0.1. Fundamental Problems.** Let $D$ be a domain in the complex plane $\mathbb{C}^1$ with rectifiable boundary $\partial D$ and $f$ a complex valued function, continuous on $\bar{D}$ together with its Cauchy–Riemann derivative:

$$\frac{\partial f}{\partial \bar{z}} = \frac{1}{2}\left(\frac{\partial f}{\partial x} + i\frac{\partial f}{\partial y}\right), \qquad z = x + iy.$$

The formula of Cauchy–Green–Pompéiu (1904, see [33], [40], [45]) has the form

$$f(z) = \frac{1}{2\pi i}\int_{\partial D}\frac{f(\zeta)\,d\zeta}{\zeta - z} - \frac{1}{2\pi i}\int_{D}\frac{\frac{\partial f}{\partial \bar{\zeta}}(\zeta)\,d\bar{\zeta}\wedge d\zeta}{\zeta - z}, \qquad z\in D. \tag{0.1}$$

This formula becomes the classical Cauchy formula in case $f$ is a *holomorphic function*, that is when $\partial f/\partial \bar{z} = 0$ on $D$.

The Cauchy- and Cauchy–Green formulae are fundamental technical tools in the theory of functions of one complex variable. Examples of profound applications of these formulas are given in the works of Carleson [26] and Vitushkin [80]. In the first of these, the famous "Corona" problem for the disc in $\mathbb{C}^1$ is solved. In the second is solved the problem, going back to Weierstrass and Runge, on the uniform approximation by holomorphic functions on compact sets in $\mathbb{C}^1$.

Till the beginning of the thirties, the only multidimensional integral formula was the Cauchy formula for a polydomain $D = D_1 \times \ldots \times D_n$ in $\mathbb{C}^n$, where each $D_j$ is a bounded domain in $\mathbb{C}^1$ with rectifiable boundary.

Let $f$ be a function continuous on $\bar{D}$ and holomorphic in $D$, i.e., $\bar{\partial} f = 0$, where

$$\bar{\partial} = \sum \frac{\partial}{\partial \bar{z}_j}\,d\bar{z}_j$$

is the *Cauchy–Riemann* operator. Then the *Cauchy formula* (1841, see [4]) holds

$$f(z) = \frac{1}{(2\pi i)^n}\int_{\partial D_1\times\ldots\times\partial D_n}\frac{f(\zeta)\,d\zeta_1\wedge\ldots\wedge d\zeta_n}{(\zeta_1 - z_1)\ldots(\zeta_n - z_n)}. \tag{0.2}$$

This formula allows one to prove the fundamental properties of holomorphic functions of several variables, for example, the local representation of holomorphic functions by power series, the property of uniqueness of analytic continuation, etc.

Using the classical Cauchy formula, Hartogs (1906, see [81], [23]) showed that in $\mathbb{C}^n$, for $n > 1$, there is a domain $D$ such that each function holomorphic on $D$ necessarily has a holomorphic continuation to some large domain $\Omega \supset D$.

For example, any function $f$, holomorphic in the domain

$$\Omega_1^\varepsilon \cup \Omega_2^\varepsilon = \{z: |z_1| < 1 + \varepsilon,\ 1 - \varepsilon < |z_2| < 1 + \varepsilon\} \cup$$
$$\cup \{z: |z_2| < 1 + \varepsilon,\ |z_1 - 1| < \varepsilon$$

contained in an $\varepsilon$-neighbourhood of the boundary of the bidisc $\Omega = \{z: |z_1| < 1,\ |z_2| < 1\}$, has a holomorphic continuation to the bidisc $\Omega^\varepsilon = \{z: |z_1| < 1 + \varepsilon,\ |z_2| < 1 + \varepsilon\}$, given by the formula

$$F(z_1, z_2) = \frac{1}{2\pi i}\int_{|\zeta_2| = 1}\frac{f(z_1, \zeta_2)\,d\zeta_2}{(\zeta_2 - z_2)}$$

The function $F$ is holomorphic in $\Omega^\varepsilon$ by construction, agrees with $f$ in $\Omega_2^\varepsilon$ by the Cauchy formula, and finally, agrees with $f$ in $\Omega_1^\varepsilon$ by the uniqueness theorem.

Poincaré (1907, see [83]) using the expansion of a function on a sphere by spherical harmonics, showed that each function, holomorphic in a neighbourhood of the boundary of a ball in $\mathbb{C}^2$, extends holomorphically to the interior of this ball.

The more general theorem of Hartogs asserts that if $D$ is a bounded domain in $\mathbb{C}^n$ $(n \geq 2)$ with connected complement $\mathbb{C}^n \setminus D$, then any function holomorphic in a neighbourhood of $\partial D$ uniquely extends to a holomorphic function on $D$. However, a rigourous proof of this assertion was obtained only in 1936 by Brown (see [23]).

These phenomena led Hartogs to the following definition. A domain $D$ in $\mathbb{C}^n$ is called a *domain of holomorphy* if there does not exist a larger domain $\Omega \supset D$ to which every holomorphic function on $D$ extends holomorphically.

Using the classical Cauchy formula as in the above examples, Hartogs showed that for any domain of holomorphy in $\mathbb{C}^n$, the following *continuity principle* holds (see [81], [23]).

**Theorem 0.1** (Hartogs, 1909). *Let $\Omega$ be a domain of holomorphy. If $\{D_\nu\}$ is an arbitrary sequence of analytic discs whose closures are contained in $\Omega$ and such that $\lim_{\nu \to \infty} D_\nu = D_0$, $\partial D_0 \subset \Omega$, where $D_0$ is an analytic disc, then $D_0 \subset \Omega$.*

Here, an *analytic disc* means the holomorphic imbeding in $\mathbb{C}^n$ of the unit disc in $\mathbb{C}^1$.

E. Levi (1910) introduced a notion of domain of meromorphy. He proved that the Hartogs continuity principle is valid for domains of meromorphy as well (see [23], [10]). Following Hartogs and Levi a domain in $\mathbb{C}^n$ is called *pseudoconvex* if the continuity principle from theorem 0.1 is valid for it.

E. Levi (1911) formulated a natural problem: to show that each pseudoconvex domain is a domain of holomorphy. This problem turned out to be exceedingly difficult and was solved by Oka only in 1942 (see §3).

The other principal problem of multidimensional complex analysis was the problem, going back to Weierstrass and Poincaré, of representing an arbitrary meromorphic function in a domain $\Omega \subset \mathbb{C}^n$ as a quotient of functions holomorphic in $\mathbb{C}^n$. It is known (see [65]) that the solution to this problem would have been most welcome to Weierstrass in constructing the theory of Abelian functions. In order to solve this problem, Weierstrass and Poincaré thought it was necessary to have a complete description for the structure of the zeros and poles of meromorphic functions in $\mathbb{C}^n$.

The local properties of the zero sets of holomorphic functions are described by the following result of Weierstrass (see [33]).

**Theorem 0.2** (Weierstrass, 1879). *Let $f = f(z_1, \ldots, z_n)$ be a function holomorphic in a neighbourhood of zero in $\mathbb{C}^n$ and suppose the function $f(0, z_n) \cdot z_n^{-k}$ is* *holomorphic and different from zero at zero. Then, the function $f$ can be represented uniquely in the form $f = g \cdot h$, where $h$ is holomorphic in a neighbourhood of zero, $h(0) \neq 0$, and the function $g$ (a Weierstrass polynomial) is of the form*

$$g(z) = z_n^k + \sum_{j=0}^{k-1} g_j(z') \cdot z_n^j,$$

*where the functions $g_j(z')$, $z' = (z_1, \ldots, z_{n-1})$, are holomorphic in a neighbourhood of zero and $g_j(0) = 0$.*

Let $\Omega$ be a domain in $\mathbb{C}^n$, $f$ a function meromorphic in $\Omega$, and $M_+$ and $M_-$ the sets of zeros and poles respectively of $f$ in $\Omega$. From Theorem 0.2 it follows that the sets $M_\pm$ are the closure of the set of their regular points and the regular points split into irreducible complex $(n-1)$-dimensional components $\{M_\nu^\pm\}$. Let us denote by $\gamma_\nu^\pm$ the *multiplicity of the zeros* ($\gamma_\nu^+ > 0$) or *poles* ($\gamma_\nu^- < 0$) of the function $f$ on a component $M_\nu^\pm$ (see Article III of the present volume).

From Theorem 0.2 follows the following integral representation of Poincaré (1883, 1898) for $\log|f|$:

$$\log|f| = G + H, \tag{0.3}$$

where $H$ is a harmonic functions on $\Omega$ and $G$ is a potential of the form

$$G = C_n \sum_\nu \gamma_\nu^\pm \int_{M_\nu^\pm} |\zeta - z|^{2-2n} dV_{2n-2}(\zeta),$$

where $dV_k$ is the $k$-dimensional volume form on $k$-dimensional submanifolds of $\mathbb{C}^n$ (see [73], [32]).

The representation of a meromorphic function $f$ in $\mathbb{C}^n$ as a quotient of entire functions was accomplished by Weierstrass for $n = 1$ (in 1876), by Poincaré for $n = 2$ (in 1883), and in the general case by Cousin (in 1895) as a consequence of the following result.

**Theorem 0.3** (Weierstrass, Poincaré, Cousin, 1895, see [23], [32], [33]). *Let $M$; be an $(n-1)$-dimensional analytic set in $\mathbb{C}^n$, $\{M_\nu^+\}$ its regular $(n-1)$-dimensional components, and $\{\gamma_\nu^+\}$ the multiplicities of these components. Then, there exists an entire function $f_+$ whose zeros are on $M^+$ with multiplicity $\gamma_\nu^+$, $\nu = 1, 2, \ldots$.*

The approach of Poincaré, which allowed him to obtain only the case $n = 2$, consisted in constructing an entire function satisfying equation (0.3).

The original approach of Cousin consisted in the following successful generalization of the Mittag-Leffler Theorem (1877).

**Theorem 0.4** (Cousin 1895, see [23], [32], [33]). *Let $\{\Omega_j\}$, $j \in J$, be a locally finite cover of a polydomain $D = D_1 \times \ldots \times D_n \subset \mathbb{C}^n$, where each domain $D_k$, with the possible exception of one, is simply connected in $\mathbb{C}^1$. Let $\{\varphi_j\}$ be a family of functions meromorphic on $\{\Omega_j\}$ and such that the functions $\{\varphi_i - \varphi_j\}$ (respectively $\varphi_i \varphi_j^{-1}$) are holomorphic on $\{\Omega_i \cap \Omega_j\}$, $i$, $j \in J$. Then there exists a meromorphic* *function $\varphi$ on $D$ such that all functions $\{\varphi - \varphi_j\}$ (respectively $(\varphi/\varphi_j)^{\pm 1}$ are holomorphic on $\{\Omega_j\}$, $j \in J$.*

For the proof of Theorem 0.4, using the classical Cauchy formula, Cousin established the following fundamental result on the splitting of singularities of holomorphic functions of several complex variables.

**Theorem 0.5** (Cousin, 1895, see [23], [32], [33]). *Let $D$ be a polydomain in $\mathbb{C}^n$ containing the origin. Set*

$$\Omega_1 = \{z \in D : \operatorname{Im} z_1 > -\varepsilon\};$$

$$\Omega_2 = \{z \in D : \operatorname{Im} z_1 < \varepsilon\}$$

$$S = \{z \in D : \operatorname{Im} z_1 = 0\}, \qquad \varepsilon > 0.$$

*Then, any function $f$, holomorphic in the domain $\Omega_1 \cap \Omega_2$ can be represented on $S$ in the form $f = f_1 - f_2$, where $f_1$ and $f_2$ are functions holomorphic in the domains $\Omega_1$ and $\Omega_2$ respectively.*

The problems of extending the assertions of Theorems 0.3 and 0.4 to general domains of holomorphy in $\mathbb{C}^n$ came to be called the *problems of Poincaré* and *Cousin* respectively (see [10]). Just as in the case of the Levi problem, no progress was made on these problems until the works of A. Weil, H. Cartan and Oka (see [10]).

In the end, the principal difficulties arose (see [10]) in attempts to extend to several complex variables, the result of Runge (1885) on the representation of an arbitrary analytic function on a domain $\Omega \subset \mathbb{C}^1$ as the limit of a sequence of rational functions uniformly converging in $\Omega$.

**0.2. A Survey of Results.** The reason for difficulties in proving the general Hartogs Theorem, solving the Levi problem, the Cousin and Poincaré problems, obtaining a multidimensional analog of Runge's approximation theorem, and solving a series of other problems in the theory of functions of several complex variables was tied to the absence of a natural analog in several variables of the Cauchy formula for a more general class of domains than polydiscs.

Finally, in the thirties two sufficiently general formulas were found which played a remarkable role in the development of multidimensional complex analysis (A. Weil, 1932; Bergman 1934; Martinelli, 1938; Bochner, 1943).

A simple formula with non-holomorphic kernel, found by Martinelli and Bochner, allowed one to give a simple proof of the Hartogs–Brown Theorem and to obtain a generalization of this theorem leading to the theory of *CR*-functions. These results are dealt with in section §1.

A deeper formula, with holomorphic kernal, found by A. Weil and Bergman for functions holomorphic in analytic polyhedra allowed one: first of all, to obtain a multidimensional Runge approximation theorem (A. Weil, 1932); secondly, to solve the problems of Cousin, Poincaré and Levi (H. Cartan, 1934;

Oka, 1936, 1953; Norguet, 1954; Bremermann, 1954); and thirdly, to give a fundamental formulation for the Oka and H. Cartan's, theory of ideals of analytic functions, 1940–1950.

In this connection Oka (1953) remarked that in the theory of functions of several complex variables, for a long time problems were only accumulated; A. Weil made the first step in the opposite direction, i.e. in the direction of solving these problems. We remark that the majority of the following steps were made by Oka himself. An account of the results related to Weil's formula is given in sections §2 and §3.

In the fifties H. Cartan (1951) and Grauert (1958) using the sheaf theory of Leray not only obtained far reaching generalizations of Oka's theorems but also completely banished the constructive method of integral representations from multidimensional complex analysis (see [33]).

In the sixties further developments in the Oka–Cartan–Grauert theory took place essentially thanks to the methods of a priori $L^2$-estimates for the $\bar{\partial}$-Neuman–Spencer problems, developed in the works of Morrey (1958), Kohn (1963) and Hörmander (1965) (see [22], [45], and below §6).

The constructive method of integral representations was revived anew essentially in the seventies. Here the source of fundamental new ideas was the works of Leray (1956) and Lelong (1953).

In particular, Leray starting from an idea of Fantappiè found a completely general integral formula for functions holomorphic on arbitrary domains, which as became clear later, contains in itself the cases of the formulas of Bochner, Martinelli, Weil, and others.

The formula of Cauchy–Fantappiè–Leray allowed one to construct (Grauert, G.M. Khenkin, Lieb, Ramirez, 1969) a good integral formula, in strictly pseudo-convex domains $D$ of $\mathbb{C}^n$, for holomorphic functions and for solutions $f$ of the non-homogeneous Cauchy–Riemann equation: $\bar{\partial} f = g$, where $g$ is a (0, 1)-form in $D$ (see [19], [40]).

These integral representations led to the theorem on uniform approximation by functions holomorphic on the compact set $\bar{D}$ (G.M. Khenkin, 1969, 1974); to the formula for the continuation, with uniform estimates, of functions holomorphic on submanifolds of $D$ (G.M. Khenkin, Leiterer, Amar, 1972, 1980); to decompositions of functions in an ideal in terms of generators (Berndtsson, 1982) etc. (see [11] and [40]). These results are explained in §§4, 5, 6.

Lelong transformed the Poincaré equation (0.3) to a $\bar{\partial}$-equation of the form:

$$\bar{\partial} \frac{df}{f} = i[M],$$

called the *Poincaré–Lelong equation*, and he found for it a fundamental integral formula, analogous to the canonical Weierstrass product, for entire functions with finite order of growth and having zeros on a given $(n-1)$-dimensional analytic set in $\mathbb{C}^n$. A similar construction was obtained independently by Stoll (1953), based on the work of Kneser (1938) (see [73], [32] and [78]).

In the elaboration of these works certain integral representations were found (G.M. Khenkin, Skoda, Sh.A. Dautov, P.L. Polyakov, Charpentier, 1975–1984) for solutions of the $\bar{\partial}$-equations on strictly pseudoconvex domains and analytic polyhedra which allowed a complete description of the zero sets of functions of finite order on these domains (see [49] and [41]). A survey of these results is given in §6.

Integral representation is strictly pseudoconvex domains were also successfully applied (Kerzman, Stein, Ligocka, Lieb, Range, 1978–1984) towards constructing a "parametrix" for the $\bar{\partial}$-Neuman–Spencer problem and to obtain the asymptotic expansion of Fefferman (1974) for the Bergman and Szegö kernels in such domains (see §6).

The method of integral representations turned out to be particularly fruitful in the theory of *CR*-functions, i.e., functions on real submanifolds of $\mathbb{C}^n$ satisfying the tangential Cauchy–Riemann equations. This theory, which has various applications to differential equations and mathematical physics, previously contained (see [82]) disparate although effective results (S.N. Bernstein, Bochner, H. Lewy, Bishop, Rossi, Harvey, Wells, Lawson, Sibony, 1912–1977).

In recent years, thanks to the construction of integral representations for *CR*-functions (Baouendi, G.M. Khenkin, Trèves, 1980, 1981), the theory of *CR*-functions has received significant growth: holomorphic approximation [8]; analytic representation [38]; uniqueness theorems [2], [8]; propagation of singularities [34]; holomorphic continuation [39], [2], [15], [7]; and theorems on the "edge of the wedge" for *CR*-functions [2], [3].

A survey of these results is given in §7 of the present article (see also [3]).

In sections 1–7 of the present work, then, the method of integral representations is set forth only in applications to problems of the theory of functions on domains in the space $\mathbb{C}^n$ and on submanifolds thereof. In this connection primary attention is given to those classical problems of the theory of functions which in the first place were solved or arose in the context of integral representations of holomorphic functions and which once again have developed in recent years.

However, the method of integral representations works effectively also outside the framework of classical functions theory in $\mathbb{C}^n$.

First of all, the above mentioned results on holomorphic functions and *CR*-functions have meaningful generalizations to the case of differential forms of arbitrary type.

In particular, explicit formulae were found (Lieb, Øvrelid, P.L. Polyakov, Range, Siu, Sh.A. Dautov, A.B. Romanov, 1970–1975) for solutions of the equation $\bar{\partial}f = g$, where $f$ and $g$ are differential forms of respective types $(0, q)$ and $(0, q+1)$ on piecewise pseudoconvex domains, analytic polyhedra, or their boundaries. An essential role here was played by the analog, for differential forms, of the Bochner–Martinelli formula, constructed earlier by Koppelman (1967) (see [1], [54], [49], [40], and §8 of the present article).

Certain problems in complex integral geometry (Andreotti, Norguet, Penrose, 1967–1976) and in the theory of tangential Cauchy–Riemann equations (H. Lewy, Kohn, Andreotti, Hill, Naruki, 1957–1972) required integral representations for $(0, q)$-forms not on pseudoconvex domains in $\mathbb{C}^n$ but rather on $q$-pseudoconcave domains in $\mathbb{CP}^n$ and on $q$-pseudoconcave $CR$-manifolds.

The construction of integral formulae on such domains and manifolds (G.M. Khenkin, S.G. Gindikin, P.L. Polyakov, 1977–1984) led to an exact criterion for the local and global solvability of the inhomogeneous Cauchy–Riemann equations on $q$-concave domains and on $q$-concave $CR$-manifolds, and also led to the description of the Radon–Penrose transform and its kernel (see [49], [28], [52], [49] and [51]). Section 8 of the present article is devoted to these questions.

The author extends his profound gratitude to P. Lelong, A.G. Vitushkin, S.G. Gindikin, and P.L. Polyakov for their interest in the present work and their help in its realization.

## §1. The Bochner–Martinelli Formulas and Their Applications

**1.1. The Bochner–Martinelli Formula and the Hartogs Theorem.** Let $D$ be a bounded domain in $\mathbb{C}^n$ with "rectifiable" boundary $\partial D$ (i.e., the $(2n-1)$-dimensional Hausdorff measure of $\partial D$ is finite). We choose the orientation of the space $\mathbb{C}^n$ such that

$$(-i)^n \int_{\zeta \in D} \omega(\bar{\zeta}) \wedge \omega(\zeta) > 0, \quad \text{where } \omega(\zeta) = d\zeta_1 \wedge \ldots \wedge d\zeta_n.$$

For vectors $\zeta, \eta \in \mathbb{C}^n$, we set

$$\langle \eta, \zeta \rangle = \sum_{j=1}^{n} \eta_j \cdot \zeta_j.$$

In most integral representations, a differential form $\omega'$ on $\mathbb{C}^n$ is used which has the form

$$\omega'(\eta) = \sum_{k=1}^{n} (-1)^{k-1} \eta_k \, d\eta_1 \wedge \ldots \wedge d\eta_{k-1} \wedge d\eta_{k+1} \wedge \ldots \wedge d\eta_n,$$

and satisfies the equation $d\omega'(\eta) = n \cdot \omega(\eta)$.

We consider the spaces $L^p(\partial D)$ and $L^p(D)$, consisting of functions whose $p$th powers are integrable with respect to the respective Lebesgue measures on $\partial D$ and $D$.

Let $H^p(\partial D)$ (respectively $H^p(D)$) be the subspaces of $L^p(\partial D)$) (respectively $L^p(D)$) having a holomorphic continuation to (respectively, which are holomorphic in) $D$.

Let $\mathscr{H}(D)$ denote the space of all holomorphic functions on $D$ and $C(D)$ the space of all continuous functions on $D$.

By $C^{(\alpha)}(\bar{D})$ we denote the space of all continuous functions on $\bar{D}$ which have continuous derivatives up to order $[\alpha]$ inclusively, satisfying a Lipschitz condition of order $\alpha-[\alpha]$.

We set $A^{(\alpha)}(\bar{D}) = C^{(\alpha)}(\bar{D}) \cap \mathscr{H}(D)$ and $A(\bar{D}) = C(\bar{D}) \cap \mathscr{H}(D)$.

**Theorem 1.1** (Martinelli, Bochner, 1943). *For any bounded domain $D$ with rectifiable boundary and for every function $f$ such that $f$ and $\bar{\partial} f$ are continuous on $\bar{D}$, we have the equality*

$$f(z) = \frac{(n-1)!}{(2\pi i)^n}\left[\int_{\partial D} f(\zeta)\frac{\omega'(\bar{\zeta}-\bar{z})\wedge\omega(\zeta)}{\langle\bar{\zeta}-\bar{z},\zeta-z\rangle^n} - \int_{D} \bar{\partial} f(\zeta)\frac{\omega'(\bar{\zeta}-\bar{z})\wedge\omega(\zeta)}{\langle\bar{\zeta}-\bar{z},\zeta-z\rangle^n}\right], \tag{1.1}$$

*for $z \in D$.*

This formula was proved, first of all, by Martinelli (1938) for the ball in $\mathbb{C}^n$, and then for general domains, by Martinelli and Bochner (1943), for the case $\bar{\partial} f = 0$, and by Koppelman (1967), when $\bar{\partial} f \neq 0$ (see [4] and [40]).

To prove formula 1.1 it is sufficient to apply the Cauchy–Green formula (0.1) to $f$ on each complex line passing through $z$, and then to average the equalities so obtained (see [31]).

If $f$ is a holomorphic function on $D$, then formula (1.1) becomes the classical integral representation of Martinelli–Bochner

$$f(z) = \frac{(n-1)!}{(2\pi i)^n}\int_{\partial D} f(\zeta)\frac{\omega'(\bar{\zeta}-\bar{z})\wedge\omega(\zeta)}{|\zeta-z|^{2n}}, \qquad z \in D. \tag{1.2}$$

We shall consider that a domain $D$ has a smooth boundary if

$$D = \{z \in \mathbb{C}^n : \rho(z) < 0\}, \tag{1.3}$$

where $\rho$ is a function of class $C^1(\mathbb{C}^n)$ with the property grad $p \neq 0$ on $\partial D$.

For domains with smooth boundary, Bochner obtained the equality (1.2) from the Green formula for a function $F$ harmonic on $D$:

$$F(z) = \frac{(n-1)!}{\pi^n}\int_{\partial D}\frac{F(\zeta)}{|\zeta-z|^{2n}}\left\langle \bar{\zeta}-\bar{z}, \frac{\partial\rho}{\partial\zeta}\right\rangle\frac{dV_{2n-1}(\zeta)}{|\operatorname{grad}\rho(\zeta)|} +$$

$$+\frac{(n-2)!}{\pi^n}\int_{\partial D}\frac{\left\langle \frac{\partial F}{\partial\bar{\zeta}}(\zeta), \frac{\partial\rho}{\partial\zeta}(\zeta)\right\rangle}{|\zeta-z|^{2n-2}}\frac{dV_{2n-1}(\zeta)}{|\operatorname{grad}\rho(\zeta)|}1$$

where

$$\frac{\partial\rho}{\partial\zeta} = \left(\frac{\partial\rho}{\partial\zeta_1},\ldots,\frac{\partial\rho}{\partial\zeta_n}\right), \qquad z \in D.$$

From this formula, it follows that the Martinelli–Bochner formula (1.2) holds if and only if $F$ is harmonic on $D$ and satisfies the boundary $\bar{\partial}$-condition of

Neuman–Spencer

$$\left\langle \frac{\partial F}{\partial \bar{\zeta}}(\zeta), \frac{\partial \rho}{\partial \zeta}(\zeta) \right\rangle = 0 \text{ on } \partial D. \tag{1.4}$$

The $\bar{\partial}$-condition of Neuman (1.4) forces an arbitrary harmonic function to be holomorphic, since

$$\int_D \left|\frac{\partial F}{\partial \bar{\zeta}}\right|^2 dV_{2n} = \int_{\partial D} \bar{F} \left\langle \frac{\partial F}{\partial \bar{\zeta}}(\zeta), \frac{\partial \rho}{\partial \zeta}(\zeta) \right\rangle \frac{dV_{2n-1}}{|\operatorname{grad} \rho|} = 0$$

(in this connection, see [22]).

As a consequence we have the following (see [4]).

**Theorem 1.1′** (A.M. Aronov, A.M. Kytmanov, L.A. Ajzenberg, 1975). *If for some function $f \in C(\bar{D})$, the Martinelli–Bochnner formula* (1.2) *is satisfied, then $f \in A(\bar{D})$.*

The Martinelli–Bochner formula was used by the authors to prove a more general and sharper version of the Hartogs Theorem (see [4] and [19]).

**Theorem 1.2** (Bochner, 1943; Weinstock, 1970). *Let D be a bounded domain with rectifiable boundary and with connected complement in $\mathbb{C}^n$, $n > 1$, and let $f \in L^1(\partial D)$. Then a necessary and sufficient condition in order that f extends holomorphically to D is that f satisfy the tangential Cauchy–Reimann equations: $\bar{\partial}_\tau f = 0$.*

In case $f$ is a smooth function and the domain $D$ has a smooth boundary, then the equation $\bar{\partial}_\tau f = 0$ signifies the equation

$$\bar{\partial}\tilde{f} \wedge \bar{\partial}\rho = 0 \text{ on } \partial D, \tag{1.5}$$

where $\tilde{f}$ is any smooth extension of $f$ to a neighbourhood of $\partial D$.

In the general situation, the equation $\bar{\partial}_\tau f = 0$ is to be understood in the generalized sense:

$$\int_{\partial D} f \wedge \bar{\partial}\varphi = 0 \tag{1.6}$$

for any smooth differential form $\varphi$ of order $(2n-2)$.

Theorem 1.2 is proved by the following scheme. Let $f$ be a fixed integrable function on $\partial D$ and set

$$\frac{(n-1)!}{(2\pi i)^n} \int_{\zeta \in \partial D} f(\zeta) \frac{\omega'(\bar{\zeta}-\bar{z}) \wedge \omega(\zeta)}{|\zeta - z|^{2n}} = \begin{cases} f_+, & \text{if } z \in D, \\ f_-, & \text{if } z \in \mathbb{C}^n \setminus D. \end{cases} \tag{1.7}$$

We have the following analog of the classical theorem of Sokhotskij (see [4], [49], [35]).

$$f = f_{+|\partial D} - f_{-|\partial D}. \tag{1.8}$$

Moreover, from the condition $\bar{\partial} f = 0$ on $\partial D$ it follows that the functions $f\pm$ are holomorphic in $z$ despite the non-holomorphicity of the Martinelli–Bochner kernel. Indeed,

$$\bar{\partial} f_{\pm} = \frac{(n-1)!}{(2\pi i)^n} \int_{\partial D} f(\zeta)\bar{\partial}_z \omega'\left(\frac{\bar{\zeta}-\bar{z}}{|\zeta - z|^2}\right) \wedge \omega(\zeta) =$$

$$-\frac{(n-1)!}{(2\pi i)^n} \int_{\partial D} f(\zeta)\bar{\partial}_\zeta \omega'\left(\frac{\bar{\zeta}-\bar{z}}{|\zeta - z|^2}\right) \wedge \omega(\zeta) = 0.$$

Finally, to complete the proof of sufficiency in Theorem 1.2, we must verify that $f_- \equiv 0$ in $\mathbb{C}^n \backslash D$. Because of the connectedness of the domain $\mathbb{C}^n \backslash D$, it is sufficient to establish this equality for $z \in \mathbb{C}^n \backslash \Omega$, where $\Omega$ is some polydisc containing the domain $D$.

Noting that for $z \in \mathbb{C}^n \backslash \Omega$, the form $\omega'(\bar{\zeta} - \bar{z})/|\zeta - z|^2) \wedge \omega(\zeta)$ is $\bar{\partial}$-exact in the polydisc $\Omega$, i.e. has the form $\bar{\partial}_\zeta \varphi_z(\zeta)$, where $\varphi_z(\zeta)$ is some $(2n-2)$-form, we deduce from equations (1.6) and (1.7) that $f_-(z) = 0$. The assertion of Theorem 1.2 follows from this and from equation 1.8.

Theorem 1.2 has an explicitly geometrical reformulation: under the hypotheses of this theorem, the $(2n-1)$-dimensional graph

$$\Gamma_f = \{(w, z) \in \mathbb{C}^{n+1} : w = f(z), z \in \partial D\}$$

is the boundary of an $n$-dimensional manifold

$$G_f = \{(w, z) \in \mathbb{C}^{n+1} : w = f_+(z), z \in D\}$$

if and only if $\bar{\partial}_\tau f = 0$ on $\partial D$.

A very careful geometric analysis together with the Bochner–Martinelli formula allowed Harvey and Lawson to reformulate the above version of the Bochner–Hartogs Theorem as the following brilliant result on bounding complex manifolds by odd-dimensional cycles (see [35]).

**Theorem 1.2′** (Harvey, Lawson, 1974). *Let $M$ be a compact $(2k-1)$-dimensional $(k > 1)$ smooth manifold in $\mathbb{C}^n$. Then $M$ is the boundary of some $k$-dimensional analytic set if and only if $M$ is a maximally-complex manifold, i.e. for each point $\zeta \in M$, the real tangent space $T_\zeta M$ contains a $(k-1)$-dimensional complex subspace.*

A more expanded formulation of this result can be found in the article III by Chirka.

We have enunciated the most classical applications of the Martinelli–Bochner formula. In recent years the Martinelli–Bochner formula has been successfully used in a series of questions in the theory of multidimensional residues. On this basis, for example, one can give an elementary proof for the multidimensional logarithmic residue formula (see [4]) and for the Lefschetz formula for the number of fixed points of a holomorphic mapping (see [31]), (see also article V by Dolbeault and Vol. 8, article I by L.A. Ajzenberg, A.P. Yuzhakov, and A.K. Tsikh on the theory of multidimensional residues).

Although the formula of Martinelli–Bochner is completely universal and for $n = 1$ coincides exactly with the Cauchy–Green formula, it has an essential draw-back compared to the Cauchy formula: the kernel in (1.1), (1.2), for $n > 1$, is not holomorphic in the $z$ variable.

Nevertheless, for the ball in $\mathbb{C}^n$, Bochner deduced immediately from (1.2) the following formula with holomorphic kernel (see [22]).

**Theorem 1.3** (Bochner, 1943). *Let $D$ be the unit ball in $\mathbb{C}^n$*, i.e. $D = \{z \in \mathbb{C}: |z| < 1\}$, *and let $f \in A(\bar{D})$. Then, for any $z \in D$, we have the equalities*

$$f(z) = K^0 f(z) = \frac{(n-1)!}{(2\pi i)^n} \int_{\zeta \in \partial D} f(\zeta) \frac{\omega'(\bar{\zeta}) \wedge \omega(\zeta)}{(1 - \bar{\zeta} \cdot z)^n}, \tag{1.9}$$

$$f(z) = K^1 f(z) = \frac{n!}{(2\pi i)^n} \int_{\zeta \in D} f(\zeta) \frac{\omega(\bar{\zeta}) \wedge \omega(\zeta)}{(1 - \bar{\zeta} \cdot z)^{n+1}}. \tag{1.10}$$

For the proof of formula (1.9) let us consider a function of the type

$$F(z, w) = \frac{(n-1)!}{(2\pi i)^n} \int_{\partial D} f(\zeta) \frac{\omega'(\bar{\zeta} - w) \wedge \omega(\zeta)}{\langle \bar{\zeta} - w, \zeta - z \rangle^n}$$

holomorphic on the product of balls $\{(z, w) \in \mathbb{C}^{2n}: |z| < 1, |w| < 1\}$. On account of the Martinelli–Bochner formula (1.2), this function coincides with the function $f(z)$ on the real subspace $w = \bar{z}$.

From an elementary uniqueness theorem, we have $F(z, w) \equiv f(z)$, for $|z| < 1$, $|w| < 1$. In particular, $F(z, 0) \equiv f(z)$, for $z \in D$. The last equality is precisely (1.9). Formula (1.10) follows from (1.9) with the help of Stokes formula.

Bochner and Bergman showed that the operators $f \mapsto K^0 f$ and $f \mapsto K^1 f$, given by the integrals (1.9) and (1.10), are orthogonal projections of the spaces $L^2(\partial D)$ and $L^2(D)$ respectively onto the spaces $H^2(\partial D)$ and $H^2(D)$ (see [4] and [23]).

**1.2. The Integral Representations of Bochner and Hua Loo-keng on Classical Domains.** Let $D$ be an arbitrary bounded domain in $\mathbb{C}^n$. For a point $z \in D$ we consider the functional $z \mapsto f(z)$ on $H^2(D)$. By the Riesz Theorem we have

$$f(z) = \int_D f(\zeta) B(\zeta, z)\, dV_{2n}(\zeta) = Bf, \tag{1.11}$$

where $\bar{B}(\cdot, z) \in H^2(D)$ and $B(\zeta, z) = \overline{B(z, \zeta)}$. Bergman (1933, see [24] and [58]) showed that the operator $f \mapsto Bf$ is the orthogonal projection of $L^2(D)$ onto $H^2(D)$. The operator $B$ is called the *Bergman projection* and the kernel $B(\zeta, z)$ is called the *Bergman kernel function* for the domain $D$.

For a bounded domain $D$ in $\mathbb{C}^n$ we denote by $S(D)$ the smallest closed subset on $\partial D$ on which each function holomorphic in $D$ and continuous in $\bar{D}$ assumes its maximum (*the Bergman–Shilov boundary*).

If $D$ is the ball in $\mathbb{C}^n$, then $S(D) = \partial D$. If $D = D_1 \times \ldots \times D_n$ is a polydomain in $\mathbb{C}^n$, then

$$S(D) = (\partial D_1) \times \ldots \times (\partial D_n) \subset \partial D.$$

Let $L^2(S(D), d\mu)$ denote the space of functions $f$ square integrable with respect to the measure $d\mu$ on $S(D)$ and $H^2(S(D), d\mu)$ the subspace of functions in $L^2(S(D), d\mu)$ admitting a holomorphic continuation to $D$.

The kernel in the Cauchy formula (0.2) for the polydisc $D = \{z \in \mathbb{C}^n : |z_j| < 1, j = 1, \ldots, n\}$ gives the orthogonal *Szegö projection* from $L^2(S(D), dV_n(\zeta))$ onto $H^2(S(D), dV_n(\zeta))$, where $dV_n$ is Lebesgue measure on $S(D)$. Gleason (1962, see [4]) showed that for any domain $D$, there exists a positive measure $d\mu$ on $S(D)$, such that the kernel $K(\zeta, z)$ of the orthogonal projection of $L^2(S(D), d\mu)$ onto $H^2(S(D), d\mu)$ is holomorphic in $z \in D$ and $\mu$-integrable in $\zeta \in S(D)$. Thus,

$$f(z) = \int_{\zeta \in S(D)} f(\zeta) K(\zeta, z)\, d\mu(\zeta), \quad z \in D, \tag{1.12}$$

for any function $f \in H^2(S(D), d\mu)$.

In applications, the Cauchy formulas of type (0.2), (1.9) or (1.10) work effectively, not so much because their kernels give Szegö or Bergman projections, but rather because these kernels are holomorphic in the outer variable and have an explicit formula allowing a precise analysis of its singularities.

In the abstract formulas of Gleason (1.12) and Bergman (1.11), there is not enough information about the singularities of the kernels and so it is very difficult to apply these formulas.

For classical homogeneous domains of holomorphy, including the ball and the polydisc, the kernels of Szegö and Bergman have been explicitly calculated by Bochner (1944) and Hua Loo-keng (1958, see [46]).

A *bounded domain* $D \subset \mathbb{C}^n$ is said to be *classical* if the entire group of its analytic automorphisms (one-to-one mappings of the domain onto itself) is a classical Lie group acting transitively on $D$.

According to the classification of E. Cartan (1936) there are four types of irreducible classical domains (see [71]).

A classical *domain of the first type* is a domain $\Omega^1_{p,q}$ in $\mathbb{C}^{pq}$, consisting of all complex $p \times q$ matrices $Z$ such that the matrix $I - Z^* Z$ is positive definite ($\gg 0$), where $p \geq q \geq 1$, $I$ is the unit matrix, and $Z^* = \bar{Z}'$ is the conjugate matrix of $Z$. The Bergman–Shilov boundary of a domain $\Omega^1_{p,q}$ has real dimension $q(2p-1)$ and consists of matrices of the form

$$S(\Omega^1_{p,q}) = \{Z : Z^* Z = 1\}.$$

A classical domain of the *second* (respectively *third*) type is a domain $\Omega^2_{p,p}$ respectively $\Omega^1_{p,p}$) in the space $\mathbb{C}^{p(p+1)/2}$ (respectively $\mathbb{C}^{p(p-1)/2}$), consisting of complex symmetric (respectively skew-symmetric) matrices of order $p$, satisfying the condition $I - Z^* Z \gg 0$ (respectively $I + Z^* Z \gg 0$). The manifold $S(\Omega^2_{p,p})$ has real dimension $p(p+1)/2$ and consists of all symmetric unitary matrices of order $p$. The manifold $S(\Omega^3_{p,p})$ has dimension $(p-1)(p+1)/2$ if $p$ is even and consists of all skew-symmetric unitary matrices of order $p$. If $p$ is odd, then $S(\Omega^3_{p,p})$ has dimension $p(p-1)/2$ and consists of matrices of the form $UDU'$, where $U$ is an arbitrary unitary matrix and

$$D = \begin{pmatrix} 0 & 0 \\ -1 & 0 \end{pmatrix} \oplus \ldots \oplus \begin{pmatrix} 0 & 1 \\ -1 & 0 \end{pmatrix} \oplus 0.$$

Finally, a classical *domain of the fourth type* $\Omega^4_n$ consists of vectors $z = (z_1, \ldots, z_n) \in \mathbb{C}^n$, $n > 2$, satisfying the condition

$$|\langle z, z\rangle|^2 + 1 - 2\langle \bar{z}, z\rangle > 0; |\langle z, z\rangle| < 1.$$

The manifold $S(\Omega^4_n)$ has dimension $n$ and consists of vectors of the form $e^{i\theta}\cdot x$, with $\theta \in [0, \pi], x \in S_n$, where $S_n$ is the unit sphere in $\mathbb{R}^n$.

**Theorem 1.4** (Bochner, 1944; Hua Loo-keng, 1958). *Let $\Omega$ be a simple classical domain and $f \in H^2(S(\Omega), dV)$, where $dV$ is the Haar measure on $S(\Omega)$. Then, in the Cauchy–Szegö formula* (1.12), *the kernel $K(\zeta, z)$ has the following form:*

*for domains $\Omega^1_{p,q}$*

$$K(\zeta, Z) = [\det(I - \bar{\zeta}\cdot Z)]^{-p};$$

*for domains $\Omega^2_{p,p}$*

$$K(\zeta, Z) = [\det(I - \bar{\zeta}\cdot Z)]^{-(p+1)/2};$$

*for domains $\Omega^3_{p,p}$*

$$K(\zeta, Z) = [\det(I + \bar{\zeta}\cdot Z)]^{[p/2 - 1/4 - (-1)^p/4]};$$

*for domains $\Omega^4_n$*

$$K(\zeta, z) = \langle x - e^{i\theta} z, x - e^{-i\theta} z\rangle^{-n/2};$$

*where $\zeta = e^{i\theta}\cdot x$.*

**Theorem 1.5** (Hua Loo-keng, 1958). *Let $\Omega$ be an irreducible classical domain and $f \in H^2(\Omega)$. Then the kernel $B(\zeta, Z)$ in the Cauchy-Bergman formula* (1.11) *has the following form:*

*for domains $\Omega^1_{p,q}$,*

$$B(\zeta, Z) = V^{-1}(\Omega^1_{p,q})\cdot[\det(I - \bar{\zeta}\cdot Z)]^{-(p+q)};$$

*for domains $\Omega^2_{p,p}$,*

$$B(\zeta, Z) = V^{-1}(\Omega^2_{p,p})[\det(I - \bar{\zeta}\cdot Z)]^{-(p+1)};$$

*for domains $\Omega^3_{p,p}$,*

$$B(\zeta, Z) = V^{-1}(\Omega^3_{p,p})[\det(I + \bar{\zeta}\cdot Z)]^{-(p-1)};$$

*for domains* $\Omega_n^4$,

$$B(\zeta, z) = V^{-1}(\Omega_n^4)(1 + |\langle\zeta, z\rangle|^2 - 2\langle\bar{\zeta}, z\rangle)^{-n},$$

*where* $V(\Omega)$ *is the volume of* $\Omega \subset \mathbb{C}^n$ *in the Bergman metric* $\sum_{j,k} b_{j,k}(z)\, dz_j \cdot d\bar{z}_k$, *with*

$$b_{j,k} = \frac{\partial^2 \ln B(z, z)}{\partial z_j \cdot \partial \bar{z}_k}.$$

The formulas of Bochner–Hua Loo-keng allow one to prove the following significantly more precise form of the Hartogs–Bochner Theorem for classical domains (see [75]).

**Theorem 1.6** (Schmid, 1970; Naruki, 1970; Rossi, Vergne, 1976). *Let* $\Omega$ *be an irreducible classical domain in* $\mathbb{C}^n$ *with the property* $\dim_{\mathbb{R}} S(\Omega) > n$. *In order for a function* $f$ *in* $L^1(S(\Omega))$ *to satisfy the tangential Cauchy–Riemann equations* $\bar{\partial}_\tau f = 0$ *on* $S(\Omega)$, *it is necessary and sufficient that* $f \in H^1(S(\Omega))$.

We remark that if a classical domain $\Omega$ is such that $\dim_{\mathbb{R}} S(\Omega) = n$, then $S(\Omega)$ has no complex tangent vectors and consequently, there are also no tangential Cauchy–Riemann equations. Theorems 1.4 and 1.5 were used by Hua Loo-keng [46] in order to solve the Dirichlet problem on an arbitrary classical domain $\Omega$ for functions satisfying the Laplace equation with respect to the Bergman metric on $\Omega$.

**Theorem 1.7** (Hua Loo-keng, 1958). *Let* $\Omega$ *be an irreducible classical domain in* $\mathbb{C}^n$ *and* $\Delta_\Omega$ *the Laplace operator on* $\Omega$ *for the Bergman metric, i.e.*

$$\Delta_\Omega = \sum_{j,k} b^{j,k}(z) \frac{\partial^2}{\partial z_j \partial \bar{z}_k},$$

*where* $\{b^{j,k}\}$ *is the inverse matrix of* $\{b_{j,k}\}$. *Let* $K(\zeta, z)$ *be the Cauchy–Szegö kernel for* $\Omega$. *Then, if* $f$ *is any continuous function on* $S(\Omega)$, *there exists a function (unique moreover)* $F \in C(\bar{\Omega})$ *such that* $\Delta_\Omega F = 0$ *and* $F|_{S(\Omega)} = f$. *In addition, the function* $F$ *can be represented by the following Poisson formula:*

$$F(z) = \int_{\zeta \in S(\Omega)} f(\zeta) P(\zeta, z)\, dV(\zeta),$$

*where* $P(\zeta, z) = (K(z, z))^{-1}\, |K(\zeta, z)|^2$.

For further properties of Cauchy and Poisson type integrals, see the papers of volume 8, III, IV.

There are fundamental problems of multi-dimensional complex analysis connected with non-homogeneous domains or manifolds. In this situation it is possible to construct effective analogs of the Cauchy formula which as a rule let go of the requirement that the kernel yield an orthogonal projection of Szegö or Bergman.

## §2. The Weil Formula and the Oka–Cartan Theory

**2.1. Integral Representations in Analytic Polyhedra.** Among (non-homogeneous) domains of holomorphy, analytic polyhedra present particular interest.

A domain $D$ in $\mathbb{C}^n$ is called an *analytic polyhedron* if it can be represented in the form

$$D = \{z \in \Omega: |F_j(z)| < 1, \quad j = 1, 2, \ldots, N\}, \tag{2.1}$$

where $\{F_j\}$ are functions holomorphic in some domain $\Omega \supset \supset D$.

The role of analytic polyhedra is explained by the following important assertion (see [23], [33] and [40]).

**Theorem 2.1** (H. Cartan, Thullen, 1932; Behnke, Stein, 1938). *A domain $\Omega$ in $\mathbb{C}^n$ is a domain of holomorphy if and only if $\Omega$ can be approximated from the interior by analytic polyhedra.*

The principal advance in multidimensional complex analysis was the construction (A. Weil (1932. 1935), Bergman (1934, 1936)) of an authentic analog to the Cauchy formula for analytic polyhedra (see [81], [23] and [40]).

The boundary of an analytic polyhedron of the type (2.1) is the union of the hypersurfaces

$$\sigma_j = \{z \in \partial D: |F_j| = 1\}, \quad j = 1, 2, \ldots, N.$$

An analytic polyhedron is called a *Weil Polyhedron* if the intersection of any $k$ hypersurfaces $\sigma_J = \sigma_{j1} \cap \sigma_{j2} \cap \ldots \sigma_{jk}$ has dimension no greater than $2n - k$.

We give to the real analytic manifold (possibly with singularities) $\sigma_j$ the orientation induced by the orientation of the domain $D$. Further, by induction, we give to the manifold $\sigma_J$ the orientation induced by the orientation of $\sigma_j$. Weil (1935) considered the hypothesis that an expansion

$$F_j(\zeta) - F_j(Z) = \langle P_j(\zeta, z), \zeta - z \rangle, \tag{2.2}$$

holds, where $P_j = (P_j^1, \ldots, P_j^n)$ is a holomorphic vector function of the variables $\zeta, z \in \Omega$.

**Theorem 2.2** (A. Weil, 1935). *Let $D$ be an analytic polyhedron of the form* (2.1) *with the property* (2.2). *Then any function $f \in A(\bar{D})$ can be represented in the form*

$$f(z) = \sum_{|J| = n}' f_J(z), \tag{2.3}$$

*where*

$$f_J(z) = \int_{\sigma_J} f(\zeta)\Phi_J(\zeta, z)\omega(\zeta),$$

$$\Phi_J(\zeta, z) = \frac{(-1)^{n(n+1)/2}}{(2\pi i)^n} \frac{\det[P_{j^1}, \ldots, P_{j^n}]}{\prod_{r=1}^{n} (F_{j^r}(\zeta) - F_{j^r}(z))},$$

*the sum $\Sigma'$ in* (2.3) *being taken over strictly monotonic multiindices $J$ of length $n$.*

Bergman obtained a result which is similar but less effective. We remark that the integration in (2.3) is not over the entire boundary of the polyhedron $D$, but only on the $n$-dimensional part $S \subset \partial D$ consisting of the union of those $\sigma_j$ on which the form $\omega(\zeta) \neq 0$.

The set $S$ coincides (Hoffman, Rossi, 1962) with the Bergman–Shilov boundary $S(D)$ of the polyhedron $D$ (see [25]).

A *Weil polyhedron* (2.1) is said to be *complex non-degenerate* if for each monotone multi-index $J = (j_1, \ldots, j_n)$, we have

$$dF_{j_1} \wedge \ldots \wedge dF_{j_n} \neq 0 \text{ on } \sigma_J. \tag{2.4}$$

If $D$ is a complex non-degenerate polyhedron, then

$$S(D) = \bigcup_{|J|=n}' \sigma_J,$$

We call the $\delta$-extension of the distinguished boundary $S(D)$ the set

$$S_\delta(D) = \bigcup_{|J|=n}' \sigma_{J,\delta},$$

where

$$\sigma_{J,\delta} = \{\zeta \in \Omega;\ |F_j(\zeta)| = t, \forall j \in J; |F_j(\zeta) \leq t, \forall j \notin J, 1 \leq t \leq 1 + \delta\}.$$

A.G. Vitushkin (1968) suggested the following effective modification of the Weil formula for non-degenerate polyhedra (see [69]).

**Theorem 2.2′.** *Let $D$ be a non-degenerate Weil polyhedron of type* (2.1), (2.2) (2.4), *and $f$ a continuous function, holomorphic on $D$, and with support in the domain*

$$D_\delta = \{z \in \Omega\colon\ |F_j(z)| < 1 + \delta, \quad j = 1, 2, \ldots, N\},$$

*where $\delta > 0$ is sufficiently small. Then, for $z \in D$ the formula*

$$f(z) = \sum_{|J|=n}' f_{J,\sigma}(z) \tag{2.3′}$$

*holds, where*

$$f_{J,\delta}(z) = \int_{\sigma_{J,\sigma}} \bar{\partial} f(\zeta) \wedge \Phi_J(\zeta, z)\omega(\zeta).$$

Formula (2.3′) is obtained from formula (2.3) with the help of the Stokes Formula.

The following result on the separation of singularities of functions holomorphic in Weil polyhedra is an immediate consequence of the Weil formula.

**Proposition 2.3** (A. Weil, 1935). *Any function holomorphic in a polyhedron $D$ of type* (2.1), (2.2) *can be represented as a sum of functions $f_J$ holomorphic in the big* *domains*

$$D_J = \{z \in \Omega \colon |F_j(z)| < 1, j \in J\},$$

*where* $|J| = n$.

A less simple consequence of the Weil formula is the following stronger version for analytic polyhedra of the Hartogs Theorem on the automatic continuation of holomorphic functions.

**Proposition 2.4** ([40]). *Any function f, holomorphic in the neighbourhood of the* $(n+1)$*-dimensional distinguished boundary*

$$\bigcup_{|J| = n+1} \sigma_J$$

*of a Weil polyhedron, has a single-valued continuation to a holomorphic function on the domain D.*

With respect to the necessity of the decomposition (2.2) for Theorem 2.2, A. Weil (1935) remarked that if the functions $\{F_j\}$ are rational, then the decomposition clearly holds and stated the *conjecture* that the decomposition 2.2 is verified for any function $F$ holomorphic in a domain of holomorphy $\Omega$.

It seems that this technical problem led to the fundamental results of Oka and H. Cartan on the theory of ideals of analytic functions.

In its original form, this problem was solved through the efforts of Oka, Hefer and H. Cartan (see Theorem 2.11).

The original proof of A. Weil for Theorem 2.2 is based on the use of the classical Cauchy formula and the fact that the kernels $\Phi_J$ appearing in (2.3) form a *holomorphic cocycle*. More precisely, for any multi-index $I = (i_1, \ldots, i_{n+1})$ of length $n+1$, we have

$$\sum_{r=1}^{n+1} (-1)^r \Phi_{i_1, \ldots, i_{r-1}, i_{r+1}, \ldots, i_{n+1}}(\zeta, z) = 0, \tag{2.5}$$

where $\zeta, z$ are such that $F_i(\zeta) \neq F_i(z)$, $\forall i \in I$.

Formulas (2.3) and (2.5) essentially give an analytic representation for the evaluation functional at $z \in D$ on the space $\mathscr{H}(\bar{D})$ in terms of a holomorphic cocycle on $\Omega \backslash D$. Besides, these formulas allow us to obtain an analogous analytic representation for an arbitrary functional on $\mathscr{H}(D)$.

**Proposition 2.5** (Martineau, 1962). *Let D be an analytic polyhedron in the domain of holomorphy* $\Omega \subset \mathbb{C}^n$, $n > 1$. *Then, the Fantappiè transformation*

$$F \colon \varphi \to \Phi_J(\zeta) = \langle \varphi, K_f(\zeta, \cdot) \rangle,$$

*where*

$$\varphi \in (\mathscr{H}(\bar{D}))^*,$$

$$\zeta \in \Omega_J = \{\zeta \in \Omega \colon |F_j(\zeta)| > 1, \quad j \in J\},$$

*realizes a cannonical isomorphism between the space of functionals* $\mathscr{H}(\bar{D})^*$ *and the space of holomorphic cocycles* $\{\Phi_J\}$, $|J| = n$, *defined on the domains* $\{\Omega_J\}$, *satisfying* (2.5) *and factored by the coboundary, that is, by the subspace of cocycles* $\{\Phi_J\}$ *of the form*

$$\Phi_{j_1,\ldots,j_n} = \sum_{k=1}^{n} (-1)^k \Psi_{j_1,\ldots,j_{k-1},j_{k+1},\ldots,j_n},$$

*where the functions* $\{\Psi_{J'}\}$ *are holomorphic in the domain* $\{\Omega_{J'}\}$, $|J'| = n-1$. *Here, the holomorphic cocycle* $\{\Phi_J\}$ *corresponds to the functional* $\varphi$ *given by*

$$(\varphi, f) = \sum_{|J|=n}' \int_{\sigma_J} f(\zeta)\Phi_J(\zeta)\omega(\zeta), \qquad f \in \mathscr{H}(\bar{D}).$$

The idea of Proposition 2.5 goes back to Fantappiè who essentially obtained it for the case when $D$ is a polydisc. From Proposition 2.5 Martineau (see [28]) deduced the following

**Corollary 2.5.′** *Let K be a holomorphically convex compact subset of a domain of holomorphy* $\Omega \subset \mathbb{C}^n$, $n > 1$. *Then, the space of functionals* $(\mathscr{H}(K))^*$ *is canonically isomorphic to* $H^{n-1}(\Omega \setminus K, \mathcal{O})$, *the* $n-1$*-dimensional cohomology group on the domain* $\Omega \setminus K$ *with coefficients in the sheaf* $\mathcal{O}$ *of germs of holomorphic functions.*

### 2.2. Solution of "Fundamental Problems" in Domains of Holomorphy.

A. Weil (1932), used his own formula to obtain a multi-dimensional analog of Runge's approximation theorem (see [24]).

A *compact* set $K$ in a domain $\Omega$ is said to be *holomorphically* (respectively polynomially) *convex* with respect to this domain if for each point $z \in \Omega \setminus K$ there exists a function holomorphic in $\Omega$ (respectively a holomorphic polynomial) such that $\max_{\zeta \in K} |f(\zeta)| < |f(z)|$.

**Theorem 2.6** (A. Weil, 1932; Oka, 1937). *Let K be a holomorphically (respectively polynomially) convex compact subset of a domain* $\Omega$. *Then, each function f holomorphic in some neighbourhood* $U(K)$ *of the compact set K can be uniformly approximated on K by functions holomorphic (respectively holomorphic polynomials) on* $\Omega$.

This theorem was proved first of all by A. Weil for polynomially convex compacta and later by Oka in the general case.

The proof of Theorem 2.6 starts out by showing that because of the holomorphic convexity of the compact set $K$ with respect to the domain $\Omega$, there exists a Weil polyhedron $D$ of type (2.1) such that $K \subset D \subset U(K)$.

Then we represent the function $f$ by the Weil formula (2.3) in which we make the substitution

$$\left[\prod_{r=1}^{n} (F_{j_r}(\zeta) - F_{j_r}(z))\right]^{-1} = \sum_{k_1,\ldots,k_n \geq 0} \frac{(F_{j_1}(z))^{k_1} \ldots (F_{j_n}(z))^{k_n}}{(F_{j_1}(\zeta))^{k_1+1} \ldots (F_{j_n}(\zeta))^{k_n+1}},$$

where the series converges for $z \in D$. We obtain the formula

$$f(z) = \sum_{j_1 < \ldots < j_n} \sum_{k_1, \ldots, k_n \geq 0} f_{J,k}(z)(F_{j_1}(z))^{k_1} \ldots (F_{j_n}(z))^{k_n}, \tag{2.6}$$

where

$$f_{J,k}(z) = \frac{1}{(2\pi i)^n} \int_{\sigma_J} \frac{\det |P^s_{jr}|^n_{r,s=1} \omega(\zeta)}{(F_{j_1}(\zeta))^{k_1+1} \ldots (F_{j_n}(\zeta))^{k_n+1}}, \qquad z \in D.$$

The series in the right member of (2.6) consists of functions holomorphic on the domain $\Omega$ and converges uniformly on the compact set $K \subset D$ to the function $f$. After the appearance of Weil's work, it was remarked (H. Cartan, 1934; Oka, 1936), that if in the proof of Theorem 0.4 (Cousin, 1895) we use an integral of Weil type (compare Proposition 2.3 and Theorem 0.5) instead of a Cauchy-type integral, we obtain a generalization to analytic polyhedra on which the Weil formula holds. In particular we have (see [24]) the following.

**Proposition 2.7** (H. Cartan, 1934; Oka, 1936). *The first additive Cousin problem is solvable on any polynomially convex domain in* $\mathbb{C}^n$.

In order to show the solvability of the first Cousin problem in an arbitrary analytic polyhedron $D$, Oka represented such a polyhedron as an analytic submanifold $M_D$ of a polydisc $G$ in $\mathbb{C}^{n+N}$:

$$M_D = \{(z, w) \in G : w_j = F_j(z), \qquad j = 1, 2, \ldots, N\}, \tag{2.7}$$

where

$$G = \prod_{i=1}^{n} \{|z_i| < R\} \times \prod_{j=1}^{N} \{|w_j| < 1\}.$$

Moreover, Oka succeeded in proving (see [24] and [33])

**Proposition 2.8** (Oka, 1937). *Any analytic submanifold $M$ of type* (2.7) *in a polydisc $G$ is polynomially convex.*

We limit ourselves here to the formulation of only those criteria which guided Oka in the proof of Proposition 2.8 (see [24]).

**Proposition 2.9** (Oka, 1937). *A compact set $K \subset \mathbb{C}^n$ is polynomially convex if and only if for each point $z \in \mathbb{C}^n \setminus K$, we can construct a family of holomorphic polynomials $P_t(z)$, depending continuously on the parameter $t \in [0, \infty)$, such that none of these polynomials have zeros on $K$, $P_0(z) = 0$, and the distance from $K$ to the surface $\{P_t(z) = 0\}$ tends to infinity as $t \to \infty$.*

From Propositions 2.7 and 2.8 follows the following fundamental result (see [24], [33] and [40]).

**Theorem 2.10** (Oka, 1937). *The first Cousin problem is solvable on any domain of holomorphy in* $\mathbb{C}^n$.

We remark that for domains in $\mathbb{C}^2$, H. Cartan showed the converse: if in a domain $D \subset \mathbb{C}^2$, the first Cousin problem is solvable then $D$ is a domain of holomorphy (see [24]).

The solution of the first Cousin problem with the help of the Weil formula made it possible in turn to solve the problem of the Weil factorization (2.2) on arbitrary analytic polyhedra (see [24] and [40]).

**Theorem 2.11** (Oka, Hefer, H. Cartan, 1941–1944). *Let $\Omega$ be a domain of holomorphy in $\mathbb{C}^n$ and $g_1, \ldots, g_k$ holomorphic functions in $\Omega$, $1 \leq k \leq n$, with the property $dg_1 \wedge \ldots \wedge dg_n \neq 0$ on the set of their common zeros*

$$M = \{z \in \Omega: g_1(z) = \ldots = g_k(z) = 0\}.$$

*Then, for any function $\varphi$, holomorphic in $\Omega$ and zero on $M$, there exists functions $\varphi_1, \ldots, \varphi_k$ holomorphic in $\Omega$ such that*

$$\varphi(z) = \sum_{j=1}^{k} g_j(z) \cdot \varphi_j(z), \quad z \in \Omega.$$

The following converse also holds: if a domain $\Omega$ in $\mathbb{C}^n$ is such that for any holomorphic functions $g_1, \ldots, g_n$ having no common zeros in $\Omega$, there exist holomorphic functions $\varphi_1, \ldots, \varphi_n$ in $\Omega$ such that $\sum_j \varphi_j g_j \equiv 1$, then $\Omega$ is a domain of holomorphy (see [33]).

To obtain the factorization (2.2) from Theorem 2.11, it is sufficient to consider the functions $g_j = \zeta_j - z_j; j = 1, \ldots, n$, in the domain $\Omega \times \Omega$ and to apply Theorem 2.11 to the function $\varphi(z, \zeta) = f(z) - f(\zeta)$. Theorem 2.11 is proved in parallel with the following result which is no less important (see [24], [40]).

**Theorem 2.12** (Oka, Hefer, H. Cartan, 1941–1944). *Under the hypotheses of Theorem 2.11, for any function $f$ holomorphic on $M$, there exists a function $\tilde{f}$ holomorphic on $\Omega$, such that $\tilde{f} = f$ on $M$.*

If the domain $\Omega$ is not a domain of holomorphy, then the assertion of Theorem 2.12 also fails.

Theorems 2.11 and 2.12 are proved by induction on $k$. Let $k = 1$ and $M_1 = \{z \in \Omega: g_1(z) = 0\}$. In order to obtain Theorem 2.11 in this case, it is sufficient to set $\varphi_1 = \varphi / g_1$. To obtain Theorem 2.12, we consider a neighbourhood $\Omega_1$, of the manifold $M_1$ in $\Omega$, which admits a holomorphic retraction $z \mapsto R(z)$ onto $M_1$. The function $F(z) = f(R(z))$ is well defined in $\Omega_1$. Set $\Omega_2 = \Omega \setminus M_1$ and consider the first Cousin problem in $\Omega$ with Cousin data $F/g_1$ in $\Omega_1$ and zero in $\Omega_2$. By the Theorem of Oka 2.10, there exists a meromorphic function $\Phi$ in $\Omega$ such that $\Phi$ is holomorphic in $\Omega_2$ and $(\Phi - F/g_1)$ is holomorphic in $\Omega_1$. Consider the function $\tilde{f} = \Phi \cdot g_1$. By construction, this function is holomorphic in $\Omega$ and coincides with $f$ on $M_1$.

In order to prove Theorem 2.12 in the general case, it is necessary to consider manifolds $M_r = \{z \in \Omega: g_1(z) = \ldots = g_r(z) = 0\}$ and, using the solvability of the Cousin problem on such manifolds, successively continue the function $f$, given on $M = M_k$, to a holomorphic function $\tilde{f}_{k-1}$ on $M_{k-1}$; then to continue $\tilde{f}_{k-1}$ to a function $\tilde{f}_{k-2}$ on $M_{k-2}$ etc. until we obtain a function $\tilde{f} = \tilde{f}_0$ on $M_0 = \Omega$.

To prove Theorem 2.11 in the general case we suppose it has been verified for the case of $(k-1)$ functions $g_2, \ldots, g_k$ on the manifold $M_1$; that is,

$$\varphi(z) = \sum_{j=2}^{k} g_j(z) \cdot \tilde{\varphi}_j(z), \qquad z \in M_1.$$

Then, by Theorem 2.12, there exist holomorphic functions $\{\varphi_j\}$ on $\Omega$ agreeing with $\{\tilde{\varphi}_j\}$ on $M_1, j = 2, \ldots, k$. To complete the proof of Theorem 2.11, it is sufficient now to set

$$\varphi_1(z) = (\varphi(z) - \sum_{j=2}^{k} g_j(z) \cdot \varphi_j(z))/g_1(z), \qquad z \in \Omega.$$

Subsequently, H. Cartan using the works of Leray (1945) and Oka (1950) significantly strengthened the assertions of Theorems 2.10, 2.11 and 2.12 getting rid, in particular, of the condition that $M$ be a *complete intersection* in $\Omega$ (see [33] and [40]).

**Theorem 2.13** (H. Cartan, 1951). *Let $M$ be a closed analytic submanifold of codimension $k$ in a domain of holomorphy $\Omega$. Then,*

*a) any holomorphic function $f$ on $M$ can be continued to a holomorphic function on $\Omega$;*

*b) in any subdomain $D \Subset \Omega$, the manifold $M$ can be represented in the form*

$$D \cap M = \{z \in D : g_1(z) = \ldots = g_N(x) = 0\}, \tag{2.8}$$

*where the rank of the matrix* (grad $g_1, \ldots,$ grad $g_N$) *is $k$ everywhere on $M \cap D$;*

*c) for any function $\varphi$ holomorphic in $D$ and equal to zero on the manifold $M \cap D$ of the form* (2.8), *there exist holomorphic functions in $D$, $\varphi_1, \ldots, \varphi_N$ such that*

$$\varphi(z) = \sum_{j=1}^{N} \varphi_j(z) \cdot g_j(z), \qquad z \in \Omega;$$

*d) on $M$ the first Cousin problem is solvable.*

Part a) of Theorem 2.13 was stated earlier as a conjecture by A. Weil. The works of Oka and H. Cartan, arising in connection with the Cauchy–Weil formula, thus led to the solution of several long-standing problems of multidimensional complex analysis.

## §3. Integral Formulae and the Problem of E. Levi

**3.1. Pseudoconvex Domains. Theorems of E. Levi and H. Lewy.** E. Levi (1911) showed that applying Hartog's pseudoconvexity (see Theorem 0.1) to domains of holomorphy with smooth boundary leads to the more intuitive pseudoconvexity in the sense of Levi (see [24], [33] and [40]).

Let $D$ be a domain in $\mathbb{C}^n$ with smooth boundary. That is,

$$D = \{z \in \Omega : \rho(z) < 0\}, \tag{3.1}$$

where $\rho$ is a real-valued function in the class $C^2(\Omega)$ in some neighbourhood $\Omega$ of the compact set $\bar{D}$ and

$$d\rho(z) \neq 0, \qquad z \in \partial D.$$

The *Levi form* of a real-valued function $\rho \in C^2(\Omega)$ at a point $\zeta \in \Omega$ is the Hermitian form

$$L_{\rho,\zeta}(w) = \sum_{\alpha,\beta} \frac{\partial^2 \rho(\zeta)}{\partial z_\alpha \partial \bar{z}_\beta} w_\alpha \cdot \bar{w}_\beta, \qquad w \in \mathbb{C}^n. \tag{3.2}$$

We denote by $T_\zeta^c(\partial D)$ the *complex tangent space* to $\partial D$ at the point $\zeta \in \partial D$, i.e.

$$T_\zeta^c(\partial D) = \{z \in \mathbb{C}^n : \sum_j \frac{\partial \rho}{\partial z_j}(\zeta)(z_j - \zeta_j) = 0\}.$$

A domain $D$ given by (3.1) is called weakly (respectively strongly) pseudoconvex at the point $\zeta \in \partial D$, if it's Levi form $L_{\rho,\zeta}(w)$ is non-negative (respectively positive) for each non-zero vector

$$w \in T_\zeta^c(\partial D) - \zeta.$$

The *domain D* is called *Levi pseudoconvex* if it is weakly pseudoconvex at each point $\zeta \in \partial D$.

The *domain D* is called *strongly pseudoconvex* if it is strongly pseudoconvex at each point $\zeta \in \partial D$.

If for a domain $D$, given by (3.1), and some point $\zeta^* \in \partial D$, the E. Levi form at $\zeta^*$ has at least one non-positive (respectively negative) eigen-value, then the domain $D$ is said to be weakly (respectively strongly) *concave* at the point $\zeta^*$.

**Theorem 3.1** (E. Levi, 1911, see [33]). *If a domain D is Levi pseudoconvex, then it is Hartogs pseudoconvex. Conversely, if D given by* (3.1) *is not Levi pseudoconvex, i.e. is strongly concave at some point $\zeta^* \in \partial D$, then for some neighbourhood $U_{\zeta^*}$ of the point $\zeta^*$, every function holomorphic in $U_{\zeta^*} \cap D$ continues holomorphically to $U_{\zeta^*}$.*

H. Lewy significantly strengthened Theorem 3.1 (see [45]) in the following way.

**Theorem 3.1′** (H. Lewy, 1956). *Suppose the domain D is given by* (3.1) *and is strongly concave at some point $\zeta^* \in \partial D$. Then there exists a neighbourhood $U_{\zeta^*}$ of the point $\zeta^*$ such that any function f of class $C^1$ and satisfying the tangential Cauchy–Riemann equations $\bar{\partial} f \wedge \bar{\partial} \rho = 0$ on $U_{\zeta^*} \cap \partial D$ extends holomorphically to the domain $U_{\zeta^*} \cap (\mathbb{C}^n \setminus D)$.*

Rossi (1966) generalized the assertion of Theorem 3.1′ to the case of integrable functions satisfying the tangential Cauchy–Riemann equations in the sense of distributions on $U_{\zeta^*} \cap \partial D$.

The simplest proof of Theorem 3.1′ is to reduce it to Theorem 3.1 by representing (see Theorem 5.8 below) $f$ as the difference of functions $f_+$ and $f_-$ holomorphic respectively in the domains $U_{\zeta^*} \cap D$ and $U_{\zeta^*} \cap (\mathbb{C}^n \setminus D)$.

For a real function $\rho \in C^2(\Omega)$, we introduce the so-called *Levi polynomial*

$$F_\rho(\zeta, z) = \sum_j \frac{\partial \rho}{\partial z_j}(\zeta)(z_j - \zeta_j) + \frac{1}{2} \sum_{j,k} \frac{\partial^2 \rho}{\partial z_j \partial \bar{z}_k}(\zeta)(z_j - \zeta_j)\overline{(z_k - \zeta_k)}. \tag{3.3}$$

E. Levi, in proving Theorem 3.1 and in posing the problem of whether each pseudoconvex domain in $\mathbb{C}^n$ is a domain of holomorphy , based himself on the following result (see [40]).

**Theorem 3.2** (E. Levi, 1911). *Let $D$ be a strongly pseudoconvex domain. Then, for any ball $B$ of sufficiently small diameter, the domain $D \cap B$ is a domain of holomorphy. In addition for each point $\zeta \in B \cap \partial D$, the null-set of the Levi polynomial $F_\rho(\zeta, z)$ is a strong barrier for the domain $B \cap D$, that is,*

$$\{z\colon F_\rho(\zeta, z) = 0\} \cap B \cap \bar{D} = \{\zeta\}. \tag{3.4}$$

*If, on the other hand, the domain $D$ is strongly pseudoconcave at a point $\zeta^* \in \partial D$, then for some neighbourhood $U_{\zeta^*}$ of the point $\zeta^*$ and some two-dimensional plane $M_{\zeta^*}$ passing through $\zeta^*$, the holomorphic curve*

$$\{z \in M_{\zeta^*} \cap U_{\zeta^*}\colon F_\rho(\zeta^*, z) = 0\}$$

*lies in the domain $D$ and is tangent to $\partial D$ from within at the point $\zeta^*$.*

For the proof of Theorem 3.2 we use the Taylor formula to write

$$\rho(z) = \rho(\zeta) + 2\operatorname{Re} F_\rho(\zeta, z) + L_\rho(\zeta, z - \zeta) + o(|\zeta - z|^2). \tag{3.5}$$

If the domain $D$ is strongly pseudoconvex, then

$$L_\rho(\zeta, z - \zeta) \geq \gamma |\zeta - z|^2 \text{ on } T_\zeta^c(\partial D),$$

where $\gamma > 0$. From this and from (3.5), it follows that if the diameter of the ball $B$ is sufficiently small, then (3.4) holds.

To complete the proof of the first part of the theorem, it is sufficient now to notice that for any point $\zeta \in B \cap \partial D$, the function $[F_\rho(\zeta, z)]^{-1}$ is holomorphic in $B \cap D$ and does not continue holomorphically to the point $\zeta$.

The second part of the theorem is proved analogously.

For weakly pseudoconvex (pseudoconvex, but not strongly) domains $D$ (even with analytic boundary), there are difficulties, related to the construction of barrier functions for $\partial D$, which are well illustrated by the following surprising result (see [58]).

**Theorem 3.2′** (Kohn, Nirenberg, 1973). *Let D be the following pseudoconvex domain in* $\mathbb{C}^2$:

$$D = \{(z_1, z_2) \in \mathbb{C}^2 \colon \operatorname{Re} z_2 + |z_1 \cdot z_2|^2 + (z_1|^8 + 15/7|z_1|^2 \cdot \operatorname{Re} z_1^6 < 0\}.$$

*Let h be a function holomorphic in a neighbourhood of the point* $(0, 0) \in \partial D$ *and equal to zero at this point. Then the set* $\{(z_1, z_2) \colon h(z_1, z_2) = 0\}$ *necessarily has both some points in the interior as well as in the exterior of the domain D.*

**3.2. Oka's Solution to the Levi Problem.** From Theorem 3.2 it follows that the solution of the Levi problem for strongly pseudoconvex domains rests on clearing up the conditions under which the union of two domains of holomorphy is again a domain of holomorphy. The following principle for the joining of domains of holomorphy was brought forth by Oka in 1942 to solve the Levi problem in $\mathbb{C}^2$ (see [24]).

**Theorem 3.3** (Oka). *Let D be a domain in* $\mathbb{C}^n$ *which is the union of two domains of holomorphy of the type*

$$D_1 = \{z \in D \colon \operatorname{Im} z_1 > -\varepsilon\}, \; D_2 = \{z \in D \colon \operatorname{Im} z_1 < \varepsilon\}.$$

*Then, the domain* $D = D_1 \cup D_2$ *is also a domain of holomorphy.*

In order to prove Theorem 3.3, it is sufficient to show that the first Cousin problem is solvable in the domain $D = D_1 \cup D_2$. The proof of the latter reduces to the following assertion of Oka, which is a profound generalization of Theorem 0.5 (see [24]).

**Proposition 3.4** (Oka). *Under the hypotheses of Theorem 3.3, any function f holomorphic in a neighbourhood of the set* $S = \{z \in D \colon \operatorname{Im} z_1 = 0\}$ *can be represented on S in the form* $f = f_+ - f_-$, *where* $f_\pm$ *are functions holomorphic respectively in the closures of the domains*

$$D_\pm = \{z \in D \colon \pm \operatorname{Im} z_1 > 0\}.$$

This proposition was proved by Oka once more by making use of the Weil integral in an interesting way. The original reasoning of Oka (see [24]) is as follows.

By Theorem 2.1, without loss of generality, we may assume that the domain $D_0 = D_1 \cap D_2$ is an analytic polyhedron, that is

$$D_0 = \{z \colon |F_j(z)| < 1, \quad j = 1, 2, \ldots, N\},$$

where the functions $\{F_j\}$ are holomorphic in a neighbourhood of $\bar{D}_0$. By Theorem 2.11 we have

$$F_j(\zeta) - F_j(z) = \sum_{k=1}^{n} P_j^k \cdot (\zeta_k - z_k),$$

where $P_j^k \in \mathscr{H}(\bar{D}_0 \times \bar{D}_0)$. We consider, in the domains $D_0^\pm = D_0 \cap D_\pm$, the following functions, given by Weil-type integrals

$$f_{\pm}(z) = \sum_{|J|=n-1} \int_{\sigma_J} f(\zeta)\varphi_J^{\pm}(\zeta, z)\omega(\zeta), \tag{3.6}$$

where $\sigma_J = \{\zeta \in \partial S: |F_j| = 1, j \in J = (j_2, \ldots, j_n)\}$,

$$\varphi_J^{\pm}(\zeta, z) = \frac{1}{(\zeta_1 - z_1)} \frac{\det|P_{jr}^s|_{r,s=2}^n}{\prod_{j\in J}(F_j(\zeta) - F_j(z))}, \qquad z \in D_0^{\pm}. \tag{3.7}$$

The kernels $\varphi_J^{\pm}$ are meromorphic functions of the variables $(\zeta, z)$ in a neighbourhood $U(\sigma_J \times D_0^{\pm})$ of the compact set $\sigma_J \times \bar{D}_0^{\pm}$. By the Sokhotskij formula we have

$$f(z) = f_+(z) - f_-(z), \qquad z \in S. \tag{3.8}$$

From Oka's Theorem 2.10, there exist meromorphic functions $\Phi_J^{\pm}(\zeta, z)$ in the domains $U(\sigma_J \times \bar{D}^{\pm})$ which are holomorphic for $z \in D^{\pm} \setminus D_0^{\pm}$ and which for $z \in D_0^{\pm}$ have the same poles as the functions $\varphi_J^{\pm}$. Moreover, by the Oka–Weil Theorem 2.6, for each $\varepsilon > 0$, there are functions $G_J^{\pm}$ holomorphic in the domain $U(\sigma_J \times \bar{D}^{\pm})$ such that

$$|\Phi_J^{\pm} - \varphi_J^{\pm} - G_J^{\pm}| < \varepsilon$$

for $(\zeta, z) \in \sigma_J \times \bar{D}_0^{\pm}$. Consider the kernel $\tilde{\varphi}_J^{\pm} = \Phi_J^{\pm} + G_J^{\pm}$. Then, on $\sigma_J \times D_0^{\pm}$ we have

$$\tilde{\varphi}_J^{\pm} = \varphi_J^{\pm} + R_J^{\pm}, \tag{3.9}$$

where $|R_J^{\pm}(\zeta, z)| < \varepsilon$ for $(\zeta, z) \in \sigma_J \times \bar{D}_0^{\pm}$. Let us define holomorphic functions in the domain $\bar{D}^{\pm}$ by the formula

$$\tilde{f}_{\pm}(z) = \sum_{|J|=n-1} \int_{\sigma_J} f(\zeta)\tilde{\varphi}_J^{\pm}(\zeta, z)\omega(\zeta). \tag{3.10}$$

We set

$$\tilde{f}(z) = \tilde{f}_+(z) - \tilde{f}_-(z). \tag{3.11}$$

From (3.8), (3.9) and (3.10) follows the equality

$$\tilde{f} = f + Rf, \tag{3.12}$$

where

$$Rf = \sum_{|J|=n-1} \int_{\sigma_J} f(\zeta)(R_J^+ - R_J^-)\omega(\zeta).$$

Since the operator $R$ has small norm, equation (3.12), seen as an integral equation for $f \in \mathscr{H}(\bar{S})$, can be solved with the help of iteration, for any $\tilde{f} \in \mathscr{H}(\bar{S})$. The assertion of Proposition 3.4 follows from this and from (3.10) and (3.11).

From Theorems 3.2 and 3.3 it follows that any strongly pseudoconvex domain in $\mathbb{C}^n$ is a domain of holomorphy.

In order to solve the Levi problem in complete generality, it suffices, by Theorem 2.1 to show that any pseudoconvex domain can be approximated from within by strongly pseudoconvex domains. Such an approximation was obtained by Lelong and Oka via the characterization of pseudoconvex domains in terms of plurisubharmonic functions (see [33], [40] and [45]).

Lelong (1942) defined an upper-semicontinuous real function $\rho$ in a domain $D$ to be a *plurisubharmonic* if for any complex line $L \subset \mathbb{C}^n$, the restriction of $\rho$ to $D \cap L$ is a subharmonic function.

A typical example of a plurisubharmonic function on $D$ is furnished by the function

$$\rho(z) = \sup_{j \in J} \log |f_j(z)|,$$

where $\{f_j : j \in J\}$ is a finite collection of functions holomorphic on $D$.

A function $\rho \in C^2(\Omega)$ is called *strictly plurisubharmonic* in $\Omega$ if the Levi form $L_{p,\zeta}(\omega)$ is positive definite for each $\zeta \in \Omega$.

**Theorem 3.5** (Lelong, 1945; Oka, 1953). *In order for a domain $D \subset \mathbb{C}^n$ to be Hartogs pseudoconvex, it is necessary and sufficient that there exist a plurisubharmonic function $\rho$ which is an exhausting function for $D$, that is, for any $\alpha \in \mathbb{R}$, we have*

$$D_\alpha = \{z \in D : \rho(z) < \alpha\} \Subset D. \tag{3.13}$$

*In addition, if $D \neq \mathbb{C}^n$, then a criterion of pseudoconvexity is the plurisubharmonicity of the function $-\ln \operatorname{dist}(z, \partial D)$. Moreover, a criterion for the pseudoconvexity of $D \subset \mathbb{C}^n$ is the existence in $D$ of a strictly plurisubharmonic function in $D$ with the property* (3.13).

For bounded domains with smooth boundary, the following stronger version of Theorem 3.5 (see [54]) has also be proven.

**Theorem 3.5′** (Diederich, Fornaess, 1977). *Let $D$ be any bounded pseudoconvex domain with boundary of class $C^2$ in $\mathbb{C}^n$. Then, there exists a defining function $\rho$ for $D$ of class $C^2(\mathbb{C}^n)$ ($D = \{z : \rho(z) < 0\}$) such that the function $\tilde{\rho} = -(-p)^{1/N}$ is not only exhausting for $D$ but also strictly plurisubharmonic in $D$ for all sufficiently large $N$.*

From Theorem 3.5 it follows in particular that any pseudoconvex domain can be approximated from the interior by domains of the form

$$D = \{z \in \Omega : \rho(z) < 0\}, \tag{3.14}$$

where $\rho$ is a strictly plurisubharmonic function in the domain $\Omega \supset \bar{D}$.

By an elementary lemma of Kohn (see [32]), a domain in $\mathbb{C}^n$ with smooth boundary is strictly pseudoconvex if and only if it can be represented in the form (3.14). For this reason, in the sequel, we shall call a domain in $\mathbb{C}^n$ of type (3.14) *strongly* (or strictly) *pseudoconvex* irrespective of whether it has a smooth boundary or not.

From Theorems 3.2, 3.3 and 3.5 we have the following fundamental result.

**Theorem 3.6** (Oka, Norguet, Bremermann, 1953, see [33], [40], and [45]). *Every pseudoconvex domain in $\mathbb{C}^n$ is a domain of Holomorphy.*

**3.3. Applications and Generalizations of Oka's Theorems.** An $n$-dimensional complex manifold $\Omega$ is called a *Riemannian domain* if on $\Omega$ there is a system of $n$ holomorphic functions which form a local coordinate system in the neighbourhood of each point of $\Omega$. Oka (1953) showed that the Levi problem and the first Cousin problem are solvable for any pseudoconvex Riemannian domain. Oka's solution to the Levi problem immediately led to several remarkable results. One of them was the description of the envelope of holomorphy of an arbitrary domain in $\mathbb{C}^n$ (see [33] and [45]).

**Theorem 3.7** (Oka). *The envelope of holomorphy of an arbitrary domain $\Omega$ in $\mathbb{C}^n$ (or even of an arbitrary Riemannian domain over $\mathbb{C}^n$) is a Riemannian domain of holomorphy $\tilde{\Omega}$ over $\mathbb{C}^n$.*

This fact, that the domain of holomorphy of a domain in $\mathbb{C}^n$ can be many-sheeted, was first discovered by Thullen (1932), (see [81] and [23]).

The proof of Theorem 3.7 is based on the fact that, with the help of the Hartogs Continuity Principle (Theorem 0.1), it is relatively simple to construct a Riemannian domain $\tilde{\Omega}$ to which all functions holomorphic in $\Omega$ extend holomorphically. That there is no larger domain, to which all functions holomorphic on $\tilde{\Omega}$ extend, follows from Theorem 3.6.

Another application of the solution of the Levi problem on pseudoconvex Riemannian domains is the solution of the Poincaré problem for arbitrary domains in $\mathbb{C}^n$.

**Theorem 3.8** (Oka). *For an arbitrary domain $\Omega$ in $\mathbb{C}^n$, any meromorphic function $f$ in $D$ is the quotient of two holomorphic functions in $D$.*

We recall that for the case $\Omega = \mathbb{C}^n$, the assertion of Theorem 3.8 is a classical result of Poincaré and Cousin.

To prove Theorem 3.8, we extend the function $f$, using the continuity principle of Levi, to a function $\tilde{f}$ meromorphic on a pseudoconvex Riemannian domain $\tilde{\Omega}$. Now, using the solvability of the first Cousin problem on $\tilde{\Omega}$, we represent $\tilde{f}$ as the quotient of two holomorphic functions on $\Omega$.

One more beautiful consequence of Theorem 3.6 is the following result concerning the representation of a plurisubharmonic function in terms of modules of holomorphic functions (see [73]).

**Theorem 3.9** (Lelong (1941), $n = 1$; Bremermann (1954), $n \geq 1$). *For any continuous plurisubharmonic function $\rho$ in a pseudoconvex domain $D \subset \mathbb{C}^n$, there exist holomorphic funtions $\{f_m\}$ in $D$ such that*

$$\rho(z) = \lim_{j\to\infty} \sup_{m \leq j} \frac{1}{m} \ln|f_m(z)|, \qquad z \in D.$$

For the proof of Theorem 3.9, we notice that the domain

$$\tilde{D} = \{(z, w): z \in D, w \in \mathbb{C}^1, |\rho(z)| < \ln|w|\}$$

is pseudoconvex in $\mathbb{C}^{n+1}$. Thus, by Theorem 3.6, there exists a holomorphic function $f(z, w)$ in $\tilde{D}$ which cannot be holomorphically continued to a larger domain. Now we may take, as functions $\{f_m\}$, the coefficients in the power series:

$$f(z, w) = \sum_{m \geq 0} f_m(z) w^m.$$

For further results on plurisubharmonic functions and their role in complex analysis see [81], [73], [23], [24], [32], [33], [45] and the article Vol. 8, II.

From Oka's Theorems 2.10, 2.11 and 3.5 one can deduce the existence of global holomorphic barriers for strongly pseudoconvex domains, thus generalizing Theorem 3.2 where such barriers are constructed locally.

**Theorem 3.10** (G.M. Khenkin, 1969, 1974; Ramires, 1970; Øvrelid, 1971; Fornaess, 1974). *Let $D$ be a strongly pseudoconvex domain given by* (3.14). *Then, for some neighbourhood $U(\bar{D})$ of the compact set $\bar{D}$, there exists a smooth function $\Phi = \Phi(\zeta, z)$ of the variables $(\zeta, z) \in U(\bar{D}) \times U(\bar{D})$ such that $\Phi$ is a holomorphic function of $z \in U(\bar{D})$,*

$$2\operatorname{Re}\Phi(\zeta, z) \geq \rho(\zeta) - \rho(z) + \gamma|\zeta - z|^2 \tag{3.15}$$

*for some $\gamma > 0$, and*

$$\Phi(\zeta, z) = \langle P(\zeta, z), \zeta - z \rangle, \tag{3.16}$$

*where $P = (P_1, \ldots, P_n)$ is a smooth vector-function of the variables $(\zeta, z) \in U(\bar{D}) \times U(\bar{D})$, holomorphic in $z \in U(\bar{D})$.*

Theorem 3.10 improves results of Bremermann (1959) and Rossi (1961, see [33]) stating that for any strongly pseudoconvex domain $D$, the Bergman–Shilov boundary $S(D) = \partial D$ and each point of $\partial D$ is a peak point for the algebra $A(\bar{D})$.

The barrier functions constructed in Theorem 3.10 allow one (see [39] and here §4) to obtain analogues of the Cauchy formula for arbitrary strongly pseudoconvex domains and these formulas are as convenient as the Cauchy–Bochner formulas (1.9) and (1.10) for the ball in $\mathbb{C}^n$.

On the basis of Oka's theorems, the following description of the Bergman–Shilov boundary for an arbitrary pseudoconvex domain with smooth boundary was also obtained (see [58]).

**Theorem 3.11** (Rossi, 1961; Hakim, Sibony, 1975; Pflug, 1975; Debiard, Gaveau, 1976; Basener, 1977). *Let $D$ be a pseudoconvex domain in $\mathbb{C}^n$ with smooth boundary of class $C^2$. Then the Bergman–Shilov boundary $S(D)$ is the closure of the points of strong pseudoconvexity on $\partial D$.*

Grauert obtained the fundamental generalization of Oka's Theorem 3.6 to the case of arbitrary strongly pseudoconvex complex manifolds (see [33], [40] and [45]).

An abstract *complex manifold D* is called *strongly pseudoconvex* if there exists on $D$ an exhausting function (in the sense of (3.13) $\rho$ of class $C^2$ and a compact set $K \subset D$ such that $\rho$ is strongly plurisubharmonic on the domain $D \setminus K$.

A *complex manifold D* is called *holomorphically convex* if for each compact set $K$, its holomorphic hull

$$\hat{K} = \{ z \in D : |f(z)| \leq \sup_{\zeta \in K} |f(\zeta)|, \forall f \in \mathscr{H}(D) \}$$

is also compact in $D$.

**Theorem 3.12** (Grauert, 1958). *Every strongly pseudoconvex complex manifold is holomorphically convex.*

In contrast to the situation in $\mathbb{C}^n$, the assertion of Theorem 3.12 does not carry over to arbitrary weakly pseudoconvex complex manifolds. Counterexamples were constructed by Grauert and Narasimhan (1963).

Theorem 3.12 was first proved by Grauert based on the cohomology theory of coherent analytic sheaves (see [33]). Another proof of Theorem 3.12 follows from the fundamental works of Morrey (1958) and Kohn (1963) on the $\bar{\partial}$-problem of Neumann–Spencer (see [22]).

However, the most elementary proof of Theorem 3.12 was obtained in [49] based on an elaboration of the original proof of Theorem 3.6 by Oka and using only elementary integral formulas in strongly pseudoconvex domains (see also [4], [40] and [72]).

## §4. The Cauchy–Fantappiè Formulas

**4.1. The Formulas of Cauchy–Leray and Cauchy–Waelbroeck.** Leray (see [59] and [60]), while developing the theory of residues on complex analytic manifolds, found a general method for constructing integral representations for functions of several complex variables.

Let $\Omega$ be a domain in $\mathbb{C}^n$ and $z$ a fixed point in $\Omega$. Consider, in the domain $Q = \mathbb{C}^n \times \Omega$, with coordinates $\eta = (\eta_1, \ldots, \eta_n) \in \mathbb{C}^n$ and $\zeta = (\zeta_1, \ldots, \zeta_n) \in \Omega$, the hypersurface

$$P_z = \{(\eta, \zeta) \in Q : \langle \eta, \zeta - z \rangle = 0\}.$$

Let $h_z$ be a $(2n-1)$-dimensional cycle in the domain $Q - P_z$ whose projection on $\Omega \setminus \{z\}$ is homologous to $\partial\Omega$. Also, let $H_z$ be the class of compact homologies, on the domain $Q - P_z$, containing the cycle $h_z$.

**Theorem 4.1** (Leray, 1956, see [59]). *For any holomorphic function in the domain* $\Omega$, *we have*

$$f(z)=\frac{(n-1)!}{(2\pi i)^n}\int_{h_z\in H(Q-P_z)} f(\zeta)\frac{\omega'(\eta)\wedge\omega(\zeta)}{\langle\eta,\zeta-z\rangle^n}. \tag{4.1}$$

For the proof of Theorem 4.1, we notice that the form

$$f(\zeta)\frac{\omega'(\eta)\wedge\omega(\zeta)}{\langle\eta,\zeta-z\rangle^n}$$

is closed in the domain $Q-P_Z$. Thus, it is sufficient to prove (4.1) for any one cycle $h_Z\in H_Z$. Let $D$ be a neighbourhood of the point $z$ with rectifiable boundary, where $\bar{D}\subset\Omega$. As $h_Z$ we take the graph of the mapping $\zeta\mapsto\eta(\zeta)=\bar{\zeta}-\bar{z}$, for $\zeta\in\partial D$. For this choice of $h_Z$, formula (4.1) becomes precisely the Martinelli–Bochner formula (1.2). As an immediate corollary of Theorem 4.1 we obtain (see [60]) the following formula of Cauchy–Waelbroeck, generalizing the Bergman type integral representation.

**Theorem 4.2** (Waelbroeck, 1960; Leray, 1961). *Let D be a domain with rectifiable boundary in* $\mathbb{C}^n$ *and* $z\in D$. *Let* $\eta=\eta(\zeta,z)=(\eta_1,\ldots,\eta_n)$ *be a smooth* $\mathbb{C}^n$*-valued function of the variable* $\zeta\in\bar{D}$ *such that* $\langle\eta(\zeta,z),\zeta-z\rangle=1$, *for* $\zeta\in\partial D$. *Then, for any function f, holomorphic in D and continuous on* $\bar{D}$, *and any integer* $s\geq 0$, *we have*

$$f(z)=K^sf(z),\qquad z\in D, \tag{4.2}$$

*where*

$$K^sf(z)=\frac{(n+s-1)!}{(s-1)!(2\pi i)^n}\int_{\zeta\in D}(1-\langle\eta(\zeta,z),\zeta-z\rangle)^{s-1}f(\zeta)\omega(\eta(\zeta,z))\wedge\omega(\zeta), \tag{4.3}$$

if $s\geq 1$, *and*

$$K^0f(z)=\frac{(n-1)!}{(2\pi i)^n}\int_{\zeta\in\partial D} f(\zeta)\omega'(\eta(\zeta,z))\wedge\omega(\zeta), \tag{4.4}$$

*for* $s=0$.

We remark that Waelbroeck used formula (4.2) for the construction of a functional calculus on elements of linear topological algebras (see [25]).

For the proof of (4.2) with $s=1$, we notice that by (4.4) and Stokes formula, we have

$$f(z)=\frac{(n-1)!}{(2\pi i)^n}\int_{\zeta\in D} d[f(\zeta)\omega'(\eta(\zeta,z))\wedge\omega(\zeta)]=$$

$$=\frac{n!}{(2\pi i)^n}\int_D f(\zeta)\omega(\eta(\zeta,z))\wedge\omega(\zeta). \tag{4.5}$$

Further, from Stokes formula and the equalities

$$d[(1-\langle \eta(\zeta,z),\zeta-z\rangle)^{s-1}f(\zeta)\omega'(\eta(\zeta,z))\wedge\omega(\zeta)]=$$
$$=(s+n-1)(1-\langle \eta(\zeta,z),\zeta-z\rangle)^{s-1}f(\zeta)\omega(\eta(\zeta,z))\wedge\omega(\zeta)-$$
$$-(s-1)(1-\langle \eta(\zeta-z),\zeta-z\rangle)^{s-2}f(\zeta)\omega(\eta(\zeta,z))\wedge\omega(\zeta),$$

we deduce that the right side of (4.2) is independent of $s$. Thus, (4.2) follows from (4.5).

**4.2. Multidimensional Analogues of the Cauchy–Green Formula.** The formulas of Cauchy–Fantappié and Cauchy–Waelbroeck (4.1) and (4.2) together with the Martinelli–Bochner formula (1.2) allow (see [20]) us to obtain the following multi-dimensional analogues of the Cauchy–Green formulas.

**Theorem 4.3** (G.M. Khenkin, 1970; $s=0$; G.M. Khenkin, A.V. Romanov, 1971; $s=1$; G.M. Khenkin, 1978, $s\geq 1$). *Under the condition of Theorem* 4.2, *for any function $f$ such that both $f$ and $\bar\partial f$ are continuous on $\bar D$, the following integral representation holds:*

$$f(z)=K^s f(z)-H^s(\bar\partial f)(z), \tag{4.6}$$

*where*

$$H^0(\bar\partial f)(z)=\frac{(n-1)!}{(2\pi i)^n}\Bigg[\int_{(\zeta,\lambda_0)\in\partial D\times[0,1]}\bar\partial f(\zeta)\wedge\omega'\Big((1-\lambda_0)\frac{\bar\zeta-\bar z}{|\zeta-z|^2}+$$
$$+\lambda_0\eta(\zeta,z)\Big)\wedge\omega(\zeta)+\int_{\zeta\in D}\bar\partial f(\zeta)\wedge\omega'\left(\frac{\bar\zeta-\bar z}{|\zeta-z|^2}\right)\wedge\omega(\zeta)\Bigg], \tag{4.7}$$

*if $s=0$, and*

$$H^s(\bar\partial f)(z)=\frac{(n+s-1)!}{(s-1)!(2\pi i)^n}\int_{(\zeta,\lambda_0)\in D\times[0,1]}\bar\partial f(\zeta)\times$$
$$\times[\lambda_0(1-\langle\eta(\zeta,z),\zeta-z\rangle)]^{s-1}\times$$
$$\times\,\omega\Big((1-\lambda_0)\frac{\bar\zeta-\bar z}{|\zeta-z|^2}+\lambda_0\eta(\zeta,z)\Big)\wedge\omega(\zeta), \tag{4.8}$$

*if $s\geq 1$.*

The following extension of Theorems 4.2 and 4.3 was obtained in the recent works [11] and [12].

**Theorem 4.4** (Berndtsson, Andersson, 1982). *Under the hypotheses of Theorems* 4.2 *and* 4.3, *let $Q(\zeta,z)=(Q_1(\zeta,z),\dots,Q_n(\zeta,z))$ be an arbitrary smooth $\mathbb{C}^n$-valued function of the complex variable $\zeta\in\bar D$, and $G(t)$ a function of $t\in C^1$, analytic in a domain containing the image of $D\times D$ by the mapping $(\zeta,z)\mapsto\langle Q(\zeta,z),\zeta-z\rangle$ and such that $G(0)=1$. Then, for any function $f$ holomor-*

*phic in D and continuous in $\bar{D}$, we have*

$$f = K^G f - H^G \bar{\partial} f, \tag{4.9}$$

*where*

$$K^G f(z) = C(n)\left[\int_{\partial D} f(\zeta) \sum_{k=0}^{n-1} (-1)^k \frac{1}{k!} G^{(k)}(\langle Q, \zeta - z\rangle)\omega_n(Q, \eta) - \right.$$

$$\left. - \frac{(-1)^n}{n!} \int_D f(\zeta) G^{(n)}(\langle Q, \zeta - z\rangle)\omega_n(Q)\right], \tag{4.10}$$

$$H^G \bar{\partial} f(z) = c(n)\left[\int_D \bar{\partial} f(\zeta) \wedge \sum_{k=0}^{n-1} (-1)^k \frac{1}{k!} G^{(k)}(\langle Q, \zeta - z\rangle)\omega_k\left(Q, \frac{\bar{\zeta} - \bar{z}}{|\zeta - z|^2}\right) + \right.$$

$$+ \int_{(\zeta, \lambda_0) \in \partial D \times [0,1]} \bar{\partial} f(\zeta) \wedge \sum_{k=0}^{n-1} (-1)^k \frac{1}{k!} G^{(k)}(\langle Q, \zeta - z\rangle) \wedge$$

$$\left. \wedge\, \omega_k\left(Q, (1 - \lambda_0)\frac{\bar{\zeta} - \bar{z}}{|\zeta - z|^2} + \lambda_0 \eta(\zeta, z)\right)\right], \tag{4.11}$$

*where*

$$\omega_k(Q, \eta) = \langle \eta, d\zeta\rangle \wedge \langle \bar{\partial}\eta \wedge d\zeta\rangle^{n-1-k} \wedge \langle \bar{\partial} Q \wedge d\zeta\rangle^k,$$

$$\omega_n(Q) = \langle \bar{\partial} Q \wedge d\zeta\rangle^n, \; G^{(k)}(t) = \frac{d^k G}{dt^k}, \; c(n) = \frac{(-1)^{n(n-1)/2}}{(2\pi i)^n}.$$

In the particular case when $Q = \eta$ and $G(t) = (1 - t)^{n+s-1}$, the assertion of Theorem 4.4 reduces to Theorem 4.3.

**4.3. Integral Representations in Strictly Pseudoconvex Domains.** For concrete domains $D$ in $\mathbb{C}^n$, the formulas (4.1), (4.2), (4.6) and (4.9) yield not one integral representation but a whole class of such representations. The choice of an appropriate representation with holomorphic kernel among these is a problem requiring special consideration. For the case of a convex domain with smooth boundary: $D = \{\zeta \in \mathbb{C}^n : \rho(\zeta) < 0\}$, already Leray himself (1956, see [59]) found an appropriate realization of formula (4.1). Namely, in this case, for the cycle $h_z$ in formula (4.1) (or (4.2) with $s = 0$) we may choose the graph of the mapping

$$\zeta \mapsto \eta(\zeta, z) = \frac{\partial p}{\partial z}(\zeta) = \left(\frac{\partial p}{\partial z_1}(\zeta), \ldots, \frac{\partial p}{\partial z_n}(\zeta)\right), \qquad \zeta \in \partial D.$$

Then, formula (4.1) takes on the form

$$f(z) = \frac{(n-1)!}{(2\pi i)^n} \int_{\zeta \in \partial D} f(\zeta) \frac{\omega'\left(\frac{\partial \rho}{\partial \zeta}(\zeta)\right) \wedge \omega(\zeta)}{\left\langle \frac{\partial \rho}{\partial \zeta}(\zeta), \zeta - z\right\rangle^n}, \tag{4.12}$$

for $f \in A(\bar{D})$, $z \in D$.

L.A. Ajzenberg (1967, see [1]) remarked that formula (4.12) holds also for arbitrary *linearly convex domains D* with smooth boundary, i.e. domains such that $T_\zeta^{\mathbb{C}}(\partial D) \subset \mathbb{C}^n \setminus D$ for each point $\zeta \in \partial D$.

The most appropriate realization of formulas (4.1) and (4.2) for strictly pseudoconvex domains (G.M. Khenkin, 1969, 1970; Ramirez, 1970; G.M. Khenkin, A.V. Romanov, 1971, see [40]) of the form (3.14) is obtained by setting

$$\eta(\zeta, z) = \frac{P(\zeta, z)}{\Phi(\zeta, z) - \rho(\zeta)} \tag{4.13}$$

in (4.2), where $\Phi(\zeta, z) = \langle P(\zeta, z), \zeta - z \rangle$ is the barrier function from Theorem 3.10. In this case, formula (4.2) acquires the form

$$f = K^0 f(z) = \frac{(n-1)!}{(2\pi i)^n} \int_{\zeta \in \partial D} f(\zeta) \frac{\omega'(P(\zeta, z)) \wedge \omega(\zeta)}{[\Phi(\zeta, z)]^n} \tag{4.14}$$

for $s = 0$, and

$$f = K^s f(z) = \frac{(n+s-1)!}{(s-1)!(2\pi i)^n} \int_{\zeta \in D} f(\zeta) \left( \frac{-\rho(\zeta)}{\Phi(\zeta, z) - \rho(\zeta)} \right)^{s-1} \times$$
$$\times\, \omega\left( \frac{P(\zeta, z)}{\Phi(\zeta, z) - \rho(\xi)} \right) \wedge \omega(\zeta),$$

for $s > 0$, where $f \in A(\bar{D})$.

The more general formulas (4.6) and (4.9) also turn out to be most useful in applications to strictly pseudoconvex domains when the vector-function $\eta$ has the form (4.13). Namely, Theorems 4.3 and 4.4 were first formulated for this case in the cited works.

We remark that in the particular case where $D$ is the ball $\{\zeta : |\zeta| < 1\}$ in $\mathbb{C}^n$ and

$$\eta(\zeta, z) = \frac{\bar{\zeta}}{\langle \bar{\zeta}, \zeta - z \rangle - (\langle \zeta, \bar{\zeta} \rangle - 1)} = \frac{\bar{\zeta}}{1 - \langle \bar{\zeta}, z \rangle},$$

the operator $K^0$ is precisely the Szegö–Bochner projection (1.9); the operator $K^1 f$ is the Bergman projection (1.10); and the operators $K^s f$, $s \geq 1$, are the Bergman projections in the $L^2$-spaces with weighted measures $(1 - |\zeta|^2)^{s-1}\, dV_{2n}(\zeta)$ according to the work of Rudin and Forelli (1974, see [76]).

**4.4. The Theorem of Fantappiè–Martineau on Analytic Functionals.** Having obtained formulas (4.1) and (4.12), Leray made the following remark [59]: "Fantappié a, plus généralement, exprimé $f(z)$ comme somme de puissances $p$-ièmes de fonctions linéaires de $z$ ($p$: entier négative); d'où l'un des résultats essentiels de sa théorie des fonctionnelles linéaires analytiques: une telle fonctionnelle $F[f]$ est connue quand on connait les valeurs qu'elle prend lorsque $f$ est la puissance $p$-ième d'une fonction linéaire."

For a linearly convex domain $D$ in $\mathbb{C}^n$, containing the origin and having smooth boundary, the more general Cauchy–Fantappié formula, to which Leray refers, can be expressed (S.G. Gindikin, G.M. Khenkin [29]) in the form

$$f(z) = \frac{(p-1)!}{(2\pi i)^n} \int_{\zeta \in \partial D} (D_p)^{n-p} f(\zeta) \frac{\omega'\left(\frac{\partial \rho}{\partial \zeta}(\zeta)\right) \wedge \omega(\zeta)}{\left\langle \frac{\partial \rho}{\partial \zeta}(\zeta), \zeta - z \right\rangle^p \left\langle \frac{\partial \rho}{\partial \zeta}(\zeta), \zeta \right\rangle^{n-p}}, \tag{4.15}$$

where

$$D_p f = pf + \sum_{k=1}^n \zeta_k \frac{\partial f}{\partial \zeta_k}, \qquad f \in C^{(n-p)}(\bar{D}) \cap \mathscr{H}(D).$$

In the special case when $D$ is a convex and circular domain with smooth boundary, a formula equivalent to (4.15) was obtained first by A.A. Temlyakov (1954, see [4]).

We now formulate a result from the theory of analytic functionals which is easily proved with the help of formulas (4.12) and (4.15).

Let $\mathbb{CP}^n$ be $n$-dimensional projective space with homogeneous coordinates $z = (z_0, z_1, \ldots, z_n)$. A domain $D$ in $\mathbb{CP}^n$ is called *strictly linearly concave* if for each point $z \in D$ there exists an $(n-1)$-dimensional projective plane $\zeta_z$, depending continuously on $z$, with the property $z \in \zeta_z \subset D$.

The set of all $(n-1)$-dimensional hyperplanes $\{z \in \mathbb{CP}^n : \langle \zeta, z \rangle = 0\}$, contained in $D$, forms a domain $D^*$ in the dual projective space with homogeneous coordinates $(\zeta_0, \zeta_1, \ldots, \zeta_n)$.

A compact set $G$ in $\mathbb{CP}^n$ is called *strictly linearly convex* if the domain $\mathbb{CP}^n \setminus G$ is strictly linearly concave.

Let $\mathscr{H}^*(G)$ denote the space of linear functionals on the space $\mathscr{H}(G)$.

Let $h \in \mathscr{H}^*(G)$. The *Fantappié indicator* $F_p h$, $p < 0$, of the functional $h$ is a holomorphic function of the variable $\zeta = (\zeta_0, \zeta_1, \ldots, \zeta_n) \subset (\mathbb{CP}^n \setminus G)^*$ defined by $F_p h(\zeta) = (h, \langle \zeta, z \rangle^p)$. We have $F_p h(\lambda \zeta) = \lambda^p F_p h(\zeta)$. Functions with this property are called sections of the line bundle $O(p)$ over $(\mathbb{CP}^n)^*$. The space of all holomorphic sections of $O(p)$ over $D^*$ is denoted by $\mathscr{H}(D^*, O(p))$.

**Theorem 4.5** (Martineau, 1962, 1966; L.A. Ajzenberg, 1966; S.G. Gindikin, G.M. Khenkin, 1978; S.V. Znamenskij, 1979). *Let $G$ be a strictly linearly convex compact set in $\mathbb{CP}^n$. Then for any integer $p < 0$, the Fantappié mapping $h \mapsto F_p h$ yields an isomorphism between the space $\mathscr{H}^*(G)$ and the space $\mathscr{H}((\mathbb{CP}^n \setminus G)^*, O(p))$.*

In the original work of Fantappié (1943, see [68]) this result was proved essentially for the case when $G$ is a polydisc. The case of convex compacta was examined in the works of A. Martineau and L.A. Ajzenberg; the general case was considered in [28].

**4.5. The Cauchy–Fantappiè Formula in Domains with Piecewise-Smooth Boundaries.** The Cauchy–Fantappié–Leray formula (4.1) allows one also to obtain an effective integral representation in domains with piecewise smooth boundary and with several barrier functions, for example, in Weil polyhedra or in classical domains.

The general scheme for obtaining such representations is as follows. Suppose the boundary $\partial G$ of a bounded domain $G$ in $\mathbb{C}^n$ has a regular decomposition into smooth oriented $(2n-1)$-dimensional pieces $\Gamma_j$, $j = 1, 2, \ldots, N$, such that

$$\partial G = \bigcup_{j=1}^{N} \Gamma_j, \quad \partial \Gamma_J = \bigcup_{j=1}^{N} \Gamma_{J,j}, \tag{4.16}$$

where $J = (j_1, \ldots, j_r) \subset \{1, \ldots, N\}$, and each $\Gamma_{J,j}$ is an oriented $(2n-r-1)$-dimensional (smooth) piece of the boundary of the $(2n-r)$-dimensional manifolds $\Gamma_J$. Here, the manifolds $\Gamma_J$ and $\Gamma_{J'}$ are disjoint if $J$ is not a permutation of $J'$ and coincide in the opposite case. The orientations of the manifolds $\Gamma_J$ are, on account of (4.16), antisymmetric with the respect to permutations of the multi-indices $J$.

Now let us consider the simplex

$$\Delta = \left\{ t = \{t_0, t_1, \ldots, t_N\} \in \mathbb{R}^{N+1} : t_j \geq 0, \sum_{j=0}^{N} t_j = 1 \right\}$$

endowed with the standard orientation. For each ordered subset $J = (j_1, \ldots, j_k)$ of $(0, 1, \ldots, N)$ we set

$$\Delta_J = \{t \in \Delta : \sum_{j \in J} t_j = 1\}.$$

The orientation of each subsimplex is chosen such that

$$\partial \Delta_J = \sum_{\nu=1}^{k} (-1)^{\nu+1} \Delta_{J\hat{\nu}},$$

where $J\hat{\nu} = (j_1, \ldots, j_{\nu-1}, j_{\nu+1}, \ldots, j_k)$.

For each ordered subset $I = (i_1, \ldots, i_k)$ of $(1, \ldots, N)$ with strictly increasing components, let $\eta_I = \eta_I(\zeta, z, t)$ be a given $\mathbb{C}^n$-valued vector-function of class Lip 1 in the variables $\zeta \in \Gamma_I$, $z \in G$, $t \in \Delta_I$ with the properties

$$\langle \eta_I(\zeta, z, t), \zeta - z \rangle \neq 0 \tag{4.17}$$

for $(\zeta, z, t) \in \Gamma_I \times G \times \Delta_I$; and

$$\eta_{I\hat{\nu}} = \eta_I(\zeta, z, t) \tag{4.17'}$$

for

$$(\zeta, z, t) \in \Gamma_I \times G \times \Delta_{I\hat{\nu}}.$$

For fixed $z \in G$ and $I \subset (1, 2, \ldots, N)$ we denote by $h_z^I$ the manifold

$$\bigcup_{t \in \Delta_I} \{(\zeta, \eta) : \zeta \in \Gamma_I, \quad \eta = (\zeta, z, t)\}$$

in $\mathbb{C}^n \times \mathbb{C}^n$ with orientation induced by the orientation of the manifold $\Gamma_I \times \Delta_I$.

By (4.17) the chain

$$h_z = \sum_{|I|=1}^{N} h_z^I$$

is a cycle in the domain

$$\{(\zeta, \eta) \in \mathbb{C}^n \times \mathbb{C}^n : \langle \eta, \zeta - z \rangle \neq 0\}$$

and satisfies the conditions of Theorem 4.1.

Formula (4.1) leads to the following integral representation.

**Theorem 4.6** (Norguet, 1961; G.M. Khenkin, 1971; R.A. Ajrapetyan, G.M. Khenkin [3]). *Let $G$ be a domain with boundary of type* (4.16) *and $\{\eta_I\}$ vector functions with the properties* (4.17). *Then for any $z \in G$ and any function $f$ such that $f$ and $\bar\partial f$ are continuous on $\bar G$, we have the equality:*

$$f = Kf - H\bar\partial f, \tag{4.18}$$

*where*

$$Kf = \frac{(n-1)!}{(2\pi i)^n} \sum_{|J|=1}^{n}{}' K_J f, \tag{4.19}$$

$$K_J f(z) = \int_{(\zeta, t) \in \Gamma_J \times \Delta_J} f(\zeta) \omega' \left( \frac{\eta_J(\zeta, z, t)}{\langle \eta_J(\zeta, z, t), \zeta - z \rangle} \right) \wedge \omega(\zeta),$$

$$H(\bar\partial f) = \frac{(n-1)!}{(2\pi i)^n} \sum_{|J|=1}^{n-1}{}' H_J(\bar\partial f) + H_0(\bar\partial f), \tag{4.20}$$

$$H_J(\bar\partial f)(z) = (-1)^{|J|} \int_{(\zeta, t_0, t) \in \Gamma_J \times \Delta_{0,J}} \bar\partial f(\zeta) \wedge$$

$$\omega' \left( (1 - t_0) \frac{\bar\zeta - \bar z}{|\zeta - z|^2} + t_0 \frac{\eta_J(\zeta, z, t)}{\langle \eta_J(\zeta, z, t), \zeta - z \rangle} \right) \wedge \omega(\zeta),$$

$$H_0(\bar\partial f)(z) = \int_{\zeta \in D} \bar\partial f(\zeta) \wedge \omega' \left( \frac{\bar\zeta - \bar z}{|\zeta - z|^2} \right) \wedge \omega(\zeta).$$

The assertion of Theorem 4.6 was first obtained by Norguet (1961) for holomorphic functions and by G.M. Khenkin (1971) for smooth functions with the following additional assumption:

$$\eta_J(\zeta, z, t) = \sum_{j \in J} t_j \eta_j(\zeta, z). \tag{4.21}$$

In [4] it is remarked that the hypothesis (4.21) can be successfully replaced by the more general hypothesis (4.17).

In the situation when (4.21) holds it is possible in formulas (4.19) and (4.20) to integrate explicitly with respect to the parameter $t \in \Delta_J$.

**Proposition 4.7** ([74], [3]). *For a vector-function $\eta_J$ of type* (4.21) *we have for fixed $\zeta$ and $z$*

$$\int_{t\in\Delta_J} \omega'\left(\frac{\eta_J(\zeta,z,t)}{\langle \eta_J(\zeta,z,t),\zeta-z\rangle}\right) =$$

$$= \frac{(n-k)!}{(n-1)!}(-1)^{k(k-1)/2}\sum_{\beta\in J}{}' \frac{\det(\eta_{j_1},\ldots,\eta_{j_k},\bar\partial_\zeta P_{\beta_1},\ldots,\bar\partial_\zeta P_{\beta_{n-k}})}{\prod_{\nu=1}^{k}\langle \eta_{j_\nu},\zeta-z\rangle \prod_{\nu=1}^{n-k}\langle \eta\beta_\nu,\zeta-z\rangle}, \tag{4.22}$$

*where the sum $\Sigma'$ runs over all monotone multi-indices $\beta \in J$.*

Formula (4.22) generalizes an assertion of Fantappié (1943, see [63]): The kernel in the Cauchy formula (0.2) for the polydisc can be expressed in the form

$$\frac{1}{(\zeta_1-z_1)\ldots(\zeta_n-z_n)} = (-1)^{\frac{n(n-1)}{2}}(n-1)!\int_{t\in\Delta}\frac{\omega'(t)}{\langle t,\zeta-z\rangle^n} \tag{4.23}$$

**4.6. Integral Representations in Pseudoconvex Polyhedra and Siegel Domains.** In order to obtain, for example, from formula (4.18), the Weil formula (2.3) for analytic polyhedra of type (2.1), (2.2), it is sufficient in (4.18) to set

$$\eta_J(\zeta,z,t) = \sum_{j\in J} t_j \frac{P_j(\zeta,z)}{F_j(\zeta)-F_j(z)} \tag{4.24}$$

and to notice that by Proposition 4.7, in this situation the summands $K_J f$ in (4.18) are zero if $|J|<n$, and

$$K_J f(z) = \frac{(-1)^{\frac{n(n-1)}{2}}}{(n-1)!}\int_{\Gamma_J} f(\zeta)\frac{\det(P_{j_1}(\zeta,z),\ldots,P_{j_n}(\zeta,z))\,\omega(\zeta)}{\prod_{\nu=1}^{n}(F_{j_\nu}(\zeta)-F_{j_\nu}(z))},$$

*if* $|J|=n$.

The integral representation (4.18) works particularly well in pseudoconvex polyhedra, a natural class of domains which includes both analytic polyhedra and piecewise-strictly pseudoconvex domains.

We call a bounded domain $D$ in $\mathbb{C}^n$ (see [19], [54]) a *strictly pseudoconvex polyhedron* if there are given holomorphic mappings $\chi_j$, $j=1,2,\ldots,N$, from some domain $\Omega \supset \bar D$ onto a domain $\Omega_j$ in the space $\mathbb{C}^{m_j}$ of dimension $m_j(\le n)$ and strictly plurisubharmonic functions $\rho_j \in C^2(\Omega)$ such that the domain $D$ has the form

$$D = \{\zeta\in\Omega: \rho_j(\chi_j(\zeta)) < 0, \qquad j=1,2,\ldots,N\}. \tag{4.25}$$

If each $\chi_j$ is the identity mapping of $D$ onto itself, then definition (4.25) yields an arbitrary piecewise strictly pseudoconvex domain. If all $m_j$ are 1, then definition (4.25) yields arbitrary analytic polyhedra.

The boundary of a pseudoconvex polyhedron (4.25) consists of pieces

$$\Gamma_j = \{\zeta \in \partial D: \rho_j(\chi_j(\zeta)) = 0\}, \qquad j = 1, 2, \ldots, N,$$

joined together by the sets $\Gamma_J = \Gamma_{j_1} \cap \ldots \cap \Gamma_{j_k}$, where $J = \{j_1, \ldots, j_k\} \subset \{1, 2, \ldots N\}$.

Let us denote by $D_j$ the domain in $\mathbb{C}^{m_j}$ of the form

$$D_j = \{\xi \in \Omega_j : \rho_j(\xi) < 0\}.$$

Let

$$\tilde{\Phi}_j(\xi, w) = \sum_{\alpha=1}^{m_j} \tilde{P}_j^\alpha(\xi, w)(\xi_\alpha - w_\alpha),$$

be a barrier function for the domain $D_j$ with the properties in Theorem 3.10, where $\xi \in U(\partial D_j)$ and $z \in U(\bar{D}_j)$.

By Theorem 2.11 the holomorphic mappings $\chi_j = (\chi_{j,1}, \ldots, \chi_{j,m_j})$ admit the representations

$$\chi_{j,\alpha}(\zeta) - \chi_{j,\alpha}(z) = \sum_{\beta=1}^{n} Q_{j,\alpha}^\beta(\zeta, z)(\zeta_\beta - z_\beta),$$

where $\zeta, z \in \Omega, \alpha = 1, \ldots, m_j$, and the functions $Q_{j,\alpha}^\beta$ are holomorphic in the domain $\Omega \times \Omega$.

We may take the barrier functions $\Phi_j(\zeta . z)$, for the surfaces $\Gamma_j$, of the form

$$\Phi_j(\zeta, z) = \tilde{\Phi}_j(\chi_j(\zeta), \chi_j(z)).$$

We have

$$\Phi_j(\zeta, z) = \sum_{\beta=1}^{n} P_j^\beta(\zeta, z)(\zeta_\beta - z_\beta), \tag{4.26}$$

where

$$P_j^\beta(\zeta, z) = \sum_{\alpha=1}^{m_j} Q_{j,\alpha}^\beta(\zeta, z) \cdot \tilde{P}_j^\alpha(\chi_j(\zeta), \chi_j(z)).$$

For pseudoconvex polyhedra of type (4.25), it is natural to take the vector-functions $\{\eta_J(\zeta, z, t)\}$ of the form

$$\eta_J(\zeta, z, t) = \sum_{j \in J} t_j P_j(\zeta, z), \tag{4.27}$$

where $t \in \Delta_J, \zeta \in \Gamma_J, z \in D$.

These vector-functions satisfy conditions (4.17) and (4.21). From this and Theorem 4.6 and Proposition 4.7, it follows that in a pseudoconvex polyhedron of type (4.25) satisfying the non-degeneracy condition:

$$d\rho_{j_1}(\chi_{j_1}(\zeta)) \wedge \ldots \wedge d\rho_{j_k}(\chi_{j_k}(\zeta)) \neq 0 \text{ on } \Gamma_J,$$

where $J = (j_1, \ldots, j_k)$, $|J| \leq n$, the integral formula (4.18) for holomorphic functions has the form:

$$f(z) = Kf(z) = \frac{1}{(2\pi i)^n} \sum_{k=|J|=1}^{n} (n-k)! \sum_{\beta \in J}' (-1)^{\frac{k(k-1)}{2}} \times$$
$$\times \int_{\zeta \in \Gamma_J} f(\zeta) \frac{\det(P_{j_1}, \ldots, P_{j_k}, \bar{\partial}_\zeta P_{\beta_1}, \ldots, \bar{\partial}_\zeta P_{\beta_{n-k}}) \wedge \omega(\zeta)}{\Phi_{j_1}(\zeta, z) \ldots \Phi_{j_k}(\zeta, z) \Phi_{\beta_1}(\zeta, z) \ldots \Phi_{\beta_{n-k}}(\zeta, z)}, \tag{4.28}$$

where $J = (j_1, \ldots, j_k)$.

We now show how the integral representation of Cauchy–Bochner–Gindikin follows from formula (4.18) for an important class of convex, domains which, generally speaking, have non-smooth boundaries, namely, Siegel domains. For example, by a theorem of I.I. Piatetski–Shapiro (1961) every bounded homogeneous domain in $\mathbb{C}^n$ can be realized as a Siegel domain (see [71]).

Let $V$ be an open cone in $\mathbb{R}^k$ with vertex at the origin and not containing a single entire line, i.e. $V$ is an acute cone. Let $\Phi$: $\mathbb{C}^{n-k} \times \mathbb{C}^{n-k} \to C^k$ be an Hermitian $V$-positive form, that is,

$$\Phi(z, w) = \sum_{\alpha,\beta} \Phi_{\alpha,\beta} z_\alpha \cdot \bar{w}_\beta; \qquad \Phi_{\alpha,\beta} = \bar{\Phi}_{\beta,\alpha} \in \mathbb{C}^k,$$

$\Phi(w, w) \in V$ for all $w \in \mathbb{C}^{n-k}$ and $\Phi(w, w) = 0$, only if $w = 0$.

A *Siegel domain* is a domain in $\mathbb{C}^n$ of the form

$$D = \{z = (z', z'') \in \mathbb{C}^k \times \mathbb{C}^{n-k} \colon \rho(z) = \operatorname{Im} z' - \Phi(z'', z'') \in V\}. \tag{4.29}$$

We introduce for consideration the dual cone $V^*$; this is the cone in $(\mathbb{R}^k)^*$ given by

$$V^* = \{\lambda \in (\mathbb{R}^k)^* \colon \langle \lambda, x \rangle > 0, \qquad \forall x \in V\}.$$

In terms of $V^*$ the Siegel domain (4.29) has the form:

$$D = \{z = (z', z'') \colon \langle \lambda, \rho(z', z'') \rangle > 0, \qquad \forall \lambda \in V^*\}.$$

If $k = n$, then the domain (4.29) is called a Siegel domain of the first kind or a tubular domain over the acute cone $V$. *If* $k > n$, then the domain (4.29) is called a Siegel domain of the second kind.

The manifold

$$S = \{(z', z'') \in \mathbb{C}^n \colon \operatorname{Im} z' = \Phi(z'', z'')\} \tag{4.30}$$

is called the skeleton of the Siegel domain (4.29). We shall say that a holomorphic function $f$ on a Siegel domain belongs to the class $H^p(D)$, if

$$\|f\|_{H^p} = \sup_{x \in V} \|f(z' + ix, z'')\|_{L_p(S)} < \infty.$$

**Theorem 4.7** (see [3]). *Let D be a Siegel domain* (4.29) *and* $f \in H^1(D)$. *Then, we have*

$$f(z)=\frac{(n-1)!}{(2\pi i)^n}\int_{(\zeta,\lambda)\in S\times V_1^*} f(\zeta)\frac{\omega'\left(\left\langle \lambda, \frac{\partial\rho}{\partial\zeta}(\zeta)\right\rangle\right)\wedge\omega(\zeta)}{\left\langle\left\langle \lambda, \frac{\partial\rho}{\partial\zeta}(\zeta)\right\rangle, \zeta-z\right\rangle^n}, \tag{4.31}$$

*where* $\rho(\zeta)=\rho(\zeta',\zeta'')$, $V_1^*=\{\lambda\in V^*: |\lambda|=1\}$.

For Siegel domains of the first kind, formula (4.31) is equivalent to the Bochner formula (see article III in Vol. 8). For Siegel domains of the second kind, formula (4.31) is equivalent to a formula of S.G. Gindikin and Koranyi–Stein [57], which in turn generalizes the formulas of Bochner and Hua Loo-keng introduced in §1 for classical domains (see also [63], [74]). In order to obtain formula (4.31) from formula (4.18), it is sufficient to set

$$\eta(\zeta,z)=\frac{\left\langle \lambda, \frac{\partial\rho}{\partial\zeta}(\zeta)\right\rangle}{\left\langle\left\langle \lambda, \frac{\partial\rho}{\partial\zeta}(\zeta)\right\rangle, \zeta-z\right\rangle},$$

in formula (4.19), where $\zeta\in\partial D, \lambda\in\{\lambda\in V_1^*: \langle\lambda,\rho\rangle(\zeta)=0\}$, and to convince oneself that the integral $Kf$ of type (4.19) is equal to (4.31) on $S$ and equal to zero on $\partial D\backslash S$.

An effective integral representation for smooth functions in classical domains was obtained recently by Lu Qi-Keng (Lu Qi-Keng, On the Cauchy–Fantappiè Integral and the $\bar\partial$-Problem, Preprint, Institute of Mathematics Academia Sinica, 1985).

## §5. Integral Representations in Problems from the Theory of Functions on Pseudoconvex Domains

The integral representations of Cauchy–Fantappiè from §4 allow us to obtain, anew and in a more constructive manner, fundamental facts from the theory of functions in pseudoconvex domains introduced in §§2, 3.

Particularly complete results are obtained for strictly pseudoconvex domains. In fact, for strictly pseudoconvex domains, the kernels of Cauchy-type operators are not only holomorphic in $z$ but also allow estimates analogous to the estimates for the Cauchy kernel in one variable. We introduce here typical examples of such estimates.

### 5.1. Estimates for Integrals of Cauchy–Fantappiè Type and Asymptotics of Szegö and Bergman Kernels in Strictly Pseudoconvex Domains.

**Theorem 5.1** (G.M. Khenkin 1968, 1974, see [19], [40]). *Let $D$ be a strictly pseudoconvex domain of type* (3.14); *$g$ a function in $C^\alpha(\bar{D})$, $\alpha>0$; and $K^s$ a Cauchy-type operator of the form* (4.14), $s \geq 0$. *Then, the Hankel operators given by*

$$f \mapsto g \cdot f - K^s(g \cdot f)$$

*are completely continuous operators from $H^\infty(D)$ to the space $C(\bar{D})$, while the Töplitz operators $f \mapsto K^s(g \cdot f)$ are bounded operators in the space $H^\infty(D)$ with a closed range in $H^\infty(D)$ and finite dimensional kernel. Moreover,*

$$g \cdot f - K^s(g \cdot f) = H^s(f \cdot \bar{\partial} g), \tag{5.1}$$

*where $H^s$ is given by* (4.7), (4.8), (4.13).

**Theorem 5.2** (Ahern, Schneider, Phong, Stein, 1977, see [76], $s=0$; Ligotcka, 1982, see [62], $s=1$). *Let $D$ be a strictly pseudoconvex domain with boundary of class $C^\infty$ and let $K^s$ be a Cauchy operator given by* (4.14). *Then for any integer $s \geq 0$ and any non-integer $\alpha>0$, the operator $K^s$ maps the space $C^\alpha(\bar{D})$ continuously into the space $C^\alpha(\bar{D}) \cap \mathscr{H}(D)$.*

Of course a rather direct proof of Theorems 5.1 and 5.2 can be based on the estimate (3.15) for the barrier function $\Phi(\zeta, z) = \langle P(\zeta, z), \zeta - z \rangle$. Formula (5.1) is an immediate consequence of formula (4.6).

More complete information on estimates of integrals of Cauchy–Szegö type for the ball or for strictly pseudoconvex domains with smooth boundary are set forth in the paper by A.B. Aleksandrov Vol. 8, II.

For a strictly pseudoconvex domain $D$ with smooth boundary, we may represent the operators $K^0$ and $K^1$ given by (4.14) in the form

$$\begin{aligned} K^0 f(z) &= \int_{\partial D} f(\zeta) K^0(\zeta, z) dV_{2n-1}(\zeta), \\ K^1 f(z) &= \int_{D} f(\zeta) K^1(\zeta, z) dV_{2n}(\zeta), \end{aligned} \tag{5.2}$$

where $dV_k$ is the element of $k$-dimensional volume in $\mathbb{C}^n$.

The estimates of the projections $K^s$, contained in Theorems 5.1 and 5.2 allow us to show that for an arbitrary strictly pseudoconvex domain, the Cauchy projections given by (5.2) differ respectively from the orthogonal Szegö projection $S: L^2(\partial D) \to H^2(\partial D)$ and the orthogonal Bergman projection $\mathrm{B}: L^2(D) \to H^2(D)$ by completely continuous operators. More precisely, let us introduce the operators given by

$$A^0 f(z) = \int_{\partial D} f(\zeta) A^0(\zeta, z)\, dV_{2n-1}(\zeta),$$

$$A^1 f(z) = \int_{D} f(\zeta) A^1(\zeta, z)\, dV_{2n}(\zeta), \tag{5.3}$$

where $A^s(\zeta, z) = K^s(\zeta, z) - \bar{K}^s(\zeta, z), s = 0, 1$.

**Theorem 5.3** (Kerzman, Stein [48], Ligotcka [62]). *For any strictly pseudoconvex domain $D$ with boundary of class $C^\infty$, the operators $A^0$ and $A^1$ given by (5.3) are completely continuous and the operators $(I - A^0)$ and $(I - A^1)$ are continuous and invertible in the spaces $L^2(\partial D)$ and $L^2(D)$ respectively. Moreover,*

$$S = K^0(I - A^0)^{-1}; \quad B = K^1(I - A^1)^{-1}.$$

*In addition, the operators $A^s$, $s = 0,1$, map the spaces $C^\alpha(\partial D)$ and $C^\alpha(\bar{D})$ respectively into the spaces $C^{\alpha + 1/2}(\partial D)$ and $C^{\alpha + 1/2}(\bar{D})$, while the orthogonal projections $S$ and $B$ respectively map continuously into the spaces $C^\alpha(\partial D)$ and $C^\alpha(\partial \bar{D})$ for any non-integer $\alpha > 0$.*

From Theorem 5.3 one can deduce the asymptotic representation for the Bergman and Szegö kernels originally obtained by Fefferman (1974, see [9]) by rather difficult methods.

**Theorem 5.4** (Fefferman, 1974; Boutet de Monvel, Sjöstrand, 1976, see [9]). *Let $D = \{z \in \mathbb{C}^n : \rho(z) < 0\}$ be a strictly pseudoconvex domain with boundary of class $C^\infty$. Then the kernels $S(z, w)$ and $B(z, w)$ of the orthogonal Szegö and Bergman projections respectively have the following asymptotic form:*

$$S(z, w) = \frac{\varphi_0(z, w)}{(-\rho(z, w))^n} + \tilde{\varphi}_0(z, w)\log(-\rho(z, w)),$$

$$B(z, w) = \frac{\varphi_1(z, w)}{(-\rho(z, w))^{n+1}} + \tilde{\varphi}_1(z, w)\log(-\rho(z, w)), \tag{5.4}$$

*where $\rho(z, w)$, $\varphi_s(z, w)$, $\tilde{\varphi}_s(z, w)$, $s = 0,1$, are functions of class $C^\infty(\bar{D} \times \bar{D})$ with the properties:*

*a)* $\rho(z, z) = \rho(z)$;

*b)* $\varphi_0(z, z) = \tilde{\varphi}_0(z, z) = c_n \begin{bmatrix} \rho & \dfrac{\partial \rho}{\partial \bar{z}_k} \\ \dfrac{\partial \rho}{\partial z_j} & \dfrac{\partial^2 \rho}{\partial z_j \partial \bar{z}_k} \end{bmatrix}$ *on* $\partial D$;

*c) the forms $\bar{\partial}_z \varphi_s$, $\bar{\partial}_z \tilde{\varphi}_s$, $\partial_w \varphi_s$, $\partial_w \tilde{\varphi}_s$, $s = 0,1$, together with all derivatives, vanish for $z = w$.*

For analytic polyhedra or more general pseudoconvex polyhedra of type (4.25), one can also obtain good estimates for the Cauchy-type operator (4.28), but under the following non-degeneracy conditions: for any ordered choice of indices $A=\{\alpha_1,\dots,\alpha_s\}$ and $B=\{\beta_1,\dots,\beta_t\}$ the real Jacobian of the mapping

$$z\mapsto(\rho_{\beta_1}(\chi_{\beta_1}(z)),\dots,\rho_{\beta_t}(\chi_{\beta_t}(z)),\chi_{\alpha_1}(z),\dots,\chi_{\alpha_s}(z)) \tag{5.5}$$

is constant in a neighbourhood of the set $\Gamma_{A\cup B}\subset\partial D$.

**Theorem 5.5** (A.I. Petrosyan [69], G.M. Khenkin, A.G. Sergeev [54]). *Let $D$ be a pseudoconvex polyhedron given by* (4.25), *satisfying the non-degeneracy condition* (5.5). *Let $K$ be a Cauchy-type operator of the form* (4.28). *Then, for any function $g\in C^\alpha(\bar D)$, $\alpha>0$, the Töplitz operator $f\mapsto K(g\cdot f)$ is bounded in the space $H^\infty(D)$.*

Theorem 5.5 was first proved by A.I. Petrosyan [69] for non-degenerate Weil polyhedra using the Weil–Vitushkin formula 2.3′. For general pseudoconvex polyhedra, the assertion of the theorem follows from the main results in [54].

**Theorem 5.6** (R.A. Ajrapetyan [1], Jöricke [47]). *Suppose that the functions $\rho_j(\chi_j(\zeta))$, defining the pseudoconvex polyhedron in Theorem 5.5, are of class $C^\infty$. Then for any $\alpha>\varepsilon>0$, the Cauchy-type operator $f\mapsto Kf$ operates continuously from the space $C^\alpha(\bar D)$ to the space $C^{\alpha-\varepsilon}(\bar D)$.*

Theorem 5.6 was first obtained by R.A. Ajrapetyan for piecewise strictly pseudoconvex domains, then by Jöricke for Weil polyhedra. The general case was formulated in [40].

Although the pseudoconvex polyhedra form a rather wide class of domains, they are far from exhausting the fundamental domains for which it would be useful to have estimates for Cauchy-type integrals.

In particular, for multidimensional complex harmonic analysis, Siegel domains are very important, but the only ones in the above class are balls and their direct products.

A good indication of the behaviour of Cauchy-type integrals for Siegel domains other than direct products of balls is given by the following assertion (see [47]).

**Theorem 5.7** (Jöricke, 1983). *Let $D$ be a tubular domain in $\mathbb{C}^n$* over the spherical cone $V=\{Y\in\mathbb{R}^n\colon Y_1^2>Y_2^2+\dots+Y_n^2,\ Y_1>0\}$ *and let $S$ be the Shilov boundary of this domain. Then, for any $\alpha>n/2-1$, $\varepsilon>0$, and $n\geq 2$, the Cauchy–Szegö–Bochner type integral $f\to Kf$ given by* (4.31) *is a continuous operator from $C^\alpha(\bar D)\cap L^2(S)$ into the space $C^{\alpha-n/2+1-\varepsilon}(\bar D)\cap L^2(S)$. Moreover, the result is sharp with respect to the exponent.*

More detailed information on estimates of Cauchy-type integrals for tubular domains can be found in the paper of Vol. 8, IV.

**5.2. Localization of Singularities and Uniform Approximation of Bounded Holomorphic Functions.** We pass now directly to the treatment of those questions from the theory of functions in pseudoconvex domains for whose solution the most natural apparatus turns out to be the Cauchy–Fantappié formula.

We begin with a significant generalization of the Cousin–Oka Proposition 3.4 (see [35]).

**Theorem 5.8** (Andreotti, Hill, 1972; E.M. Chirka, 1975). *Let $D = \{z \in \Omega: \rho(z) < 0\}$ be a strictly pseudoconvex domain and $S$ a smooth hypersurface given by*

$$S = \{z \in \Omega: \rho_1(z) = 0\},$$

*where $\rho_1 \in C^1(\Omega)$ and $d\rho_1 \wedge d\rho \neq 0$ on $S \cap \partial D$.*

*Set*

$$D_\pm = \{z \in D: \pm\rho_1 < 0\}.$$

*Then, for any $\alpha > 0$, any function of class $C^\alpha$ on $\overline{S \cap D}$, satisfying the tangential Cauchey–Riemann equations on $S \cap D$, can be represented in the form*

$$f(z) = f_+(z) - f_-(z), \qquad z \in S \cap D,$$

*where $f_\pm$ are functions holomorphic in the domains $D_\pm$ and continuous on $\bar{D}_\pm$.*

This important theorem was first proved by Andreotti and Hill for the case when $f$ and $f_\pm$ are functions of class $C^\infty(S \cap D)$. The Cauchy–Fantappié formulas (see [38]) allow us to write the desired functions $f_\pm$ with the formulas

$$f_\pm(z) = \frac{(n-1)!}{(2\pi i)^n}\left[\int\limits_{S \cap D} f(\zeta)\omega'\left(\frac{\bar{\zeta} - \bar{z}}{|\zeta - z|^2}\right) \wedge \omega(\zeta) + \right.$$
$$\left. + \int\limits_{(S \cap \partial D) \times [0,1]} f(\zeta)\omega'\left((1-\lambda)\frac{\bar{\zeta} - \bar{z}}{|\zeta - z|^2} + \lambda\frac{P}{\Phi}\right) \wedge \omega(\zeta)\right],$$

where $z \in D_\pm$, $\Phi(\zeta, z) = \langle P(\zeta, z), \zeta - z\rangle$ is the barrier function from Theorem 3.10. The equality $\bar{\partial} f_\pm(z) = 0$ for $z \in D_\pm$ can be verified by an immediate calculation considering that $\bar{\partial} f = 0$ on $S \cap D$. The estimates $f_\pm \in A(D_\pm)$ follow from estimates for Cauchy-type integrals and Theorem 5.6.

The following theorem on the "localization of singularities" for bounded functions holomorphic in strictly pseudoconvex domains is a significant generalization to several variables of a well-known result of A.G. Vitushkin (1966, see [80], [25]).

**Theorem 5.9** (G.M. Khenkin, 1969, 1974, see [40]). *Let $D$ be an arbitrary strictly pseudoconvex domain given by* (3.14) *whose boundary is not necessarily smooth and let $U_j \subset \mathbb{C}^n$ $(j = 1, 2, \ldots, N)$ be open sets such that $\bar{D} \subset \bigcup_{j=1}^N U_j$. Then, any bounded holomorphic function $f$ on $D$ can be written in the form*

$$f = \sum_{j=1}^{N} f_j, \tag{5.6}$$

*where each function $f_j$ is bounded and holomorphic in some neighbourhood of the set $\bar{D}\setminus(\partial D \cap U_j)$. Moreover, if $f$ is continuous on $\bar{D}$, then each function $f_j$ is also continuous on $\bar{D}$.*

Theorem 5.9 also extends results of A. Weil and Oka on the localization of singularities of functions (without estimates) contained in Propositions 2.3 and 2.4.

To prove Theorem 5.9 we choose smooth function $\chi_j$ on $\mathbb{C}^n$ such that $\Sigma\,\chi_j = 1$ on $\bar{D}$ and $\chi_j = 0$ on $\mathbb{C}^n\setminus U_j$. Further, we set $f_j = K^1(\chi_j f)$, where $K^1$ is the Cauchy-type operator given by (4.14). The functions $\{f_j\}$: satisfy formula (5.6) by the Cauchy formula (4.14); are holomorphic on $\bar{D}\setminus(\partial D \cap U_j)$ by inequality 3.15; and are bounded on $\bar{D}$ by Theorem 5.1.

Theorems 5.9, 5.1, 5.5 allow us to obtain the following result on uniform approximation which for pseudoconvex polyhedra is a more precise version of Oka–Weil approximation theorem 2.6.

**Theorem 5.10** (G.M. Khenkin, 1968, 1974, see [40]). *Let $K$ be a strictly pseudoconvex compact set given by*

$$K = \{z \in \Omega \colon \rho(z) \leq 0\}, \tag{5.7}$$

*where $\rho$ is a strictly plurisubharmonic function of class $C^2$ defined in neighbourhood $\Omega$ of the compact set $K$. Then any function continuous on $K$ and holmorphic on the interior $K_0$ of the compact set $K$ can be uniformly approximated on $K$ by functions holomorphic in a neighbourhood of $K$.*

Theorem 5.10 contains two interesting extreme cases. If the set $K_0$ of interior points is empty, then the compact set $K$ is the zero set of a non-degenerate plurisubharmonic function, and in this case, the theorem was first proved in the works of Hörmander–Wermer, Nirenberg–Wells, and Harvey–Wells in the years 1968–1972.

On the other hand if $K_0 \neq \phi$ and $\operatorname{grad} \rho \neq 0$ on $\partial K_0$, then $K_0$ is a strictly pseudoconvex domain with smooth boundary, and in this case, the theorem was first proved in the works of G.M. Khenkin, and Lieb and Kerzman in the years 1968–1970 (see [82]).

**Theorem 5.11** (A.I. Petrosyan, 1970; G.M. Khenkin, 1974). *Let $D$ be a pseudoconvex polyhedron given by* (4.25) *and satisfying the non-degeneracy condition* (5.5). *Then, any function holomorphic on $D$ and continuous on $\bar{D}$ can be uniformly approximated by functions holomorphic in the domain $\Omega \supset \bar{D}$.*

Theorem 5.11 was first obtained by A.I. Petrosyan [69] for non-degenerate Weil polyhedra; for general non-degenerate pseudoconvex polyhedra, Theorem 5.11 was formulated in [19]. (see also [54]).

For domains in $\mathbb{C}^1$, Theorem 5.10 and 5.11 are particular cases of the classical approximation theorem of A.G. Vitushkin: for any compact set $K \subset \mathbb{C}^1$ having no inner boundary, any function in $A(K)$ can be uniformly approximated by rational functions (see [80]).

**5.3. Interpolation and Division with Uniform Estimates.** Let us now formulate two typical results on the continuation of holomorphic functions given on submanifolds of pseudoconvex domains. These results sharpen corresponding results of Oka and Cartan 2.12, 2.13 (see [40]).

**Theorem 5.12** (G.M. Khenkin, 1972; G.M. Khenkin, Leiterer, 1980; Amar, 1980). *Let $D$ be an arbitrary strictly pseudoconvex domain given by* (3.14) *and $M$ and arbitrary closed analytic submanifold of a domain $\Omega \supset \bar{D}$. Then:*

*a) for each bounded holomorphic function $f$ on $M \cap D$, there exists a bounded holomorphic function $F$ on $D$ such that $F = f$ on $M \cap D$;*

*b) if moreover $f$ is continuous on $M \cap \bar{D}$, then there exists a function $F$, continuous on $\bar{D}$, holomorphic on $D$ and agreeing with $f$ on $M \cap \bar{D}$.*

Theorem 5.12 was proved in the first place (G.M. Khenkin, 1972) under the assumption that the intersection of $M$ with $\partial D$ is transversal. We shall indicate here the outline of the proof of Theorem 5.12 for the situation when $M$ is a complete intersection in $\Omega$, i.e. when

$$M = \{z \in \Omega: g_1(z) = \ldots = g_k(z) = 0\}, \tag{5.8}$$

where $g_1, \ldots, g_k$ are holomorphic functions in $\Omega$ such that $dg_1 \wedge \ldots \wedge dg_k \neq 0$ on $M$.

Under the condition (5.8), the function $F$, satisfying the assertion a) in Theorem 5.12, can be found in the following form (G.M. Khenkin, 1972; Stout, 1975; Berndtsson, 1983 (see [40], [4], [74])):

$$F(z) = \frac{(n-k)!}{(2\pi i)^{n-k}} \int_{\zeta \in M \cap D} f(\zeta) \det\left(P_1, \ldots, P_k, \bar{\partial}_\zeta \frac{P}{\tilde{\Phi}}, \ldots, \bar{\partial}_\zeta \frac{P}{\tilde{\Phi}}\right) \wedge \omega_g(\zeta) \tag{5.9}$$

where $\omega_g(\zeta)$ is a form on $M$ with the property

$$dg_1(\zeta) \wedge \ldots \wedge dg_k(\zeta) \wedge \omega_g(\zeta) = \omega(\zeta),$$

$\tilde{\Phi}(\zeta, z) = \langle P(\zeta, z), \zeta - z \rangle - \rho(\zeta)$ is the barrier function from Theorem 3.10; $\{P_j\}, j = 1, 2, \ldots, k$ are $\mathbb{C}^n$-valued functions of the variables $\zeta, z \in \Omega$ with the properties

$$g_j(\zeta) - g_j(z) = \langle P_j(\zeta, z), \zeta - z \rangle. \tag{5.10}$$

Such functions $\{P_j\}$ exist by Theorem 2.11 (see also Theorem 5.14 below). In order to verify the equality $F|_M = f$, it is sufficient to convince oneself that, for $z \in M \cap D$, formula (5.9) yields the Cauchy–Fantappié representation for $f$. Indeed, let us consider the Cauchy–Fantappié formula (4.28) in the polyhedral domain

$$D_\varepsilon = \left\{\zeta \in D: \sup_j |g_j(\zeta)| < \varepsilon\right\}.$$

Passing to the limit as $\varepsilon \to 0$ in this formula, for $z \in M$ we obtain formula (5.9). To complete the proof of Theorem 5.12, it is necessary to show (and this is the hardest part) that

$$\sup_{z\in D} |F(z)| \leq \gamma \sup_{z\in M} |f(z)|.$$

Assertion b) in Theorem 5.12 is proved on the basis of assertion a) and the approximation Theorem 5.10. For related results see [39].

**Theorem 5.13** (G.M. Khenkin, P.L. Polyakov [42]). *Let $D = \{\zeta\in\mathbb{C}^n: |\zeta_k| < 1, k = 1, \ldots, n\}$ be the polydisc in $\mathbb{C}^n$ and $\Gamma_J = \{\zeta\in\partial D: |\zeta_j| = 1, j\in J\}$ be a component of $\partial D$ with multi-index $J = (j_1, \ldots, j_p)$. Let $M$ be a closed analytic manifold in a neighbourhood of $\bar{D}$ satisfying a transversality condition of the form: for each point $z\in M\cap\partial D$, there exists a neighbourhood $U_z$ and holomorphic functions $\{g_r\}$ in $U_z$ such that $M\cap U_z = \{\zeta\in U_z: g_r(\zeta) = 0, r = 1, \ldots, k\}$ and*

$$dg_{r_1}(z) \wedge \ldots \wedge dg_{r_q}(z) \wedge d\zeta_{j_1}(z) \wedge \ldots \wedge d\zeta_{j_p}(z) \neq 0 \tag{5.11}$$

*for $z\in M\cap\Gamma_J$, $q \leq k$, $p + q \leq n$.*

*Then, for each bounded holomorphic function $f$ on $M\cap D$, there exists a bounded holomorphic function $F$ on $D$ such that $F|_M = f$.*

We remark (Rudin, 1969) that without the transversality condition (5.11) for inner points of the boundary components, the assertion of Theorem 5.13 cannot hold (see [42]).

A scheme for proving Theorem 5.13 using a formula for the solution of the $\bar{\partial}$-equations is discussed in §6. Integral formulas also turn out to be the best tool for the problem of continuing holomorphic functions from submanifolds of $\mathbb{C}^n$ to entire functions with optimal estimates (Berndtsson [11]), (see also the paper of L.I. Ronkin, Vol. 9, I).

For strictly pseudoconvex domains, the following result significantly strengthens the division theorem of Oka–Hefer 2.11.

**Theorem 5.14** (Berndtsson [11], Bonneau, Cumenge, Zeriahi [16]). *Let $D = \{z\in\Omega: \rho(z) < 0\}$ be a strictly pseudoconvex domain with boundary of class $C^\infty$ and $M$ a submanifold of $\Omega$ given by*

$$M = \{z\in\Omega: g_1(z) = \ldots = g_k(z) = 0\}, \tag{5.12}$$

*where $g_j\in\mathscr{H}(\Omega)$, $j = 1, \ldots, k$, and $dg_1 \wedge \ldots \wedge dg_k \neq 0$ on $M$. Suppose $M$ intersects $\partial D$ transversally.*

*Then, for any function $f\in H^\infty(D)$ such that $|f(z)| = 0(|g(z)|)$, where $|g|^2 = \Sigma_j |g_j|^2$, one can give functions $g_j\in H^\infty(D)$, $j = 1, 2, \ldots, k$, using explicit formulas, such that*

$$f(z) = \sum_{j=1}^{k} \varphi_j(z)\cdot g_j(z), \qquad z\in D. \tag{5.13}$$

*If moreover $f\in C^{\alpha/2}(\bar{D})$, then for non-integral $\alpha > 1$, we have $\varphi_j\in C^{(\alpha-1)/2}(\bar{D})$, $j = 1,, \ldots, k$.*

For the case $k = n$, i.e. when $M$ consists of isolated points with $M \cap \partial D = \phi$, the assertion of Theorem 5.14 was obtained earlier (Z.L. Lejbenzon (1966) for convex domains; G.M. Khenkin (1968), Kerzman, Nagel (1971) for strictly pseudoconvex domains) as the solution of a problem of Gleason (1964, see [76]).

In the general case Theorem 5.14 was proved in [16] on the basis of explicit formulas [11] for finding the functions $\{\varphi_j\}$ satisfying (5.13).

Let us invoke formula (4.9) setting therein $G(t) = (1 - t)^k$,

$$\eta(\zeta, z) = \frac{P(\zeta, z)}{\Phi(\zeta, z)}, Q(\zeta, z) = \sum_j \frac{\bar{g}_j(\zeta)}{|g(\zeta)|^2 + \varepsilon} \cdot P_j(\zeta, z),$$

where $\Phi(\zeta, z) = \langle P(\zeta, z), \zeta - z \rangle$ is the barrier function from Theorem 9.10, and $\{P_j\}$ are functions satisfying equation (5.10).

In this situation formula (4.9): $f = K^G f$ for holomorphic functions in $D$, acquires the form

$$f(z) = c(n) \int_{\partial D} f(\zeta) \sum_{j=0}^{k} (-1)^j C_k^j \left( \frac{\langle \bar{g}(\zeta), g(z) \rangle + \varepsilon}{|g(\zeta)|^2 + \varepsilon} \right)^{k-j} \wedge$$

$$\wedge \left\langle \frac{P(\zeta, z)}{\Phi(\zeta, z)} \wedge d\zeta \right\rangle \wedge \left\langle \bar{\partial}_\zeta \frac{P(\zeta, z)}{\Phi(\zeta, z)} \wedge d\zeta \right\rangle^{n-1-j} \wedge$$

$$\wedge \langle \bar{\partial}_\zeta Q(\zeta, z) \wedge d\zeta \rangle^j.$$

If we pass to the limit as $\varepsilon \to 0$ in this formula, we obtain the equality (see [11]):

$$f(z) = F(z) + \sum_{j=1}^{k} \varphi_j(z) \cdot g_j(z), \qquad z \in D, \tag{5.14}$$

where $F$ is a function of type (5.9) and the functions $\{\varphi_j\}$, holomorphic in $D$, have the form

$$\varphi_j(z) = c(n) \int_{\partial D} f(\zeta) \sum_{j=0}^{k-1} (-1)^j \bar{g}_j(\zeta) c_k^j \frac{\langle \bar{g}(\zeta), g(z) \rangle^{k-j-1}}{|g(\zeta)|^{2(k-j)}} \wedge$$

$$\wedge \left\langle \frac{P(\zeta, z)}{\Phi(\zeta, z)} \wedge d\zeta \right\rangle \wedge \left\langle \bar{\partial}_\zeta \frac{P(\zeta, z)}{\Phi(\zeta, z)} \wedge d\zeta \right\rangle^{n-1-j} \wedge$$

$$\wedge \langle \bar{\partial}_\zeta Q(\zeta, z) \wedge d\zeta \rangle^j.$$

For the proof of Theorem 5.14, we notice that $F(z) = 0$ in $D$ since, under the hypotheses of the theorem, $f = 0$ in $M$.

Further, under the condition that $|f(\zeta)| = 0(|g(\zeta)|)$, one can immediately convince oneself of estimates for the boundedness of the functions $\{\varphi_j\}$ and also that the $\{\varphi_j\}$ belong to the class $C^{(\alpha-1)/2}(\bar{D})$ *if* $f \in C^{\alpha/2}(\bar{D})$, $\alpha > 1$.

The formula of Berndtsson (5.14) allows one to obtain an accurate theorem on division under circumstances more general than in Theorem 5.14. For example, on the basis of (5.14), the following assertion has been proven.

**Theorem 5.14′** (Passare [67]). *Let $g_1, \ldots, g_k$ be holomorphic functions in a pseudoconvex domain $\Omega$ such that the analytic set $M$ of type (5.12) has codimension $k$. Then, a function $f$ in $\mathscr{H}(\Omega)$ can be represented in the form (5.14), where $\varphi_j \in \mathscr{H}(\Omega)$ $j = 1, \ldots, k$, if and only if the current (generalized (0, k)-form) given by $f\bar{\partial}(1/g_1) \wedge \ldots \wedge \bar{\partial}(1/g_k)$ vanishes in $\Omega$.*

## §6. Formulas for Solving $\bar{\partial}$-Equations in Pseudoconvex Domains and Their Applications

**6.1 The $\bar{\partial}$-Equations. The Theorem of Dolbeault.** The Cauchy–Green formula (0.1) allows one to write down a solution of the inhomogeneous Cauchy–Riemann equation $\partial f/\partial \bar{z} = g$ in an explicit form, where $g$ is an integrable function in a bounded domain $D \subset \mathbb{C}^1$. Namely, the function given by

$$f(z) = \frac{1}{2\pi i} \int_D \frac{g(\zeta)}{\zeta - z}\, d\zeta \wedge d\bar{\zeta} \tag{6.1}$$

satisfies the required equation.

Moreover, formula (6.1) singles out the unique solution of the equation $\partial f/\partial \bar{z} = g$ which satisfies the boundary condition

$$Kf(z) = \frac{1}{2\pi i} \int_{\partial D} \frac{f(\zeta)\, d\zeta}{\zeta - z} = 0, \qquad z \in D.$$

Let us consider, in a domain $D \subset \mathbb{C}^n$, the system of Cauchy–Riemann equations given by

$$\frac{\partial f}{\partial \bar{z}_j} = g_j, \qquad j = 1, \ldots, n, \tag{6.2}$$

where $\{g_j\}$ are fixed functions in the domain $D$.

Grothendieck in 1950, using formula (6.1), gave an elementary proof (see [33]) that for any polydomain $D = D_1 \times \ldots \times D_n$ in $\mathbb{C}^n$, necessary and sufficient conditions for the solvability of the system (6.2) are the following compatibility conditions for the right member of (6.2)

$$\frac{\partial g_k}{\partial \bar{z}_j} = \frac{\partial g_j}{\partial \bar{z}_k}, \qquad j, k = 1, 2, \ldots, n. \tag{6.3}$$

The equations (6.2) and the conditions (6.3) can be written briefly in the form

$$\begin{aligned} \bar{\partial} f &= g \quad (\bar{\partial}\text{-equation for the function } f), \\ \bar{\partial} g &= 0 \quad (\bar{\partial}\text{-closure of the form } g), \end{aligned} \tag{6.4}$$

where $g$ is the (0, 1)-form given by $g = g_1 d\bar{z}_1 + \ldots + g_n d\bar{z}_n$.

This theorem of Grothendieck together with Oka's theorems 2.10 and 3.6 lead to the solvability of the $\bar{\partial}$-equation (6.4) in an arbitrary pseudoconvex domain (see [33], [45]).

**Theorem 6.1** (Dolbeault, 1953). *For any domain of holomorphy $D$ and any $\bar{\partial}$-closed $(0, 1)$-form $g$ of class $C^{(\pm\infty)}_{0,1}(D)$, there exists a function $f \in C^{(\pm\infty)}(D)$ satisfying the $\bar{\partial}$-equation (6.4) in $D$.*

Here, $C^{(-\infty)}(D)$ denotes the space of generalized functions of the domain $D$.

As a simple corollary of Theorem 6.1, we obtain the following strengthened variant of Theorem 3.4 on the splitting of singularities (see [32]).

**Theorem 6.2** (H. Cartan, 1951). *Let $\Omega$ be a domain in $\mathbb{C}^n$ and $\{\Omega_j\}$ a locally-finite covering of the domain $\Omega$ by domains $\Omega_j$, $j = 1, 2, \ldots$. Let $\{f_{i,j}\}$ be a holomorphic cocycle on $\Omega_i \cap \Omega_j$, i.e. a collection of functions, holomorphic in the domains $\Omega_i \cap \Omega_j$, with the properties*

$$f_{i,j} + f_{j,k} + f_{k,i} = 0 \tag{6.5}$$

*on $\Omega_i \cap \Omega_j \cap \Omega_k$ for each $i, j, k$. If $\Omega$ is a domain of holomorphy, then the cocycle $\{f_{i,j}\}$ is a coboundary, i.e. there exists a collection of functions $\{f_j\}$, holomorphic in the domains $\{\Omega_j\}$, and such that*

$$f_i - f_j = f_{i,j} \tag{6.6}$$

*on $\Omega_i \cap \Omega_j$ for each $i, j$.*

In the language of cohomology, Theorem 6.2 states that for any domain of holomorphy, we have $H^1(\Omega, \mathcal{O}) = 0$, where $H^1(\Omega, \mathcal{O})$ is the first cohomology group of $\Omega$ with coefficients in the sheaf $\mathcal{O}$ of germs of holomorphic functions on $\Omega$.

In order to deduce Theorem 6.2 from Theorem 6.1, we introduce for consideration a partition of unity subordinate to the covering $\{\Omega_j\}$, i.e. smooth functions $\{\chi_j\}$ with supports in the domains $\{\Omega_j\}$ such that $\Sigma\, \chi_j \equiv 1$ on $\Omega$. Further, we set $\tilde{f}_j = \Sigma\, f_{j,k}\chi_k$. On account of (6.5), we have $\tilde{f}_i - \tilde{f}_j = f_{i,j}$. In order to find holomorphic functions $\{f_i\}$ satisfying (6.6), let us consider in $\Omega$ the $\bar{\partial}$-closed form $g$ given by $g = \bar{\partial}\tilde{f}_j$ for $z \in \Omega_j$, $j = 1, 2, \ldots$. This is well-defined since $\bar{\partial}\tilde{f}_j = \bar{\partial}\tilde{f}_k$ on $\Omega_j \cap \Omega_k$. By Theorem 6.1 there exists a smooth function $f$ such that $\bar{\partial}f = g$ in $\Omega$. We set $f_j = \tilde{f}_j - f$. By construction, the functions $\{f_j\}$ are holomorphic in the domains $\{\Omega_j\}$ and satisfy (6.6).

**6.2. Problems of Cousin and Poincaré as $\bar{\partial}$-Equations. Currents of Lelong and Schwartz.** Schwartz in 1953 suggested that the first additive Cousin problem be interpreted in terms of solving $\bar{\partial}$-equations (6.4) with a special generalized $(0, 1)$-form (current) on the right side (see [42]).

Namely, let $\{\Omega_j\}$ be a locally finite cover of the domain $\Omega$ and $f_j = \varphi_j/g_j$ data for the first additive Cousin problem. Consider a generalized $(0, 1)$-form $G$ given by

$$G = \bar{\partial} f_j = \varphi_j \bar{\partial}(1/g_j), \quad z \in \Omega_j, \quad j = 1, 2, \ldots. \tag{6.7}$$

Each form $\varphi_j \bar{\partial}(1/g_j)$ yields a functional on smooth $(n, n-1)$-forms $\psi$ of compact support in $\Omega_j$ by the formula

$$\left\langle \varphi_j \bar{\partial} \frac{1}{g_j}, \psi \right\rangle = \lim_{\varepsilon \to 0} \frac{1}{2\pi i} \int_{|g_j| = \varepsilon} \frac{\varphi_j}{g_j} \wedge \psi = - \lim_{\varepsilon \to 0} \frac{1}{2\pi i} \int_{|g_j| \geq \varepsilon} \frac{f_j}{g_j} \wedge \bar{\partial}\psi. \tag{6.8}$$

The existence of the latter limit is guaranteed by the Theorem of Herrera–Lieberman–Dolbeault (1971), (see the paper by P. Dolbeault V).

Since the functions $f_i - f_j$ are holomorphic on $\Omega_i \cap \Omega_j$, the equations (6.7), (6.8) define a $\bar{\partial}$-closed (0, 1)-current on the whole domain $\Omega$, which we shall call the Schwartz current.

In the language of $\bar{\partial}$-equations, the theorem of Oka (1937, 1953) on the solvability of the first Cousin problem in pseudoconvex domains can be formulated in the following way (see [42]).

**Theorem 6.3** (Oka, 1953; Schwartz, 1953). *For any pseudoconvex Riemannian domain $\Omega$ and any Schwartz current $G$ given by* (6.7), *there exists a meromorphic function $F$ on $\Omega$ such that $\bar{\partial} F = G$.*

Theorem 6.3 leaves open the question of characterizing intrinsically those (0, 1)-currents which can be represented as Schwartz currents. Results in this direction are considered in the paper of Dolbeault.

Theorem 6.3 allows one (see [18], [5], [42]) to reformulate, in terms of solving $\bar{\partial}$-equations, the classical problem of the continuation of a function, holomorphic on a hypersurface $M = \{z \in \Omega\colon g(z) = 0\}$, to a function $\Phi$, holomorphic in the domain $\Omega$. Namely, using Theorem 6.3, Oka's procedure for constructing such a function $\Phi$ (see Theorem 2.12) acquires the form

$$\Phi = g \cdot F, \qquad \text{where } \bar{\partial} F = \varphi \bar{\partial} \frac{1}{g}. \tag{6.9}$$

Lelong (1953) found an even more fruitful and instructive interpretation of the second (multiplicative) Cousin problem in terms of solving a special $\bar{\partial}$-problem the so-called Poincaré–Lelong equation (see [32], [35] and article III).

For a given cover $\{\Omega_j\}$ of a domain $\Omega$ let us consider functions $f_j$ meromorphic in $\Omega_j$ such that for each $i$ and $j$, the function $f_i/f_j$ is holomorphic in $\Omega_i \cap \Omega_j$.

These data define a global $(n-1)$-dimensional analytic subset $M = \bigcup_\nu \gamma_\nu M_\nu$ of $\Omega$ consisting of the zeros and poles of the functions $f_j$. Here, $\{M_\nu\}$ are the irreducible regular $(n-1)$-dimensional components of the set $M$ and $\{\gamma_\nu\}$ are the multiplicities on $\{M_\nu\}$ of the zeros ($\gamma_\nu > 0$) or poles ($\gamma_\nu < 0$) of the functions $\{f_j\}$.

Let us consider the generalized (1,1)-form (current) $G_M$ given by

$$G_M = \bar{\partial} \frac{\partial f_j}{f_j}, \quad z \in \Omega_j, \quad j = 1, 2, \ldots. \tag{6.10}$$

Lelong showed that the form $G_M$ yields a functional, on the space of smooth $(n-1, n-1)$-forms $\psi$ of compact support in the domain $\Omega$, by the formula

$$\langle G_M, \psi \rangle = i \sum_{\nu} \gamma_\nu \int_{M_\nu} \psi = i([M], \psi). \tag{6.11}$$

By a *divisor in* $\Omega$ we mean any (1, 1)-current $[M]$ defined by the second equation in (6.11), where $M$ is any closed $(n-1)$-dimensional analytic subset $M = \bigcup_\nu \gamma_\nu M_\nu$ of $\Omega$.

Lelong established that any divisor is a closed (1,1)-current and hence the classical results of Oka, Stein and Serre on the solvability of the multiplicative Cousin problem as well as the Poincaré problem in pseudoconvex domains can be formulated in the following stronger form (see [35]).

**Theorem 6.4** (Oka, 1939; Stein, 1951; Serre, Lelong, 1953). *Let* $[M]$ *be any divisor given by* (6.11) *in a pseudoconvex Riemannian domain* $\Omega$. *Then, in order that there exist a meromorphic function* $F$ *satisfying the equation*

$$\bar{\partial} \frac{dF}{F} = i[M], \tag{6.12}$$

*it is necessary and sufficient that the two-dimensional cohomology class of the* (1,1)*-current* $M$ *vanish in the two-dimensional integral cohomology* $H^2(\Omega, \mathbb{Z})$.

Theorems 6.3 and 6.4 paved the way to solving the Cousin and Poincaré problems with estimates as well as related problems. For this of course it is necessary also to have solutions to $\bar{\partial}$-equations with suitable estimates.

However, the solution of the $\bar{\partial}$-equations for any pseudoconvex domain (Theorem 6.1) was obtained by such a long and difficult path (first trod by Oka) that it was difficult to pass to the estimates for the $\bar{\partial}$-equations in some metric or other necessary in applications, even for such a simple domain as the ball in $\mathbb{C}^n$.

**6.3. The $\bar{\partial}$-Problem of Neuman–Spencer.** Exploiting the method of orthogonal projection of H. Weyl, Spencer in the early fifties introduced an original approach to solving the $\bar{\partial}$-equations called the $\bar{\partial}$-problem of Neuman–Spencer (see [22]).

Let $D = \{z \in \mathbb{C}^n : \rho(z) < 0\}$ be a domain with smooth boundary. We consider $\bar{\partial}$ as an operator from $L^2(D)$ to $L^2_{0,1}(D)$ and denote by $\bar{\partial}^*$ the adjoint operator which in the Euclidean metric on $D$ has the form

$$\bar{\partial}^* g = -\sum_{j=1}^{n} \frac{\partial g_j}{\partial z_j},$$

where the form $g = \sum_j g_j d\bar{z}_j$ belongs to the domain of definition Dom $(\bar{\partial}^*)$ of the operator $\bar{\partial}^*$. *For a smooth form* $g$, as Spencer remarked, the latter is equivalent to the equation

$$\bar{\partial}\rho g \stackrel{\text{def}}{=} \sum_j g_j \frac{\partial \rho}{\partial z_j} = 0 \text{ on } \partial D.$$

The solution to the $\bar{\partial}$-problem of Neuman–Spencer based on *a priori* $L^2$-estimates for the $\bar{\partial}$-operator can be formulated as follows:

**Theorem 6.5** (Kohn [22], [56]). *Let $D$ be any strictly pseudoconvex domain with boundary of class $C^\infty$ or any weakly pseudoconvex domain with real-analytic boundary in $\mathbb{C}^n$.*

*Then there exists (and moreover uniquely) a completely continuous operator $N: L^2_{0,1}(D) \to L^2_{0,1}(D)$ such that*

$$N(L^2_{0,1}(D)) \subset \operatorname{Dom}(\bar{\partial}^*)$$

*and each function $f \in L^{2,1}(D)$ admits in $L^2(D)$ an orthogonal decomposition of the form*

$$f = Bf + \bar{\partial}^* N \bar{\partial} f, \tag{6.13}$$

*where $B$ is the Bergman projection from $L^2(D)$ onto $H^2(D)$. Moreover, $N(L^{2,k}_{0,1}(D)) \subset L^{2,k+1}_{0,1}(D)$, where $L^{2,k}(D)$ is the space of functions $L^2$-integrable in $D$ together with their derivatives up to order $k$.*

The operator $N$ satisfying (6.13) is called the *Neuman–Spencer operator*.

From Theorem 6.5 issues the following corollary.

**Theorem 6.6** (Kohn [22], [56]). *If $D$ is a domain as in Theorem 6.5 and $g$ a $\bar{\partial}$-closed form of class $L^2_{(0,1)}(D)$, then*

*a) there exists a unique solution $f \in L^2(D)$ to the equation $\bar{\partial} f = g$ which is orthogonal to the space $H^2(D)$; this solution is given by the equation $f = \bar{\partial}^* N g$;*

*b) moreover $f \in L^{2,k}(\bar{D})$ if $g \in L^{2,k}_{0,1}(\bar{D})$.*

As Hörmander has shown, assertion a) of Theorem 6.6 is valid for any bounded pseudoconvex domain $D$. Moreover, the following result holds.

**Theorem 6.7** (Hörmander [44]). *Let $D$ be any pseudoconvex domain in $\mathbb{C}^n$ and $\varphi$ any plurisubharmonic function in $D$. Then, the equation $\bar{\partial} f = g$ admits a solution with the following global $L^2$-estimate*

$$\int_D |f(z)|^2 e^{-\varphi(z)} (1 + |z|^2)^{-2} \, dV_{2n}(z) \le \int_D |g(z)|^2 e^{-\varphi(z)} \, dV_{2n}(z). \tag{6.14}$$

Kohn, using Theorems 6.6, 6.7 obtained the following result on smoothness up to the boundary for solutions of the $\bar{\partial}$-equations.

**Theorem 6.8** (Kohn, 1973, [55]). *Let $D$ be an arbitrary bounded pseudoconvex domain in $\mathbb{C}^n$ with boundary of class $C^\infty$ and $g$ a $\bar{\partial}$-closed form of class $C^\infty_{0,1}(\bar{D})$. Then, the equation $\bar{\partial} f = g$ has a solution $f \in C^\infty(\bar{D})$.*

**6.4. Formulas for Solving $\bar\partial$-Equations.** The method of *a priori* $L^2$-estimates is not suitable for obtaining the estimates in the $\bar\partial$-equations which are necessary for the applications related to the $L^\infty$ or $L^1$-metrics.

Moreover, problems from the theory of functions require explicit formulas for the solution of $\bar\partial$-equations, and also, as a rule, each problem requires its own formula.

The integral representations of type Cauchy–Fantappiè prove to be well adapted for obtaining such formulas.

Namely, from Theorem 4.3 it is easy to obtain the following result (see [19]).

**Theorem 6.9** (G.M. Khenkin, 1970, $s = 0$; G.M. Khenkin, A.V. Romanov, 1971, $s = 1$; S.A. Dautov, G.M. Khenkin, 1978, $s > 1$). *Let $D$ be an arbitrary strictly pseudoconvex domain of type (3.14), $K^s$ and $H^s$ the integral operators given by (4.7), (4.8), (4.13), and (4.14), where $s \geq 0$, and $g$ a $\bar\partial$-closed form of class $L^1_{0,1}(D)$. Then the equation $\bar\partial f = g$ has a unique solution $f$ satisfying the boundary condition $K^s f = 0$ and $f$ is given by the formula*

$$f = H^s g. \tag{6.15}$$

*Here the operator $g \to H^s g$ is completely continuous from $L^p_{0,1}(\bar D)$ to $L^p(D)$ for all $1 \leq p \leq \infty$. Moreover if the domain $D$ has a smooth boundary, then for $s > \alpha + 1$, the estimates*

$$||f|\,|\rho|^{\alpha-1}|^1_{L(D)} \leq \gamma \|\,|g|\cdot|\rho|^\alpha + |g \wedge \bar\partial\rho|\cdot|\rho|^{\alpha-1/2}\|_{L^1(D)}$$

*for $\alpha > 0$, and*

$$\|f\|_{L(D)} \leq \gamma \|\,|g| + |g \wedge \bar\partial\rho|\cdot|\rho|^{-1/2}\|_{L(D)} \tag{6.16}$$

*for $\alpha = 0$, hold.*

The latter estimate was first obtained using other formulas in the works of G.M. Khenkin and Skoda in 1975 (see [49]).

Theorem 4.4 allows one to obtain even more general global formulas and estimates for solutions of $\bar\partial$-equations. For example, if in Theorem 4.4 we set

$$G(t) = (1-t)^m; \quad Q_j(\zeta) = \frac{\zeta_j}{1+|\zeta|^2}, \quad \eta(\zeta, z) = \frac{\bar\zeta - \bar z}{|\zeta - z|^2},$$

then for $\Omega = \mathbb{C}^n$ we obtain the following result.

**Theorem 6.10** (Skoda, 1971; Berndtsson, Andersson, 1982, see [12]). *Let $g$ be a $\bar\partial$-closed (0,1) form in $\mathbb{C}^n$ such that*

$$r^{-2n} \int_{|\zeta|<r} |g(\zeta)|\,dV_{2n}(\zeta) = 0((1+r)^\alpha), \qquad \alpha \geq -2n.$$

*Then there exists a solution $f$ of the equation $\bar\partial f = g$ with estimate*

$$r^{-2n} \int_{|\zeta|<r} |f(\zeta)|\,dV_{2n}(\zeta) = 0((1+r)^{\alpha+1}(1+\log(1+r))).$$

*This solution is given by the formula*

$$f(z) = H_m g(z) = c \int_{\mathbb{C}^n} g(\zeta) \sum_{k=0}^{\min(m,n-1)} C_m^k \left( \frac{1+\bar{\zeta}z}{1+|\zeta|^2} \right)^{m-k} \times$$

$$\times \frac{\partial|\zeta - z|^2 \wedge (\bar{\partial}\partial|\zeta - z|^2)^{n-1-k} \wedge (\bar{\partial}Q(\zeta))^k}{|\zeta - z|^{2(n-k)}}, \tag{6.17}$$

*where* $m > \alpha + n \geq m + \min(n-1, m) - n$, *and has the property*

$$P_m f(z) = c \int_{\mathbb{C}^n} f(\zeta) \frac{(1+\bar{\zeta}z)^{m-n}}{(1+|\zeta|^2)^{m+1}} (\bar{\partial}\partial|\zeta|^2)^n = 0, \qquad z \in \mathbb{C}^n,$$

*where* $P_m$ *is the orthogonal projection from*

$$L^2\left(\frac{dV_{2n}(z)}{(1+|z|^2)^{m+1}}\right) \text{ onto } H^2\left(\frac{dV_{2n}(z)}{(1+|z|^2)^{m+1}}\right).$$

Returning to the assertion of Theorem 6.9, we now remark that for the ball in $\mathbb{C}^n$, $K^1 = B$, and hence in this situation, the solution $H^1 g$ is precisely the Kohn solution $\bar{\partial}^* Ng$ from Theorem 6.5.

However, the operators $H^1$ and $\bar{\partial}^* N$, which agree on $\bar{\partial}$-closed forms, are given, nevertheless, by different kernels.

An exact calculation of the kernel of the operator $N$ for the ball was carried out only recently (see [37] and [61]).

Theorems 5.3 and 6.9 allow us to obtain precise $L^p$-estimates ($1 \leq p \leq \infty$) for the Kohn solution.

**Theorem 6.11** (Øvrelid, 1976; Greiner, Stein, 1977, see [61]). *Let D be a strictly pseudoconvex domain in* $\mathbb{C}^n$ *with boundary of class* $C^\infty$. *Then the operator* $g \to f = \bar{\partial}^* Ng$, *solving the* $\bar{\partial}$-*equation* $\bar{\partial}f = g$ *in D, is for every* $p \in [1, \infty]$ *and* $k \geq 0$ *a completely continuous operator from the space* $L^{p,k}_{0,1}(\bar{D})$ *to the space* $L^{p,k}(\bar{D})$.

Further developments in the cited works led (see [9] and [61]) to an asymptotic formula for the Neuman–Spencer operator $N$ in strictly pseudoconvex domains from which the assertion of Theorem 6.11 can be derived immediately.

Obtaining solutions of $\bar{\partial}$-equations with $L^\infty$ or $L^1$-estimates in weakly pseudoconvex domains or in pseudoconvex domains with non-smooth boundary turned out to be more difficult problems which up to now have been solved only in the class of pseudoconvex polyhedra of type (4.25).

**Theorem 6.12** (G.M. Khenkin, A.G. Sergeev, 1980, see [54], [19]). *Let D be a pseudoconvex polyhedron given by* (4.25) *and satisfying the non-degeneracy condition 5.5. Let K be the Cauchy–Fantappié operator* (4.28) *and H the operator* (4.20), (4.27). *Then the solution to the equation* $\bar{\partial}f = g$ *with the property* $Kf = 0$ *is* *represented by the formula $f = Hg$. Also*

$$\|f\|_{L^p(\bar{D})} \leq \gamma \|g\|_{L^p(\bar{D})}, \qquad 1 \leq p \leq \infty, \tag{6.18}$$

*for any $\bar{\partial}$-closed form $g \in L^p_{0,1}(\bar{D})$.*

The operator $H$ in this theorem (in contrast to the operator $H^s$ in Theorem 6.9), is generally speaking, not continuous from $L^p_{0,1}(\bar{D})$ to $L^p(\bar{D})$; that is, the condition that the form $g$ be $\bar{\partial}$-closed is necessary, not only for the solvability of the $\bar{\partial}$-equation, but also for the validity of the $L^p$-estimates (6.18).

The fundamental difficulty in the proof of the estimate 6.18 consists in finding a reformulation of the formula $f = Hg$ such that the estimate becomes immediate.

In practice, for example, the following global formula for solving the $\bar{\partial}$-equation in the polydisc

$$D^n_1 = \left\{\zeta \in \mathbb{C}^n : \sup_j |\zeta_j| < 1\right\}$$

works well.

For ordered multi-indices $K$, $J$ we set

$$\gamma_K(z_J) = \{\zeta \in D^n_1 : \zeta_J = z_J;\ |z_{j_1}| \geq \ldots \geq |z_{j_r}| \geq \\ \geq |\zeta_{k_1}| = \ldots = |\zeta_{k_p}|;\ |\zeta_l| \leq |\zeta_{k_p}|.\ \forall l \notin K \cup J\},$$

where $r = |J|$, $p = |K|$,

$$H_{J,K}(\zeta, z) = \left[\frac{1}{2\pi i}\right]^{n-r} \bigwedge_{k \in K} \frac{(1 - \zeta_k\bar{\zeta}_k)^{\beta_k} d\zeta_k}{(\zeta_k - z_k)(1 - \bar{\zeta}_k z_k)^{\beta_k}} \\ \bigwedge_{l \notin K \cup J} \frac{\beta_l d\bar{\zeta}_l \wedge d\zeta_l (1 - \zeta_l\bar{\zeta}_l)^{\beta_l - 1}}{(1 - \bar{\zeta}_l z_l)^{\beta_l + 1}},$$

where $\beta_k \geq 0$, $k = 1, \ldots, n$.

**Theorem 6.13** (G.M. Khenkin, P.L. Polyakov [65], [40], [41]). *Let $D^n_1$ be the polydisc in $\mathbb{C}^n$, and $g$ a $\bar{\partial}$-closed* (0, 1) *form whose coefficients are finite measures on $D^n_1$. Then, the function*

$$f = Hg(z) = -\sum_{r=0}^{n-1} \sum_{J:|J|=r} \left( \sum_{K: K \cap J = \{\emptyset\}} (-1)^{c(n,r,p)} \int_{\gamma_K(z_J)} g(\zeta) \wedge H_{J,K}(\zeta, z) \right) \tag{6.19}$$

*is integrable on $D$ and satisfies the equation $\bar{\partial} f = g$ in $D$. Moreover the operator $H$ is continuous from $L^p_{0,1}(D^n_1)$ to $L^p(D^n_1)$ for all $1 \leq p \leq \infty$.*

Formula 6.19 generalizes a series of formulas obtained earlier for the case of the bidisc (see [19], [17]).

Integral representation of Cauchy–Fantappiè type allow one also to prove, in an elementary fashion, the following general result on $C^\infty$-estimates for solutions of the $\bar{\partial}$-equation in arbitrary convex domains thus complementing Theorem 6.8.

**Theorem 6.14** ([52]). *Let D be an arbitrary bounded convex or strictly linearly convex domain in* $\mathbb{C}^n$ *and g any* $\bar{\partial}$*-closed form of class* $C_{0,1}^{(s)}(\bar{D})$, $s > n - 2$. *Then, for each* $\varepsilon > 0$, *the equation* $\bar{\partial} f = g$ *has a solution in D of class* $C^{s-n+2-\varepsilon}(\bar{D})$.

For $s = \infty$ this theorem was first proved by A. Dufresnoy. (Ann. Inst. Fourier, 29, pp. 229–238, 1979). See also [66].

A description of those pseudoconvex (or even convex) domains $D$, for which the equation $\bar{\partial} f = g$ has a solution with a uniform estimate, $\|f\|_{L_\infty(D)} \leq \gamma \|g\|_{L_\infty(D)}$, present itself as a rather difficult problem. We remark that Sibony [77] constructed a pseudoconvex domain $D$ with boundary of class $C^\infty$ and a $\bar{\partial}$-closed form $g \in L_{0,1}^\infty(D)$ such that the equation $\bar{\partial} f = g$ has no bounded solutions in $D$.

**6.5. The Poincaré–Lelong Equation. Construction of Holomorphic Functions with Given Zeros.** Formulas and estimates for solutions of the $\bar{\partial}$-equation yield a rather flexible apparatus for many problems in the theory of functions on domains of $\mathbb{C}^n$.

All of the results in §5, for example, can be well interpreted in the language of $\bar{\partial}$-equations. In particular, Theorem 5.13 on the continuation of bounded holomorphic functions from a submanifold to the polydisc was obtained in [42] on the basis of formulas (6.9) and (6.19).

Formulas for solving the $\bar{\partial}$-equations turned out to be particularly useful in the problem of constructing holomorphic functions with finite order of growth and having given zeros.

We present now the basic results in this direction respectively for $\mathbb{C}^n$, for strictly pseudoconvex domains, and for the polydisc.

Let $F$ be a function holomorphic in $\mathbb{C}^n$. By $\operatorname{Ord} F$ we denote the infimum of those $\alpha > 0$ for which

$$\int_{z \in \mathbb{C}^n} |z|^{-\alpha-1} \ln^+ |F(z)| dV_{2n}(z) < \infty.$$

If $F$ is a function holomorphic in a strictly pseudoconvex domain $D = \{z \in \mathbb{C}^n: \rho(z) < 0\}$ or in the polydisc $D_1^n = \{z \in \mathbb{C}^n: \rho_1(z) = \sup_j \ln |z_j| < 0\}$ then we denote by $\operatorname{Ord} F$ the infimum of those $\alpha > 0$ for which $F$ belongs respectively to $N_\alpha(D)$ or $N_\alpha(D_1^n)$, where

$$N_\alpha(D) = \{F \in \mathscr{H}(D): \int_{z \in D} |\rho(z)|^{\alpha-1} \ln^+ |F(z)| dV_{2n}(z) < \infty\},$$

$$N_\alpha(D_1^n) = \{F \in \mathscr{H}(D_1^n): \int_{\{z \in D_1^n: |z_1| = \ldots = |z_n|\}} |\rho_1(z)|^{\alpha-1} \ln^+ |F(z)| dV_{n+1}(z) < \infty\},$$

where $\ln^+(t) = \sup\{\ln t, 0\}$.

Further, let $M = \sum_\nu \gamma_\nu M_\nu$ be an $(n-1)$-dimensional analytic set respectively in $\mathbb{C}^n$, $D$ or $D_1^n$, where $\{M_\nu\}$ are the irreducible $(n-1)$-dimensional components of $M$ and $\{\gamma_\nu\}$ are the multiplicities of these components.

We denote by $\operatorname{Ord} M$ the infimum of those $\alpha > 0$ for which

$$\sum_\nu \gamma_\nu \int_{z \in M_\nu} |z|^{-\alpha - 2n + 2} dV_{2n-2}(z) < \infty,$$

if $M \subset \mathbb{C}^n$;

$$\sum_\nu \gamma_\nu \int_{z \in M_\nu} |\rho(z)|^{\alpha+1} dV_{2n-2}(z) < \infty, \tag{6.20}$$

if $M \subset D$; and finally,

$$\sum_\nu \gamma_\nu \int_{M_{\nu,\gamma}} |\rho_1(z)^{\alpha+2-n} dV_{2n-2}(Z) < \infty,$$

where

$$M_{\nu,\gamma} = \{z \in M_\nu\colon \sup_j (\ln|z_j|) \le \gamma \inf_j (\ln|z_j|), \gamma \ge 0\},$$

if $M \subset D_1^n$.

From the formula of Poincaré–Lelong or the multi-dimensional formula of Jensen–Stoll, it follows immediately (see [78]) that for any function $F$ holomorphic in $\mathbb{C}^n$, $D$ or $D_1^n$, the inequality

$$\operatorname{Ord} F \ge \operatorname{Ord} M_F$$

holds, where $M_F$ is the set of zeros (counting multiplicities) of the function $F$.

The construction of a function $F$ with given zeros $M$ such that $\operatorname{Ord} F = \operatorname{Ord} M$ was carried out, in the case of one complex variable, in classical works: for entire functions, by Borel in 1900, and for functions in the unit disc, by M.M. Dzhrbashyan in 1948 (see [21]). For functions of several variables, the first precise results were obtained by Lelong and Stoll.

**Theorem 6.15** (Lelong, 1953; Stoll, 1953; see [73], [32], [78]). *For any $(n-1)$-dimensional analytic set $M$ in $\mathbb{C}^n$, there exists a holomorphic function $F$ such that $M = \{z \in \mathbb{C}^n\colon F(z) = 0\}$ and $\operatorname{Ord} F = \operatorname{Ord} M$.*

Functions $F$ with the property $\operatorname{Ord} F = \operatorname{Ord} M$ were constructed in the cited works with explicit formulas, which, in the light of Theorem 6.4, can be considered as explicit solutions to the Poincaré–Lelong equation: $\bar\partial\,(dF/F) = i[M]$. For example, we obtain Lelong's formula if we use formula (6.17) of Theorem 6.10 to solve the Poincaré–Lelong equation.

For strictly pseudoconvex domains, it is possible, with the help of Theorem 6.9, to give (see [20], [49]) a complete characterization of the zero-sets for functions of the class $N_\alpha(D)$, $\alpha \geq 0$, where

$$N_0(D) = \left\{ F \in \mathscr{H}(D): \int_{\partial D} \ln^+ |F(z)| dV_{2n-1}(z) < \infty \right\}.$$

**Theorem 6.16** (G.M. Khenkin, 1975; Skoda, 1975; S.A. Dautov, G.M. Khenkin, 1977). *Let $D$ be a strictly pseudoconvex domain in $\mathbb{C}^n$ with smooth boundary. In order that an $(n-1)$-dimensional set $M$ in $D$ be the set of zeros of a function of class $N_\alpha(D)$, $\alpha \geq 0$, it is necessary and sufficient that condition* (6.20) *hold, if $\alpha > 0$, and that the Blaschke condition hold*

$$\sum_j \gamma_j \int_{M_j} |\rho(z)| dV_{2n-2}(z) < \infty,$$

*if $\alpha = 0$.*

Although for the case of the polydisc $D_1^n$, a complete characterization of the zero sets of functions of class $N_\alpha(D_1^n)$ has not yet been found, nevertheless, the formulas of Theorem 6.13 allow us to prove the following.

**Theorem 6.17** (G.M. Khenkin, P.L. Polyakov, 1984; Charpentier, 1984; P.L. Polyakov, 1986). *For any $(n-1)$-dimensional analytic set $M$ in the polydisc $D_1^n$, there exists a holomorphic function $F$ whose zeros lie in $M$ and such that* $\operatorname{Ord} F = \operatorname{Ord} M$.

This result was obtained for the bidisc in [41], [17] and for the general case by P.L. Polyakov in [70].

One of the central unsolved problems in the theory of functions in pseudoconvex domains is the "*corona*" *problem*: let $f_1, \ldots, f_N$ be bounded holomorphic functions in a strictly pseudoconvex domain $D$ such that $|f_1|^2 + \ldots + |f_N|^2 > \delta > 0$ in $D$. Do there exist bounded holomorphic functions $h_1, \ldots, h_N$ in $D$ such that $f_1 h_1 + \ldots + f_N h_N \equiv 1$ in $D$?

Making use of formula (6.15) and a suitable generalization of the celebrated construction of Carleson and Wolf (see [26]) in the "Corona" theorem for domains in $\mathbb{C}^1$, it is possible to prove (see [49], [40]) an assertion which yields hope that a generalization of Carleson's theorem to strictly pseudoconvex domains in $\mathbb{C}^n$ could be established.

**Theorem 6.18** (G.M. Khenkin, 1977; Varopoulos, 1977; Amar, 1980). *With the hypotheses of the "corona" problem, there exist functions $h_j \in \bigcap_{p<\infty} H^p(\partial D)$, $j = 1, 2, \ldots, N$, such that*

$$\sum_{j=1}^N f_j h_j \equiv 1.$$

The fundamental difficulty in the multi-dimensional "corona" problem consists in finding precise conditions on a form $g$ in the domain $D$ under which the equation $\bar{\partial} f = g$ possesses a bounded solution in $D$.

The strongest result in this direction was obtained recently by Berndtsson (Berndtsson, B., An $L^\infty$-estimate for the $\bar{\partial}$-equation in the unit ball in $\mathbb{C}^n$. Preprint, University of Göteborg, 1983, 35 pp. and Berndtsson, B., $\bar{\partial}_b$ and Carleson type inequalities, Preprint, University of Göteborg, 1986, 17 pp.).

## §7. Integral Representations in the Theory of *CR*-Functions

Integral representations of Cauchy–Fantappié type allow us not only to make classical results (with estimates on the boundary) more precise in pseudoconvex domains but also to significantly advance the theory of functions on real submanifolds of $\mathbb{C}^n$.

Let $M$ be a real smooth submanifold of $\mathbb{C}^n$, $T_z(M)$ the real tangent space to $M$ at the point $z \in M$, and $T_z^{\mathbb{C}}(M)$ the largest complex subspace lying in $T_z(M)$.

The manifold $M$ is called a Cauchy–Riemann manifold (*CR-manifold*) if the number $\dim_{\mathbb{C}} T_\tau^{\mathbb{C}}(M) = CR \dim M$ does not depend on the point $\tau \in M$. If $CR \dim M = 0$, then the *manifold* $M$ is said to be *totally real.*

A *CR*-manifold $M$ is said to be *generic* or of general type if $T_z(M) \oplus JT_z(M) = \mathbb{C}^n$, where $J$ is the operator in $\mathbb{R}^{2n}$ obtained by multiplication by $\sqrt{-1}$ in the space $\mathbb{C}^n \simeq \mathbb{R}^{2n}$.

Such manifolds can be locally represented in the form

$$M = \{z \in \Omega: \rho_1(z) = \ldots = \rho_k(z) = 0\}, \tag{7.1}$$

where $\rho = \{\rho_1, \ldots, \rho_k\}$ is a collection of real smooth functions in the domain $\Omega$ with the property $\bar{\partial}\rho_1 \wedge \ldots \wedge \bar{\partial}\rho_k \neq 0$ on $M$.

In terms of the representation (7.1), we have

$$T_z^{\mathbb{C}}(M) = \left\{\zeta \in \mathbb{C}^n: \sum_{j=1}^n \frac{\partial \rho_\nu}{\partial z_j}(z)(\zeta_j - z_j) = 0, \quad \nu = 1, \ldots, k\right\}.$$

A smooth function $f$ on a *CR*-manifold $M$ is called a *CR-function,* in other words, a function satisfying the tangential Cauchy–Riemann equations $\bar{\partial}_M f = 0$, if for any complex vector field $\zeta(z) = (\zeta_1(z), \ldots, \zeta_n(z)) \in T_z^{\mathbb{C}}(M)$, $z \in M$, we have

$$\sum_{j=1}^n \bar{\zeta}_j(z) \frac{\partial f}{\partial \bar{z}_j}(z) = 0.$$

For generic *CR*-manifolds given by (7.1), this equation is equivalent to the equation

$$\bar{\partial}\tilde{f} \wedge \bar{\partial}\rho_1 \wedge \ldots \wedge \bar{\partial}\rho_k = 0 \text{ on } M,$$

where $\tilde{f}$ is any smooth extension of the function $f$ to the domain $\Omega \supset M$.

If $f$ is locally integrable or even a generalized function on $M$, then the equation $\bar{\partial}_M f = 0$ should be understood in the generalized sense: $\int_M f \wedge \bar{\partial}\varphi = 0$, for any smooth differential form $\varphi$ of compact support.

If the $CR$-manifold $M$ is real analytic and $f$ a real analytic function on $M$, then as Tomassini (1966, see [3]) has shown, the equation $\bar{\partial}_M f = 0$ is equivalent to the requirement that the function $f$ be the trace on $M$ of some function holomorphic in a neighbourhood of $M$.

However, for a smooth $CR$-function on $M$, generally speaking, a holomorphic continuation to a neighbourhood of $M$ does not exist.

**7.1. Approximation and Analytic Representation of *CR*-Functions.** The theory of $CR$-functions is rather far advanced for the case of real hypersurfaces in $\mathbb{C}^n$. An especially effective result here is the analytic representation of $CR$-functions on hypersurfaces given by Theorem 5.8.

For a $CR$-manifold $M$ of arbitrary codimension, it is natural to define an *analytic representation of a CR-function* $f$ of the form

$$f = \sum_{\nu=1}^{N} f_\nu, \tag{7.2}$$

where each $f_\nu$ is a $CR$-function on $M$ which admits a holomorphic extension to some domain $D_\nu$, in $\mathbb{C}^n$ such that $M \subset \partial D_\nu$.

Such a representation has been obtained thus far only for isolated cases. First of all, an analytic representation holds (see [38]) for functions on arbitrary totally real manifolds.

**Theorem 7.1** (G.M. Khenkin, 1979). *Let $M$ be a $C^s$-smooth generic totally real submanifold of a domain $\Omega \subset \mathbb{C}^n$ given by (7.1). Let $\{\Omega_\nu\}$ be a finite covering of the domain $\Omega \backslash M$ by strictly pseudoconvex domains of the form*

$$\Omega_\nu = \{z \in \Omega \colon \rho_\nu < 0\}, \qquad \nu = 1, 2, \ldots, n+1,$$

*where $\{\rho_\nu\}$ are strictly plurisubharmonic functions on $\Omega$ such that $\rho_\nu = 0$ on $M$ and $\bar{\partial}\rho_{\nu_1} \wedge \ldots \wedge \bar{\partial}\rho_{\nu_n} \neq 0$ on $M$ for each $\nu_1 < \ldots < \nu_n$. Then, for any pseudoconvex subdomain $D_0 \Subset \Omega$ and any function $f \in C^\alpha(D_0 \cap M)$, $\alpha \leq s - 2$, one can construct functions $f_{\nu_1, \ldots, \nu_n}$, holomorphic in the domains $D_0 \cap D_{\nu_1} \cap \ldots \cap D_{\nu_n}$ and of class $C^{\alpha - \varepsilon}$ on $D_0 \cap M$ such that*

$$f(z) = \sum_{\nu=1}^{n+1} (-1)^{n+\nu} f_{1, \ldots, \nu-1, \nu+1, \ldots, n+1}(z), \qquad z \in M \cap D_0. \tag{7.3}$$

In the case where $M = \mathbb{R}^n$ and $\alpha = \pm\infty$, the assertion of Theorem 7.1 is contained in the well-known theorems of Martineau (1970) and Bros-Jaglonitzer (1975) which complement the theory of hyperfunctions of Sato (see Vol. 8, article IV).

It was found [38] that it is possible to write functions satisfying the assertion of Theorem 7.1 in the explicit form

$$f_{\nu_1, \ldots, \nu_n}(z) = \frac{(-1)^{\frac{n(n-1)}{2}}}{(2\pi i)^n} \int_{\zeta \in M \cap D_0} f(\zeta) \frac{\det[P_{\nu_1}(\zeta, z), \ldots, P_{\nu_n}(\zeta, z)]\omega(\zeta)}{\Phi_{\nu_1}(\zeta, z) \ldots \Phi_{\nu_n}(\zeta, z)}, \quad (7.4)$$

where

$$z \in D_0 \cap D_{\nu_1} \cap \ldots \cap D_{\nu_n}; \ \Phi_\nu(\zeta, z) = \langle P_\nu(\zeta, z), \zeta - z \rangle$$

are the barrier functions for the strictly pseudoconvex hypersurfaces $\{z \in D_0: \rho_\nu = 0\}$ with the properties from Theorem 3.10.

The holomorphy of the functions $f_{\nu_1, \ldots, \nu_n}$ in the domains $D_0 \cap D_{\nu_1}, \ldots, \cap D_{\nu_n}$ follows immediately from the definition (7.4).

To obtain the equality (7.3) it is sufficient to consider the integral representation (4.28) in the pseudoconvex polyhedron

$$D_0^\varepsilon = \{z \in D_0: \rho_\nu(z) < \varepsilon, \ \nu = 1, 2, \ldots, n+1\}$$

and to pass to the limit as $\varepsilon \to 0$.

The smoothness of the functions $f_{\nu_1}, \ldots, {}_{\nu_n}$ on $M \cap D_0$ is a consequence of the estimates for Cauchy–Fantappié type integrals given by Theorem 5.6.

Formulas (7.3) and (7.4) can be seen as generalizations to totally real manifolds of the following Fourier–Radon type transformation

$$f(y) = \frac{(n-1)!}{(2\pi i)^n} \int_{\lambda \in S^{n-1}} f(x) \frac{[1 + (\lambda, x - y)]\omega'(\lambda) \wedge \omega(x)}{[(\lambda, x - y) + i|x - y|^2 + i0]^n}, \quad (7.5)$$

where $x, y \in \mathbb{R}^n$, found in the works of Sato–Kawai–Kashiwara, Bros–Jaglonitzer, and Bony in the years 1973–1976 (see Vol. 8, article IV).

Theorem 7.1 together with formulas (7.4) and (7.5) admit several nice corollaries. First of all, from Theorem 7.1, the well-known result on global approximation of functions on totally real submanifolds in $\mathbb{C}^n$ follows (see [19], [40], [82]).

**Theorem 7.2** (Hörmander, Wermer, 1968; Nirenberg, Wells, 1969; Harvey, Wells, 1952; Range, Siu, 1974). *On any $C^s$-smooth totally real manifold $M$ in $\mathbb{C}^n$, any function $f$ of class $C^{s-1}(M)$ can be approximated in the $C^{s-1}(K)$ norm on each compact set $K \subset M$ by functions holomorphic in a neighbourhood of $M$.*

Theorem 7.2 in turn allows one to prove immediately an important uniqueness theorem for $CR$-functions on arbitrary $CR$-manifolds.

**Theorem 7.3** (R.A. Ajrapetyan, G.M. Khenkin [2], Baounedi–Treves [8]). *Let $M$ be any $CR$-manifold of the form (7.1) and $N$ a generic $CR$-submanifold of $M$, i.e. $N = \{z \in M: \rho_{k+j} = 0, \ j = 1, \ldots, m\}$, where $m \leq n - k$, and $\{\rho_{k+j}\}$ are smooth real functions such that $\bar{\partial}\rho_1 \wedge \ldots \wedge \bar{\partial}\rho_{k+m} \neq 0$ on $N$. Then, any continuous $CR$-function $f$ on $M$ which is equal to zero on $N$ vanishes in some neighbourhood of the manifold $N$ in $M$.*

For the case when $M$ is a hypersurface in $\mathbb{C}^n$, the result of Theorem 7.3 was first obtained by S.I. Pinchuk (1974).

*Proof* of Theorem 7.3 (see [2]). Without loss of generality, we may assume that $m = n - k$, i.e. $N$ is totally real. Let $\Omega_0$ be a neighbourhood of the manifold $N$ such that $N$ as well as all totally real manifolds $N^0$ sufficiently close to $N$ in the $C^1$-topology are holomorphically convex in $\Omega_0$.

Let $N^0$ be any totally real submanifold of $M$ close to $N$ and with the property $\partial(N \cup N^0) = \varnothing$; and let $Q$ be a manifold in $M$ such that $\partial Q = N \cup N^0$.

Suppose now that $f$ satisfies the hypotheses of the theorem. In order to prove that $f$ vanishes on $N^0$, it is sufficient, by Theorem 7.2, to verify that

$$\int_{N^0} f \wedge \varphi = 0$$

for any holomorphic $(n, 0)$-form. Indeed, by Stokes' formula and by the hypotheses of the theorem, we have

$$\int_{N_0} f \wedge \varphi = \int_Q \bar{\partial} f \wedge \varphi + \int_N f \wedge \varphi = 0.$$

Baouendi and Treves [8] obtained Theorem 7.3 as a corollary of their theorem on the local approximation of $CR$-functions on arbitrary $CR$-manifolds by holomorphic functions.

**Theorem 7.4** (Baouendi–Treves, 1981). *Let $M$ be an arbitrary $CR$-manifold of class $C^s$, $s \geq 1$ in $\mathbb{C}^n$ and $f$ an arbitrary $CR$-function of class $C^{s-1}(M)$. Then, for each point $z^* \in M$, there exists a neighbourhood $\Omega_{z^*}$ such that on $M \cap \Omega_{z^*}$, the function $f$ can be approximated with arbitrary precision in the topology $C^{(s-1)}(M \cap \Omega_{z^*})$ by polynomials in $z_1, \ldots, z_n$.*

The result of Theorem 7.4 was known previously only for hypersurfaces (Nirenberg–Wells, E.M. Chirka) and also for standard $CR$-manifolds (Naruki) (see [3]).

The proof of Theorem 7.4 in [8] is based on constructing a generalization of the classical proof of the Weierstrass approximation theorem. Namely, let us fix on $M$ a totally real submanifold $M_{z^*}$ passing through the point $z^*$ and of the form

$$M_{z^*} = \{z \in M : \rho_{k+j}(z) = 0, j = 1, \ldots, n-k\},$$

where $\{\rho_{k+j}\}$ are functions of class $C^s$ such that $\rho_{k+j}(z^*) = 0$ and

$$\bar{\partial}\rho_1 \wedge \ldots \wedge \bar{\partial}\rho_n \neq 0$$

in some neighbourhood $\Omega_{z^*}$ on $M$ of the point $z^*$. Also, we set

$$M^{\varepsilon}_{z^*} = \{z \in M : \rho_{k+1} = \varepsilon_{k+1}, \ldots, \rho_n = \varepsilon_n\}.$$

Consider the entire function $f_\nu$ given by

$$f_\nu(z) = \int_{\zeta \in (M_{z^*} \cap \Omega_{z^*})} f(\zeta) \left( \frac{\nu}{\sqrt{2\pi}} \right)^n e^{-\nu^2 \langle z-\zeta,\, z-\zeta \rangle} \omega(\zeta). \tag{7.6}$$

It is a classical fact that for some neighbourhood $\Omega^0_{z^*}$ of the point $z^*$ and for any function $f \in C^{(s-1)}(M)$, we have $f_\nu | M_{z^*} \to f | M_{z^*}$ as $\nu \to \infty$ in the topology of $C^{(s-1)}(M_{z^*} \cap \Omega^0_{z^*})$. Baouendi and Treves showed that using the condition $\bar\partial_M f = 0$ on $M$ we may conclude that for any sufficiently small vector $\varepsilon = (\varepsilon_{k+1}, \ldots, \varepsilon_n)$, we also have that

$$f_\nu |_{M^\varepsilon_{z^*}} \to f|_{M^\varepsilon_{z^*}},$$

in the topology of $C^{(s-1)}(M^\varepsilon_{z^*} \cap \Omega^0_{z^*})$. From this the assertion of Theorem 7.2 follows.

Thus (7.6) can be viewed as a local approximate integral representation of a $CR$-function on $M$.

Although analytic representations of the form (7.2) for arbitrary $CR$-functions are impossible on arbitrary $CR$-manifolds (J.M. Trépreau, 1985), there is a suitable integral representation (generally with a non-holomorphic kernel) which works well in practice and which generalizes the representations (7.3) and (7.4). A $\mathbb{C}^n$-valued vector function $P = (P^{(1)}, \ldots, P^{(n)})$ of class $C^1(\Omega \times \Omega)$ is called a strictly regular barrier for the level function $\rho \in C^2(\Omega)$ if

$$P^{(j)}(\zeta, z) = \frac{\partial \rho}{\partial \zeta_j}(z) + O(|\zeta - z|)$$

and

$$2\operatorname{Re}\Phi(\zeta, z) \geq \rho(\zeta) - \rho(z) + \gamma|\zeta - z|^2,$$

where $(\zeta, z) \in \Omega \times \Omega$, $\Phi(\zeta, z) = \langle P(\zeta, z), \zeta - z \rangle$, and $\gamma$ is a positive constant.

Further, let $M$ be an arbitrary $CR$-manifold of the form (7.1),

$$\rho_{k+1} = -\sum_{j=1}^{k} \rho_j; \quad D_0 = \{z \in \Omega: \rho_0(z) < 0\}$$

a relatively compact subdomain of $\Omega$; $M_0 = M \cap D_0$; and

$$D_j = \{z \in D_0: \rho_j(z) < 0\}; \qquad j = 1, 2, \ldots, k.$$

Let $\{P_j\}$ be arbitrary strictly regular barriers for the level functions $\rho_j \in C^2(\Omega)$, $j = 0, 1, \ldots, k+1$.

For each multi-index $J = (j_1, \ldots, j_k)$, we set

$$(-1)^{\frac{k(k-1)}{2}} \frac{(2\pi i)^n}{(n-k)!} f_J(z) =$$

$$= \sum_{\beta \in J}{}' \int_{\zeta \in M_0} f(\zeta) \frac{\det[P_{j_1}, \ldots, P_{j_k}, \bar\partial_\zeta P_{\beta_1}, \ldots, \bar\partial_\zeta P_{\beta_{n-k}}] \wedge \omega(\zeta)}{\Phi_{j_1} \ldots \Phi_{j_k} \Phi_{\beta_1} \ldots \Phi_{\beta_{n-k}}}$$

$$- \sum_{\alpha \in (0, J)}{}' \int_{\zeta \in \partial M_0} f(\zeta) \frac{\det[P_0, P_{j_1}, \ldots, P_{j_k}, \bar\partial_\zeta P_{\alpha_1}, \ldots, \bar\partial_\zeta P_{\alpha_{n-k-1}}]}{\Phi_0 \Phi_{j_1} \ldots \Phi_{j_k} \Phi_{\alpha_1} \ldots \Phi_{\alpha_{n-k-1}}} \wedge \omega(\zeta) \tag{7.7}$$

where $z \in D_J = D_{j_1} \cap \ldots \cap D_{j_k}$.

**Theorem 7.5** (G.M. Khenkin, 1979). *For any integrable CR-function $f$ on $M_0$, the functions $f_J$, defined in the domains $D_J$ by equation (7.7), have generalized boundary values on $M_0$. Moreover, we have*

$$\int_{z\in M_0} f(z) \wedge \varphi(z) = \sum_{v+1}^{k+1} (-1)^{k+v} \int_{z\in M^0} f_{1,\ldots,v-1,v+1,\ldots k+1}(z) \wedge \varphi(z), \quad (7.8)$$

*where $\varphi$ is an arbitrary $(n, n-k)$-form of compact support in $D_0$ and of class $C^\alpha$, $\alpha > 0$.*

Recently, Baouendi, Rothschild and Treves (Invent. Math. *82*, 359–396, 1985) have obtained analytic representations of *CR*-functions on standard (or rigid) *CR*-manifolds of the form (7.9). This result can be obtained also as a consequence of formulas (7.7), (7.8) for *CR*-functions on such *CR*-manifolds.

**7.2. *CR*-Functions and the "Edge-of-the-Wedge" Theorems.** The following result on analytic continuation of *CR*-functions through the "edge-of-the-wedge" was obtained in [2] and [3] on the basis of the integral representations (7.7) and (7.8).

**Theorem 7.6** (R.A. Ajrapetyan, G.M. Khenkin, 1981). *Let us view a generic CR-manifold $M$ given by (7.1) as the common boundary of CR-manifolds given by*

$$M_j = \{z \in \Omega: \rho_j(z) < 0, \rho_v(z) = 0, v \neq j\}.$$

*Then, any function $f$ continuous on the compact set*

$$\bigcup_{j=1}^{k} \bar{M}_j$$

*and CR on the manifold*

$$\bigcup_{j=1}^{k} M_j$$

*has a continuous extension in $\mathbb{C}^n$ to a function holomorphic in a domain $\Omega^+$ such that for any $\varepsilon > 0$, we have*

$$\Omega^+ \supset \left\{z \in U_\varepsilon(M): \rho_j < \varepsilon\left(\sum_{v\neq j} \rho_v\right), v = 1,2,\ldots,k\right\},$$

*$M_j \cap U_\varepsilon(M) \subset \bar{\Omega}^+$, where $U_\varepsilon(M)$ is some neighbourhood of the manifold $M$.*

In the case where all of the functions $\{\rho_v\}$ which determine the *CR*-manifold $M$ given by (7.1) are pluriharmonic, then the assertion of Theorem 7.6 is true also for $\varepsilon = 0$.

**Theorem 7.6′** (R.A. Ajrapetyan, G.M. Khenkin, 1981). *Consider the CR-manifold*

$$M_0 = \{\zeta \in \Omega: \text{Re } F_0 > 0, \text{Re } F_j = 0, j = 1,2,\ldots,n\},$$

*where* $\{F_j\}$ *are functions holomorphic in the domain* $\Omega \subset \mathbb{C}^n$. *Thus,* $M_0 \subset \Omega$ *and* $M_0$ *is the common boundary of the CR-manifolds given by*

$$M_j = \{z \in \Omega\colon \operatorname{Re} F_0 > 0, \operatorname{Re} F_j > 0, \operatorname{Re} F_\nu = 0, \nu \neq j\},$$

$j = 1, \ldots, n$. *Then, any continuous CR-function* $f$ *on*

$$\bigcup_{j=1}^{k} \bar{M}_j.$$

*has a continuous extension* $F$ *which is holomorphic in the analytic polyhedron*

$$D = \{\zeta \in \Omega\colon \operatorname{Re} F_j > 0, j = 0, 1, \ldots, n\}.$$

In the case where $F_0 \equiv 1$, $F_j = z_j$, $j = 1, \ldots, n$, i.e. when $M = \mathbb{R}^n$ and

$$f \in C^\infty\left(\bigcup_{j=1}^{n} \bar{M}_j\right),$$

the result of Theorem 7.6′ was obtained first by Malgrange–Zerner (1961) whose result in turn extended the classical theorems on separate analyticity (S.N. Bernstein) and on the "edge-of-the-wedge" (N.N. Bogolyubov, Bremerman–Oehme–J.G. Taylor, V.S. Vladimirov) see [81].

From Theorems 7.6 and 7.6′ follow not only a list of classical results but also a series of generalizations thereof (Epstein, Browder, S.I. Pinchuk, E.M. Chirka, Bedford) (see [3]).

The proof of Theorem 7.6 consists in verifying that the desired holomorphic continuation $F$ of $f$ can be given by the formula of A. Weil

$$F(z) = \frac{(-1)^{\frac{n(n-1)}{2}}}{(2\pi i)^n}\left[\int_{M_0} f(\zeta)\frac{\det[P_1(\zeta, z), \ldots, P_n(\zeta, z)]}{\prod_{j=0}^{n}(F_j(\zeta) - F_j(z))}\omega(\zeta) + \sum_{\nu=1}^{n}\int_{(\partial M_\nu)\setminus M_0} f(\zeta) \times \right.$$

$$\left. \times \frac{\det[P_1(\zeta, z), \ldots, P_{\nu-1}(\zeta, z), P_0(\zeta, z), P_{\nu+1}(\zeta, z), \ldots P_n(\zeta, z)]}{\prod_{j=0}^{n}(F_j(\zeta) - F_j(z))}\omega(\zeta)\right],$$

where $F_j(\zeta) - F_j(z) = \langle P_j(\zeta, z), \zeta - z\rangle$, $j = 0, 1, \ldots, n$, and $\{P_j\}$ are holomorphic functions of the variables $\zeta, z \in \Omega$.

In applications, Theorem 7.6 works well in conjunction with the uniqueness Theorem 7.3 and the strong principle of continuity of Hartogs called the strong theorem on discs (see [81]).

**Theorem 7.7** (Bremerman, 1954). *Let* $z = z(t)$, $t \in [0, 1]$ *be a real-analytic curve in* $\mathbb{C}^n_z$ *and* $D(t) = \{(z, w)\colon z = z(t), |w| < 1\}$ *a continuous family of discs in* $\mathbb{C}^n_z \times \mathbb{C}_w$. *Then, if a function* $f = f(z, w)$ *is holomorphic in* $\bigcup_{0<t<1} D(t)$ *and at least* *in one point of the limit disc $D(0)$, then $f$ is holomorphic at all points of the limit disc $D(0)$.*

We remark that, without the assumption that the curve $z = z(t)$ be real analytic, the assertion of Theorem 7.7 is in general not true, (see S. Favorov, Funct. Anal. Appl. *12*, 90–91, 1978).

Theorem 7.7 is a corollary of a classical lemma of Oka: if $z(t)$ is a continuous curve in $\mathbb{C}^1$ and $u(z)$ is a subharmonic function in a neighbourhood of the curve, then

$$\overline{\lim_{t\to 0}}\, u(z(t)) = u(z(0)).$$

The following result [34] can be seen as a deep generalization of both the strong theorem on discs as well as the Bogolyubov theorem on the "edge-of-the-wedge".

**Theorem 7.8** (Hanges, Treves, 1983). *Let $M$ be a generic CR-manifold in $\mathbb{C}^n$; $N$ a connected complex analytic submanifold of $M$; and $f$ a (generalized) CR-function on $M$. Then, if $f$ has a holomorphic continuation in the neighbourhood of some point $z^* \in N \subset M$, then $f$ extends holomorphically in the neighbourhood of each point $z \in N$.*

The original and rather difficult proof of Theorem 7.8 in [34] uses the technique in the works [7] and [8] and the Fourier–Bros–Jagolnitzer transformation (7.5).

A simpler proof of Theorem 7.8 can be obtained by combining Theorems 7.5–7.7.

**7.3. Holomorphic Continuation of *CR*-Functions Given on Concave *CR*-Manifolds.** The theory of *CR*-functions has its origin in theorems of Bochner and H. Lewy which give conditions (respectively global and local) under which a *CR*-function defined on a real hypersurface of $\mathbb{C}^n$ necessarily has a holomorphic continuation to some domain in $\mathbb{C}^n$.

In 1960, H. Lewy found the first example of a *CR*-manifold of higher codimension in $C^n$ for which an analogous situation prevails.

The fundamental results on holomorphic continuation of *CR*-functions are formulated using a natural generalization of the Levi form for arbitrary *CR*-manifolds.

Let us denote by $N_z(M)$ the space normal to the *CR*-manifold (7.1) at the point $z \in M$. We shall assume further that the functions $\{\rho_j\}$, in the representation (7.1), are smooth of order $C^2(\Omega)$ and so chosen that the vectors $\operatorname{grad} \rho_j(z)$ form an orthogonal basis for $N_z(M)$. The *Levi form of the manifold $M$* at the point $z \in M$ is the quadratic form $L_z^M(\zeta)$ on $T_z^{\mathbb{C}}(M)$ with values in $N_z(M)$ given by

$$L_z^M(\zeta) = -\sum_{\nu=1}^{k} \left( \sum_{\alpha,\beta} \frac{\partial^2 \rho_\nu(z)}{\partial z_\alpha \cdot \partial \bar{z}_\beta} (\zeta_\alpha - z_\alpha)(\bar{\zeta}_\beta - \bar{z}_\beta) \operatorname{grad} \rho_\nu(z) \right),$$

where $\zeta \in T_z^{\mathbb{C}}(M)$.

We have the following geometric interpretation of the Levi form in terms of the Lie bracket of a complex tangent vector field

$$L_z^M(\zeta) = \frac{1}{2\pi} \sum_{v=1}^{k} ([\zeta, \bar{\zeta}], J \operatorname{grad} \rho_v(z)) \operatorname{grad} \rho_v(z),$$

where

$$\bar{\zeta} = \sum_j \bar{\zeta}_j \frac{\partial}{\partial \bar{z}_j}$$

is the Cauchy–Riemann differential operator in the direction of the complex tangent vector field $\zeta$ with the property $\zeta(z) = \zeta \in T_z^{\mathbb{C}}(M)$. Here $[\zeta, \bar{\zeta}]$ is the commutator of the differential operators $\zeta$ and $\bar{\zeta}$.

A *CR*-manifold $M$ in $\mathbb{C}^n$ is said to be a *standard manifold* if it has the form

$$M = \{z = (z', z'') \in \mathbb{C}^k \times \mathbb{C}^{n-k}: \rho(z) = \operatorname{Im} z' - F(z'', z'') = 0\}, \tag{7.9}$$

where $F = (F_1, \ldots, F_k)$ is a $\mathbb{C}^k$ valued Hermitian form on $\mathbb{C}^{n-k}$.

If we identify the space $N_0(M)$ with the space $\mathbb{R}^k$ in which the form $F$ takes its values, then $L_0^M = F(z'', z'')$, $z'' \in \mathbb{C}^{n-k}$. For an arbitrary point $\zeta$ on a standard manifold (7.9), we have

$$T_\zeta^{\mathbb{C}}(M) = \{z \in \mathbb{C}^n: z' - \zeta' = 2i(F(z'', \zeta'') - F(\zeta'', z''))\},$$

$$L_\zeta^M(z) = \sum_{j=1}^{k} F_j(z'' - \zeta'', z'' - \zeta'') \operatorname{grad} \rho_j(\zeta).$$

The importance of the role of standard *CR*-manifolds among all *CR*-manifolds is due to the fact that any *CR*-manifold is locally equivalent up to order two to a standard one. More precisely, if $M$ is an arbitrary $C^2$-smooth *CR*-manifold given by (7.1), then for any point $z^* \in M$ there is a neighbourhood $U$ and a biholomorphic mapping

$$\varphi: U \to \Omega \subset \mathbb{C}^k \times \mathbb{C}^{n-k}$$

such that $\varphi(z^*) = 0$ and

$$\varphi(M \cap U) = \{(z', z'') \in \mathbb{C}^k \times \mathbb{C}^{n-k}: \rho = \operatorname{Im} z' - F(z'', z'') + R(z) = 0\}, \tag{7.10}$$

where $F$ is a $\mathbb{C}^k$-valued Hermitian form and the remainder term admits the estimate $|R(z)| = o(|z|^2)$.

A *CR-manifold* $M$ is called *Levi-flat* if $L_z^M(\zeta) \equiv 0$ for each $z \in M$. A real-analytic Levi-flat *CR*-manifold necessarily (see below) admits locally the following form

$$M = \{\zeta \in \Omega: \rho_j = \operatorname{Re} F_j(\zeta) = 0, j = 1, \ldots, k\},$$

where $\{F_j\}$ are functions holomorphic in the domain $\Omega$.

For a standard *CR*-manifold one can precisely determine a domain to which all *CR*-functions on $M$ automatically continue holomorphically (see [3]).

**Theorem 7.9** (Naruki, 1970; R.A. Ajrapetyan, G.M. Khenkin, 1984). *Let $M$ be any standard CR-manifold of the form* (7.9) *and such that the convex cone $V$,* *spanned by the vectors* $F(z'', z'')$, *where* $z'' \in \mathbb{C}^{n-k}$, is solid in $\mathbb{R}^k$. *By* $D_M$ *we denote the domain given by*

$$D_M = \{z \in \mathbb{C}^n : \rho = \operatorname{Im} z' - F(z'', z'') \in V\}.$$

*Then,*

a) *each continuous (or generalized) CR-function on* $M$ *has a holomorphic continuation to the domain* $D_M$;

b) *if moreover the cone* $V$ *is acute, and* $M_0$ *is a relatively compact domain on* $M$ *of the form* $M_0 = \{z \in M : \operatorname{Re} \Phi_0(z) < 0\}$, *where* $\Phi_0$ *is some holomorphic function, then the maximal domain to which an arbitrary CR-function on* $M_0$ *extends holomorphically has the form*

$$D_{M_0} = \{z \in D_M : \operatorname{Re} \Phi_0(z) < 0\}.$$

Assertion a) of Theorem 7.9. for continuous $CR$-functions was proved first in the rather difficult work of Naruki. The more precise assertion b) can be proved in a remarkably more simple fashion [3] with the help of integral formulas. Namely, if $f$ is a $CR$-function on $M_0$, then a holomorphic function on $D_{M_0}$ agreeing with $f$ on $M_0$ can be given by the explicit formula

$$F = \frac{(n-1)!}{(2\pi i)^n} (F_1 + F_2),$$

where

$$F_1(z) = \int_{(\zeta, t) \in M_0 \times U} f(\zeta) \omega' \left( \frac{P(\zeta, z, t)}{\Phi(\zeta, z, t)} \right) \wedge \omega(\zeta);$$

$$F_2(z) = \int_{(\zeta, t, t^0) \in (\partial M_0) \times U \times [0,1]} f(\zeta) \omega' \left( t^0 \frac{P_0(\zeta, z)}{\Phi_0(\zeta, z)} + (1 - t^0) \frac{P(\zeta, z, t)}{\Phi(\zeta, z, t)} \right) \wedge \omega(\zeta),$$

and $U$ denotes the set of points $t$ on the unit sphere $S^k \subset (\mathbb{R}^k)^*$, for which the scalar form $\langle t, F \rangle$ is positive definite,

$$P(\zeta, z, t) = \left\langle t, \frac{\partial \rho}{\partial \zeta}(\zeta) \right\rangle,$$

$$\Phi(\zeta, z, t) = \langle P(\zeta, z, t), \zeta - z \rangle; \ \Phi_0(\zeta, z) = \langle P_0(\zeta, z), \zeta - z \rangle.$$

Only recently [2], [15], [39] was there success in obtaining an analogue to Theorem 7.9 for arbitrary $CR$-manifolds.

**Theorem 7.10** (G.M. Khenkin, 1980; R.A. Ajrapetyan, G.M. Khenkin, 1981; Boggess, Polking, 1981). *Let* $M$ *be a CR-manifold given by* (7.1) *and such that for a point* $z \in M$, *the convex hull* $V_z$ *of the vectors* $L_z^M(w)$, *where* $w \in T_z^{\mathbb{C}}(M)$, *is a solid cone in* $N_z(M)$. *Then, for any open subcone* $V_z^0 \subset V_z$ *with the property* $\partial V_z^0 \cap \partial V_z = \{z\}$, *there exists a neighbourhood* $\Omega_0$ *of the point* $z$ *such that each continuous (or generalized) CR-function* $f$ *on* $M$ *extends to a function* $F$ *holomorphic in the domain* $(M + V_z^0) \cap \Omega_0$.

For the case $V_z = N_z(M)$, Theorem 7.10 was first obtained in [39] and solves the *Naruki problem* (1969).

In the above form, Theorem 7.10 is a rapid consequence of Theorem 7.6 on the edge-of-the-wedge for *CR*-functions and Theorem 7.16 on suspending analytic discs on *CR*-manifolds. An explicit formulation and proof of Theorem 7.10 based on the approximation Theorem 7.4 was first given in [15].

The method of integral representations ([2], [3], [39]) allows one to prove Theorem 7.10 just as we did for Theorem 7.9, with an explicit formula. In order to state this formula, it is necessary to introduce several definitions.

By the *Levi form of the CR-manifold M* (7.1) at the point $z \in M$ in the direction $t_z = \sum_j t^j \operatorname{grad} \rho_j(z)$ or simply in the direction $t = (t^1, \ldots, t^k)$, we shall mean the scalar quadratic form

$$L_z^{\rho_t}(w) = \langle t_Z, L_z^M \rangle(w),$$

where $\rho_t = \langle t, \rho \rangle$, $w \in T_z^{\mathbb{C}}(M)$.

The *manifold M* is said to be *q-concave* (respectively *weakly q-concave*) at the point $z \in M$ in the direction $t \in S^k$ if the form $L_z^{\rho_t}(w)$ has at least $q$ negative (respectively $q$ non-positive) eigen-values on $T_z^{\mathbb{C}}(M)$. We shall say that the manifold $M$ is $q$-concave (respectively weakly $q$-concave) at the point $z \in M$ if it is $q$-concave (respectively) weakly $q$-concave) in each direction $t \in S^k$.

Let us denote by $S_{q,z}$ the closed set of those points $t$ on the sphere $S^k$ for which the form $L_z^{\rho_t}(w)$, $w \in T_z^{\mathbb{C}}(M)$ has less than $q$ negative eigenvalues. In particular, $S_{1,z}$ is the set of points $t \in S^k$ for which the form $\langle t, L_z^M \rangle$ is positive semi-definite on $T_z^{\mathbb{C}}(M)$.

Integral representations of *CR*-functions (and *CR*-forms [3]) on $q$-concave *CR*-manifolds make us of the following important assertion.

**Proposition 7.11** ([3]). *Let $M$ be a $C^2$-smooth CR-manifold given by* (7.1), *$z^* \in M$, and $U(S_{q,z^*})$ a contractible neighbourhood of $S_{q,z^*}$. Then there exists a neighbourhood $\Omega_0$ of the point $z^*$ and a constant $A > 0$ such that for each $t \in S^k$, there exists a strongly regular barrier vector-function $P(\zeta, z, t)$ for the level function $\tilde{\rho}_t = \langle t, \rho \rangle + A(\rho_1^2 + \ldots + \rho_k^2)$ in the domain $\Omega_0$. Moreover, $P$ is a $C^1$-smooth function of $t$ with the property*

$$\langle \bar{\partial}_\zeta P(\zeta, z, t) \wedge d\zeta \rangle^{n-k-q+1} = 0,$$

*for each $(\zeta, z) \in \Omega_0 \times \Omega_0$ and $t \in S^k \backslash U(S_{q,z^*})$.*

Making use of the barrier function from Proposition 7.11, we can write down a function $F$ satisfying the requirements of Theorem 7.10. Indeed, we have the explicit formula [4]:

$$F = \frac{(n-1)!}{(2\pi i)^n}(K_U f + K_O f), \tag{7.11}$$

where

$$K_U f(z) = \int_{(\zeta,t)\in(M\cap\Omega_0)\times U} f(\zeta)\omega'\left(\frac{P(\zeta,z,t)}{\Phi(\zeta,z,t)}\right)\wedge\omega(\zeta),$$

$$K_0 f(z) = \int_{(\zeta,t,t^0)\in\partial(M\cap\Omega_0)\times S^k\times[0,1]} f(\zeta)\omega'\left(t^0\frac{\bar\zeta-\bar z}{|\zeta-z|}+\right.$$

$$\left.+(1-t^0)\frac{P(\zeta,z,t)}{\Phi(\zeta,z,t)}\right)\wedge\omega(\zeta),$$

$U$ is a neighbourhood of $S_{1,z^*}$ on $S^k$; and $P(\zeta,z,t)$ is the barrier vector-function satisfying Propositon 7.11 for $q=1$ and $\Phi(\zeta,z,t)=\langle P(\zeta,z,t),\zeta-z\rangle$.

**7.4. The Phenomena of Hartogs–Bochner and H. Lewy on 1-Concave *CR*-Manifolds.** Theorems 7.2, 7.3, 7.4, 7.10 and their generalizations to *CR*-forms (see [3]) lead to [50] a Hartogs–Bochner type effect for *CR*-functions on arbitrary 1-concave *CR*-manifolds.

**Theorem 7.12** (G.M. Khenkin, 1984). *Let $\Omega\subset\mathbb{C}^n$ be a pseudoconvex domain and $\rho=\{\rho_1,\ldots,\rho_k\}$ a collection of real functions of class $C^2(\Omega)$ such that $\bar\partial\rho_1\wedge\ldots\wedge\bar\partial\rho_k\neq0$ in $\Omega$ and each CR-manifold $M_\varepsilon$ given by $M_\varepsilon=\{z\in\Omega: \rho(z)=\varepsilon\}$, where $\varepsilon=(\varepsilon_1,\ldots,\varepsilon_k)$, is 1-concave. Let $\Omega_0=\{z\in\Omega:\rho_0(z)<0\}$ be a relatively compact subdomain of $\Omega$ and let $\varepsilon^0=\{\varepsilon_1^0,\ldots,\varepsilon_k^0\}$ be a collection such that the manifold $M_0=M_{\varepsilon^0}\cap\Omega_0$ has a connected complement in $M_{\varepsilon^0}$ and $\bar\partial\rho_1\wedge\ldots\wedge\bar\partial\rho_k\wedge\bar\partial\rho_0\neq0$ almost everywhere on $\partial M_0$. Then, any continuous CR-function $f$ on $\partial M_0$ can be extended to a continuous CR-function $F$ on $M_0$.*

We remark that under the conditions of Theorem 7.12, the manifold $\partial M_0$ is, in general, only an almost *CR*-manifold. Nevertheless, we say (by definition) that a function $f$ on $\partial M_0$ is a *CR*-function if

$$\int_{\partial M_0} f\wedge\bar\partial\varphi=0$$

for each smooth $(n,n-k-2)$-form $\varphi$. This definition is satisfied, for example, by the restriction to $\partial M_0$ of any *CR*-function defined in a neighbourhood of $\partial M$ on $M_{\varepsilon^0}$.

In case the diameter of the manifold $M_0$ is sufficiently small, a function $F$ satisfying Theorem 7.12 can be given by the explicit formula

$$F(z)=\frac{(n-1)!}{(2\pi i)^n}\int_{(\zeta,t,t^0)\in(\partial M_0)\times S^k\times[0,1]} f(\zeta)\omega'\left[t^0\frac{\bar\zeta-\bar z}{|\zeta-z|^2}+(1-t^0)\frac{P(\zeta.z.t)}{\Phi(\zeta,z,t)}\right]\wedge\omega(\zeta) \tag{7.12}$$

where $z\in M_0$ and the function $\Phi=\langle P,\zeta-z\rangle$ satisfies the requirement of Proposition 7.11.

From Theorem 7.10 or directly from formula 7.12 it follows that the *CR*-function $F$ constructed on $M_0$ extends further to a holomorphic function in some neighbourhood of $M_0$ in $\Omega_0$.

The result of Theorem 7.12 is probably valid for arbitrary weakly 1-concave *CR*-manifolds if the function $f$ in the formulation of the theorem is taken to be a smooth *CR*-function.

The method of integral representations allows one to prove this conjecture for those *CR*-manifolds $M$ of type (7.1) for which there exists a weakly concave regular barrier function, i.e. a smooth vector-function $P=(P^{(1)},\ldots,P^{(n)})$ of the variables $\zeta\in\Omega\backslash M$ and $z\in M$ such that the following conditions are satisfied:

$$|\langle P(\zeta,z),\zeta-z\rangle|\geq\gamma|\rho(\zeta)|x,$$
$$|\bar{\partial}_\zeta P(\zeta,z)|\leq\gamma|\rho(\zeta))|-x$$
$$\langle\bar{\partial}_\zeta(\zeta,z)\wedge d\zeta\rangle^{n-k}=0,$$

where $(\zeta,z)\in(\Omega\backslash M)\times M$; $\gamma>0$; $x>0$.

Aside from strictly 1-concave manifolds, such barrier functions can be constructed also in two important particular cases: for standard *CR*-manifolds and for *CR*-manifolds admitting folliations into holomorphic curves.

**Theorem 7.12′** ([50]). *Let $M$ be an arbitrary weakly 1-concave standard CR-manifold given by (7.9), where $\dim_{\mathbb{C}} T^{\mathbb{C}}(M)\geq 2$, and let $M_0$ be a subdomain of $M$ with smooth boundary and connected complement $M\backslash M_0$. Then, for any $C^\infty$-smooth CR-function $f$ on $\partial M_0$, there exists a $C^\infty$-smooth CR-function $F$ on $M_0$ agreeing with $f$ on $\partial M_0$.*

We remark that by a $C^\infty$-smooth *CR*-function on $\partial M_0$ here, we have in mind a function $f$ of class $C^\infty$ on $\partial M_0$ which admits a continuation $\tilde{f}$ to a neighbourhood of $\partial M_0$ with the property $\bar{\partial}\tilde{f}=0$ on $\partial M_0$ along with all its derivatives.

We shall regard that a smooth *CR*-manifold $M$ given by (7.1) admits a smooth folliation into holomorphic curves if there exists a smooth mapping of maximal rank $\pi\colon M\to\mathscr{P}\subset\mathbb{R}^{2n-k-2}$ with the property: for each $p\in\mathscr{P}$, the manifold $\pi^{-1}(p)$ is a one-dimensional complex manifold.

**Theorem 7.12″** ([50]). *Let $\Omega$ be a pseudoconvex domain and $\rho=\{\rho_1,\ldots,\rho_k\}$ be a collection of functions of class $C^2(\Omega)$ such that $\bar{\partial}\rho_1\wedge\ldots\wedge\bar{\partial}\rho_k\neq 0$ in $\Omega$ and each CR-manifold $M^\varepsilon=\{z\in\Omega:\rho(z)=\varepsilon\}$ admits a smooth folliation into holomorphic curves. Let $k\leq n-2$, i.e. $\dim_{\mathbb{C}} T^{\mathbb{C}}(M^\varepsilon)\geq 2$. Then, on each manifold $M^\varepsilon$ the assertion of Theorem 7.12 holds for smooth CR-functions $f$ and $F$.*

The classical theorem of Hartogs–Bochner 1.2 is of course a particular case of Theorems 7.12′, 7.12″.

We remark that the condition $\dim_{\mathbb{C}} T^{\mathbb{C}}(M)\geq 2$ in Theorems 7.12′ and 7.12″ may not, in general, be discarded. However from the Hartogs continuity principle (Theorem 0.1) it follows [13] that the assertion of Theorem 7.12″ is valid also for the case $\dim_{\mathbb{C}} T^{\mathbb{C}}(M)=1$ provided we suppose in addition that $\dim_{\mathbb{R}} T(M) \geq 3$, that the domain $M_0 \Subset M$ is such that the holomorphic leafs $\pi^{-1}(p) \cap M_0$ are connected and simply connected, and that the function $f$ be holomorphic in a neighbourhood of $\partial M_0$.

From Theorem 7.12 follows a beautiful maximum principle for $CR$-functions obtained first by Sibony using the Monge–Ampère equation.

**Theorem 7.13** (Sibony, 1977). *In order that an arbitrary smooth CR-function $f$ on a CR-manifold $M$ admit no local maxima on $M$, it is necessary and sufficient that the manifold $M$ be weakly 1-concave.*

If a $CR$-manifold admits a smooth folliation into holomorphic curves, then on such a manifold, not only does the Hartogs–Bochner phenomenon prevail but also the equally acclaimed effect of H. Lewy (see [50]).

**Theorem 7.14** (Hill, 1977; G.M. Khenkin, 1984). *Suppose a CR-manifold $M$ given by* (7.1) *folliates smoothly into holomorphic curves $\Gamma_p$ depending on the parameter $p \in \mathscr{P} \subset \mathbb{R}^{2n-k-2}$, $k \leq n-2$. Let the domain $\mathscr{P}$ be separated by a smooth hypersurface into subdomains $P_+$ and $P_-$ and set*

$$M_{\pm} = \bigcup_{p \in \mathscr{P}_{\pm}} \Gamma_p.$$

*Finally, let $M_0$ be a compact subdomain of $M$ whose boundary is an almost CR-manifold and let $\Gamma_p^0 = M_0 \cap \Gamma_p$, $p \in \mathscr{P}$. Then any continuous CR-function $f_{\pm}$ on $\partial M_0 \cap M_{\pm}$ extends to a continuous CR-functions $F_{\pm}$ on the CR-manifold $M_0 \cap M_{\pm}$ respectively.*

The result of Theorem 7.14, under the hypothesis that the folliation $\{\Gamma_p\}$ can be holomorphically straightened, was obtained by Hill and generalized significant works of H. Lewy. The validity of Theorem 7.14 in the general case is proved in [50] and makes use of the following simple but very convenient property of $CR$-functions (see Vol. 9, article VI).

**Proposition 7.15** (A.E. Tumanov, G.M. Khenkin, 1983). *Let $N$ be a smooth CR-manifold in $\mathbb{C}^n$ which folliates smoothly into holomorphic curves $\Gamma_p$, $p \in \mathscr{P} \subset \mathbb{R}^m$ with smooth boundaries $\partial\Gamma_p$. Suppose moreover that $\partial N$ contains a smooth CR-manifold $M$ with the property: each analytic curve $\Gamma_p$ intersects $M$ transversally in a real arc $\gamma_p \subset \partial\Gamma_p$ and also*

$$M = \bigcup_{p \in \mathscr{P}} \gamma_p.$$

*Then any continuous function $f$ on $N$ which is holomorphic on each curve $\Gamma_p$ and is a CR-function on $M$ satisfies also the tangential Cauchy–Riemann equations on $N$.*

On account of Proposition 7.15, for the proof of Theorem 7.14 it is sufficient to show that for each $p \in \mathscr{P}_+$ the function $f_+$ extends holomorphically to an analytic disc $\Gamma_p^0$. For this, since the analytic disc $\Gamma_p^0$ subtends, it is sufficient to verify that the moment condition

$$\int_{\partial\Gamma_p^0} f_+ \wedge h = 0$$

is fulfilled for each holomorphic 1-form with polynomial coefficients. Indeed, suppose the disc $\Gamma_p^0$ has the form

$$\Gamma_p^0 = \{z \in M_0 : \varphi_1(z) = \ldots = \varphi_{n-k-1}(z) = 0\},$$

where $\{\varphi_j\}$ are functions holomorphic in $\Omega$. By the Bochner–Martinelli formula (1.2), we have

$$\int_{\partial\Gamma_p^0} f_+ \wedge h = \lim_{\varepsilon \to 0} \int_{\{z \in \partial M_0 :\, |\varphi| = \varepsilon\}} f_+ \wedge h \wedge \omega_\varphi,$$

where

$$\omega_\varphi = \frac{(n-k-2)!}{(2\pi i)^{n-k-1}} \bar{\partial} |\varphi|^{4-2n+2k} \wedge \left(\sum_j d\varphi_j \wedge d\bar{\varphi}_j\right)^{n-k-2},$$

$$|\varphi|^2 = \sum_{j=1}^{n-k-1} |\varphi|^2.$$

Since the form $h \wedge \omega_\varphi$ is a smooth $\bar{\partial}$-closed form of type $(n-k, n-k-1)$ on $M_0 \backslash \Gamma_p^0$, by Stokes formula we obtain, further, the equality

$$\int_{\partial\Gamma_p^0} f_+ \wedge h = \int_{(\partial M_0) \cap (\partial M_+)} f_+ \wedge h \wedge \omega_\varphi.$$

By the well-known theorem of Andreotti–Grauert–Hörmander [44] which extends the approximation theorem of Oka–Weil to $\bar{\partial}$-closed forms, there exists in the neighbourhood of $\bar{M}_0$ a sequence of $\bar{\partial}$-closed smooth $(n-k, n-k-1)$-forms $\omega_{\varphi,\nu}$ which approximate the form $\omega_\varphi$ uniformly on $(\partial M_0) \cap (\partial M_+)$. From this and from the previous equation we have

$$\int_{\partial\Gamma_p^0} f_+ \wedge h = \lim_{\nu \to \infty} \int_{(\partial M_0) \cap (\partial M_+)} f_+ \wedge h \wedge \omega_{\varphi,\nu} = 0,$$

i.e. this proves the required moment condition.

**7.5. Analytic Discs and the Holomorphic Hull of a *CR*-Manifold.** Theorem 7.14 works well in conjunction with the following result on the local suspension of analytic discs on *CR*-manifolds with non-zero Levi form.

**Theorem 7.16** (Hill, Taiani, 1978, see [43]). *Let $M$ be a $C^s$-smooth CR-manifold given by* (7.1), *$s \geq 5$. Suppose that for a fixed point $z \in M$ and for a fixed $\zeta \in T_z^c(M)$, we have that $L_z^M(\zeta) = t_z \neq 0$. Then, there exists a neighbourhood $D$ of the point $z$ and a CR-manifold $\tilde{M}$ of real dimension greater by one with the properties:*

*a) $\partial\tilde{M} \supset M \cap D$;*

*b) $T_z(\tilde{M})$ is equal to the linear hull of the space and the vector $t_z$;*

*c) the manifold $\tilde{M}$ folliates as a family of complex one-dimensional analytic discs with boundaries on $M$ and depending smoothly on $(2n-k-1)$ real parameters;*

*d) if $M$ is a real analytic manifold, then $\tilde{M}$ is also real analytic, and moreover, the manifold $\tilde{M}$ extends to a larger real analytic manifold $\tilde{M}_\delta$ such that $M \cap \tilde{M}_\delta = M \cap D$;*

*e) if $M$ is smooth of class $C^s$, then the manifold $\tilde{M}$ is smooth of class $C^{(s-2)/3}$.*

The method of constructing (under the conditions of Theorem 7.16) the manifold $\tilde{M}$ by attaching to $M$ a family of analytic discs with boundaries on $M$ is based on the following construction.

In a neighbourhood $U_{z^*}$ of a fixed point $z^* \in M$ a $C^1$-smooth generic $CR$-manifold $M$ can be represented, by (7.10), in the form

$$M = \{(z, w) \in \mathbb{C}^k \times \mathbb{C}^{n-k} : x = h(y, w)\},$$

where $z = x + iy \in \mathbb{C}^k$ and $h$ is a $C^1$-smooth function on $\mathbb{R}^k \times \mathbb{C}^{n-k}$ with the properties $h(0) = 0$ and $dh(0) = 0$.

Let $S^1$ denote the boundary of the unit disc $D^1$ in $\mathbb{C}^1$. If we are given mappings $v: S^1 \to \mathbb{R}^k$ and $w: S^1 \to \mathbb{C}^{n-k}$, then we denote by $H(v, w)$ the mapping from $S^1$ to $\mathbb{R}^k$ given by

$$H(v, w)(e^{i\varphi}) = h(v(e^{i\varphi}), w(e^{i\varphi})).$$

By construction, any mapping of the form $g(e^{i\varphi}) = H(v, w)(e^{i\varphi}) + iv(e^{i\varphi})$ carries the circle $S^1$ to some closed contour on $M$. In order that the mapping $g$ extend to a holomorphic mapping $G$ of the disc $\bar{D}^1$ to $\mathbb{C}^n$, it is necessary and sufficient, as is well known, that the functions $H(v, w)$ and $v$ be the boundary values of conjugate harmonic functions, i.e. satisfy the Bishop equation (1965, see [14], [43])

$$v(e^{i\varphi}) = T(H(v, w))(e^{i\varphi}) + y,$$

where $T$ is the classical Hilbert transform and $y$ is a vector in $\mathbb{R}^n$.

**Theorem 7.17** (Greenfield, Wells, 1968, see [82]. Hill, Taiani, 1978, see [43]). For any smooth mapping $w: S^1 \to \mathbb{C}^{n-k}$ *with sufficiently small $C^1$-norm and any vector $y$ with sufficiently small norm in $\mathbb{R}^k$, there exists a unique continuous mapping*

$$v(w, y): S^1 \to \mathbb{R}^k,$$

*satisfying the Bishop equation. Moreover the mapping $v(w, y)$ depends continuously on $w$ and $y$ and for any $\alpha < 1$*

$$\|v(w, y)\|_{C^\alpha(S^1)} \to 0, \textit{ if } \|w\|_{C^1(S^1)} + \|y\|_{\mathbb{R}^k} \to 0.$$

From Theorems 7.14 and 7.16 an important corollary follows.

**Theorem 7.18** (Hill, Taiani, 1978, 1981, see [14]). *Under the hypotheses of Theorem 7.16, any continuous CR-function $f$ on a CR-manifold $M$ with non-zero Levi form extends to a continuous CR-function $F$ on a CR-manifold $\tilde{M} \cap D$ of dimension greater by one.*

A significant strengthening of Theorem 7.18 was obtained not long ago by Boggess and Pitts [14] for the case when a higher order Levi form of the $CR$-manifold $M$ differs from zero.

From Theorem 7.18 follow, in turn, criteria for local holomorphic convexity of sufficiently smooth $CR$-manifolds which were obtained earlier (see [82]).

**Theorem 7.19** (Greenfield, 1968; Wells, 1968). *A $C^2$-smooth CR-manifold $M$ in $\mathbb{C}^n$ is locally holomorphically convex if and only if the Levi form $L_z^M \equiv 0$ for all $z \in M$.*

Stein (1937) established this result first for hypersurfaces in $\mathbb{C}^n$.

From the Frobenius integrability theorem, it follows (Sommer, Nirenberg) that any $C^2$-smooth $CR$-manifold $M$, whose Levi form vanishes identically, folliates into complex manifolds of maximal possible complex dimension, $CR$ $\dim M$ (see [58]).

## §8. The $\bar{\partial}$-Cohomology of $p$-Convex and $q$-Concave Manifolds and the Radon-Penrose Transform

In §§1–7 the method of integral representations has been demonstrated basically on problems from the theory of functions on domains in the space $\mathbb{C}^n$. Here we introduce examples of applications of integral formuli to problems from the cohomology theory of complex manifolds which go beyond the limits of classical function theory.

**8.1. The $\bar{\partial}$-Cohomology. Theorems of Andreotti and Grauert.** To begin with we present a short survey of classical results from the theory of $\bar{\partial}$-cohomology. (see also the article of A.L. Onishchshik Vol. 10, I).

Let $\Omega$ be a complex manifold, $\{\Omega_j\}, j \in J$, an open locally finite cover of $\Omega$ and $GL(N, \mathbb{C})$; the group of invertible complex $N \times N$ matrices. Let holomorphic functions $g_{jk}$ be defined in the intersections $\Omega_j \cap \Omega_k$, taking their values in $GL(N, \mathbb{C})$, and forming multiplicative cocycles, i.e.

$$g_{jk}(z) \cdot g_{kl}(z) \cdot g_{lj}(z) = I, \qquad z \in \Omega_j \cap \Omega_k \cap \Omega_l.$$

Here $I$ denotes the identity matrix.

These compatibility conditions on the functions $\{g_{jk}\}$ allow us to introduce an equivalence relation on the disjoint union

$$\tilde{E} = \bigcup_{j \in J} (\Omega_j \times \mathbb{C}^N)$$

by setting

$$(z, v) \sim (z', v'), (z, v) \in (\Omega_j \times \mathbb{C}^N), (z', v') \in \Omega_k \times \mathbb{C}^N$$

if and only if $z = z'$ and $v' = g_{jk} v$.

The complex manifold $E$ obtained from $\tilde{E}$ by factoring by the above equivalence relation is called a *holomorphic vector bundle with base* $\Omega$ and with fibre $\mathbb{C}^N$. This definition is also closely related to the holomorphic mapping $\pi\colon E\to\Omega$ projecting a representative point $(z,v)$ of $E$ onto the point $z\in\Omega$. Here $\pi^{-1}(z) \simeq \mathbb{C}^N$. The functions $\{g_{jk}\}$ which determine this bundle $E$ are called the transition functions of this bundle with respect to the trivializations $\{\Omega_j\times\mathbb{C}^N\}$.

The bundle given by the same covering $\{\Omega_j\}$ and the multiplicative cocycle of matrix functions ${}^t(g_{\bar{j}\bar{k}}^{-1})$ is called the *dual bundle* $E^*$ to the bundle $E$.

If $D$ is a domain in $\Omega$, then a *section of the bundle* $E$ over $D$ is a system of mappings $\{h_j\}$ from the domains $D\cap\Omega_j$ to $\mathbb{C}^N$ such that $h_j=g_{jk}h_k$ on $D\cap\Omega_j\cap\Omega_k$.

The simplest non-trivial example of a holomorphic vector bundle is given by the universal line bundle $\mathcal{O}(-1)$ given by the natural projection $\pi\colon \mathbb{C}^{n+1}\setminus\{0\}\to\mathbb{CP}^n$, where $\mathbb{CP}^n$ is $n$-dimensional complex projective space. The transition matrices $g_{jk}$ for the bundle $\mathcal{O}(-1)$ have the form $g_{jk}=z_j/z_k$, where $z\in U_j\cap U_k$.

By $\mathcal{O}(l)$ we denote the line bundle over $\mathbb{CP}^n$ given by the transition functions $(z_k/z_j)^l$. The sections of the bundle $\mathcal{O}(l)$ are the functions of the variable $z^1,\dots,z^{n+1}$ which are homogeneous of degree $l$.

A *form* $f$ of type $(0,r)$ on $D\subset\Omega$ *with values in the bundle* $E$ is a section over $D$ of the bundle $E\oplus_{\mathbb{C}}\Lambda^{0,r}\,T^*(\Omega)_{\mathbb{C}}$, where $T^*(\Omega)_{\mathbb{C}}$ is the complexification of the cotangent bundle on $\Omega$. In other words, a $(0,r)$-form $f$ with values in $E$ is a system $\{f_j\}$ of $\mathbb{C}^N$-valued $(0,r)$ forms on $D\cap\Omega_j$ such that

$$f_j = g_{jk}f_k \text{ on } D\cap\Omega_j\cap\Omega_k. \tag{8.1}$$

In local coordinates on $\Omega$, the form $f_j$ can be written:

$$f_j=\sum_p f_{j,p}(z)d\bar{z}^p,$$

where
$$p=(p_1,\dots,p_r),\qquad d\bar{z}^p=d\bar{z}^{p_1}\wedge\dots\wedge d\bar{z}^{p_r}.$$

Let us denote by $C^{(s)}_{0,r}(D,E)$ the space of forms of type $(0,r)$ on $D$ whose coefficients are sections of class $C^{(s)}$ of the vector bundle $E$ over $D$. In view of (8.1) and the holomorphicity of the transition functions $g_{jk}$, for each form $f\in C^{(s)}_{0,r}(D,E)$, the form $\bar{\partial}f\in C^{(s)}_{0,r+1}(D,E)$ is well-defined.

A form $f\in C^{(s)}_{0,r}(D,E)$ is called $\bar{\partial}$-closed if $\bar{\partial}f=0$ and is called $\bar{\partial}$-exact if $f=\bar{\partial}u$, where $u\in C^{(s+1)}_{0,r-1}(D,E)$.

Let $H^{(s)}_{0,r}(D,E)$ denote the quotient group which is the group of $\bar{\partial}$-closed form in $C^{(s)}_{0,r}(D,E)$ factored by the group of $\bar{\partial}$-exact forms. The group $H^{(s)}_{0,r}(D,E)$ is called the *Dolbeault* $(0,r)$-*cohomology* group of the manifold $D$ with coefficients in $E$.

Let $H^r(D,E)$ denote the $r$-dimensional Čech cohomology group of $D$ with coefficients in the sheaf of germs of holomorphic sections of $E$. In particular $H^0(D,E)$ is the space of holomorphic sections of $E$ over $D$. The following classical result holds.

**Theorem 8.1** (Dolbeault, 1953). *For any natural number r and any s, there is a canonical isomorphism between the groups* $H^{(s)}_{0,r}(D, E)$ *and* $H^r(D, E)$.

We have already made essential use of a particular case of this isomorphism in the proof of the equivalence of the assertions in Theorems 6.1 and 6.2.

Now we formulate a fundamental result from the cohomology theory of complex manifolds (see [31], [33]).

**Theorem 8.2** (Kodaira, 1953; Grauert, 1958). *Let* $\Omega$ *be a strictly pseudo-convex complex manifold with* $C^2$*-smooth exhaustion function* $\rho$, *i.e.*

$$\Omega = \bigcup_{\alpha \geq \alpha_0} D_\alpha,$$

*where* $D_\alpha = \{z \in \Omega: \rho(z) < \alpha\} \Subset \Omega$ *and* $\rho$ *is strictly plurisubharmonic on* $\Omega \backslash D_{\alpha_0}$. *Let E be a holomorphic vector bundle over* $\Omega$. *Then*

*a) for any* $r \geq 1$ *and* $\alpha \geq \alpha_0$, *we have*

$$\dim_{\mathbb{C}} H^r(\Omega, E) = \dim_{\mathbb{C}} H^r(D_\alpha, E) < \infty;$$

*b) if* $D_{\alpha_0} = \varnothing$, *then for all* $r > 1$, *we have*

$$\dim_{\mathbb{C}} H^r(\Omega, E) = 0;$$

*c) if* $\Omega$ *is a compact manifold and* $D_{\alpha_0} = \Omega$, *then the assertion a) holds for all* $r \geq 0$;

*d) if* $\Omega$ *is non-compact, then for each* $\alpha \geq \alpha_0$, *we have*

$$\dim_{\mathbb{C}} H^0(\Omega, E) = \dim_{\mathbb{C}} H^0(D_\alpha, E) = \infty.$$

Assertion c) of Theorem 8.2, first obtained by Kodaira, can be seen as a far reaching generalization of the classical Liouville Theorem. Assertions a), b), and (d), first obtained by Grauert, are extensions of the theorems of Oka on the solvability of the problems of Cousin and Levi in pseudoconvex Riemann domains. Theorem 3.12 of Grauert is a simple consequence of assertion a). From Theorem 3.12, in turn, assertion d) of Theorem 8.2. follows quickly

We present a simple example which illustrates the result of Kodaira.

The space $H^m(\mathbb{CP}^n, \mathcal{O}(l))$ is zero-dimensional for each $m = 1, \ldots, n-1$. The space $H^0(\mathbb{CP}^n, \mathcal{O}(l))$ is zero-dimensional if $l < 0$ and consists of homogeneous holomorphic polynomials if $l > 0$. The space $H^n(\mathbb{CP}^n, \mathcal{O}(l))$ is canonically isomorphic to the space $[H^0(\mathbb{CP}^n, \mathcal{O}(-n-s-l))]^*$.

The clarification of more precise conditions on a manifold $\Omega$, under which the assertions of Theorem 8.2 hold for the cohomology $H^r(\Omega, E)$ of fixed order $r \geq 0$, led to the important notions of *p*-convex and *q*-concave manifolds.

A real *function* $\rho$ of class $C^2$ on a domain $D \subset \mathbb{C}^n$ is called *p-convex* (respectively *strongly p-convex*) if its Levi form

$$L_{\rho,\zeta}(w) = \sum_{j,k} \frac{\partial^2 p}{\partial z_j \partial \bar{z}_k}(\zeta) w_j \bar{w}_k, \qquad w \in \mathbb{C}^n,$$

has for each $\zeta \in D$ no less than $(p+1)$ non-negative (respectively positive) eigenvalues.

The property of $p$-convexity of a function $\rho$ is independent of the choice of holomorphic coordinates in the domain $D$ and hence can be well-defined for functions on complex manifolds.

A relatively compact *domain* $D_+$ (respectively $D_-$) of a complex manifold $\Omega$ is called *strongly p-convex* (respectively *strongly p-concave*) if it can be represented in the form

$$D_+ = \{z \in \Omega \colon \rho(z) < 0\} \tag{8.2}$$

respectively $D_- = \{z \in \Omega \colon \rho(z) > 0\}$, where $\rho(z)$ is a strongly $p$-convex function in a neighbourhood of $\partial D_\pm$. In particular, a strictly pseudoconvex domain is a strongly $(n-1)$-convex domain.

The notion of strong $p$-convexity of domains was introduced by Rothstein in connection with the following effective generalization of a theorem of E. Levi.

**Theorem 8.3** (Rothstein, 1955). *Let the domains $D_\pm$ given by* (8.2) *have smooth boundaries, where the domain $D_+$ is strongly p-convex and the domain $D_-$ is strongly p-concave at a point $\zeta \in D_\pm$. Then, there exists a neighbourhood $U_\zeta$ of the point $\zeta$ such that*

*a) in the domain $U_\zeta \cap \bar{D}_-$, there exists a p-dimensional analytic submanifold intersecting $\bar{D}_+ \cap U_\zeta$ only in the point $\zeta$;*

*b) any closed $(n-p+1)$-dimensional analytic subset of $U_\zeta \cap D_-$ extends to a closed analytic subset of $U_\zeta$.*

For related results on the extension of analytic sets across pseudoconcave surfaces, see article III of E.M. Chirka.

A complex *manifold* $\Omega_+$ (respectively $\Omega_-$) is called *strongly p-convex* (respectively *strongly p-concave*) if on $\Omega_+$ (respectively $\Omega_-$) there exists a $C^2$-smooth real function $\rho$ and a number $\alpha_0$ such that the domains given by

$$D_\alpha^\pm = \{z \in \Omega_\pm \colon \pm\rho < \alpha\}, \qquad \pm\alpha > \alpha_0, \tag{8.3}$$

are relatively compact in $\Omega_\pm$ and respectively strongly $p$-convex or strongly $p$-concave.

The following result significantly generalizes assertions a) and b) of Theorem 8.2 (see [22], [44]).

**Theorem 8.4** (Andreotti, Grauert, 1962). *Let $\Omega_+$ (respectively $\Omega_-$) be a strongly p-convex (respectively stongly p-concave) manifold with exhaustion function given by* (8.3). *Let E be a holomorphic vector bundle over $\Omega_\pm$. Then*

*$a^+$) for each $r \geq n-p$ and $\alpha > \alpha_0$, we have*

$$\dim_{\mathbb{C}} H^r(\Omega_+, E) = \dim_{\mathbb{C}} H^r(D_\alpha^+, E) < \infty,$$

*$a^-$) for each $r < q$ and $\alpha < -\alpha_0$, we have*

$$\dim_{\mathbb{C}} H^r(\Omega_-, E) = \dim_{\mathbb{C}} H^r(D_\alpha^-, E) < \infty,$$

*b) if $D^+_{\alpha_0}=\varnothing$, then for $r>n-p$, we have*

$$\dim_{\mathbb{C}} H^r(\Omega_+, E)=0.$$

We remark that Theorem 8.3 of Rothstein can be deduced from Theorem 8.4 if we make use of the equation of Poincaré–Lelong and Theorem 3.1 of E. Levi.

A relatively compact *domain D* in a complex manifold $\Omega$ is called *strictly q-concave* if $D$ is strongly $q$-concave and at the same time strongly $(n-q)$-convex.

The following result gives an interesting generalization of Grauert's Theorems 3.12 and 8.2 (see [30]).

**Theorem 8.5** (Andreotti, Norguet, 1966). *Let $D^-$ be an arbitrary strictly q-concave domain on a complex manifold $\Omega$ and let E be an arbitrary holomorphic vector bundle over $\Omega$. Then*

$$\dim_{\mathbb{C}} H^q(D^-, E)=\infty.$$

*Moreover, for any neighbourhood $U_\zeta$ of any point $\zeta\in\partial D^-$, there exists an element $f\in H^q(D^-, E)$ which does not extend to an element of $H^q(U_\zeta\cup D^-, E)$.*

In the language of Dolbeault cohomology, the last assertion of Theorem 8.5 signifies the following. For any point $\zeta\in\partial D^-$, there exists a $\bar\partial$-closed form $f\in C^{(\infty)}_{0,q}(D^-, E)$ such that there is no $\varphi\in C^{(\infty)}_{0,q-1}(D^-, E)$ such that $f+\bar\partial\varphi$ extends to a $\bar\partial$-closed form in the domain $U_\zeta\cup D^-$.

Theorems 8.4 and 8.5 show that on a strongly $q$-concave manifold $\Omega^-$, the role of holomorphic functions is played by the elements of the space $H^q(\Omega^-, E)$. For $q>0$ the space $H^q(\Omega^-, E)$ can have a very complicated structure, even for strictly $q$-concave manifolds. For example, Rossi (1965) constructed examples of two-dimensional strictly 1-concave manifolds $\Omega^-$ for which the *space* $H^1(\Omega^-, \mathcal{O})$ are not *separable*. In the language of Dolbeault cohomology, this means that for such $\Omega^-$, for all $s$, the set of all $\bar\partial$-exact forms in the class $C^{(s)}_{0,1}(\Omega^-, \mathcal{O})$ does not form a closed subspace of $C^{(s)}_{0,1}(\Omega^-, \mathcal{O})$.

We formulate a result giving a convenient sufficient condition in order that the space $H^q(D^-, E)$ be separable (see [22], [44]).

**Theorem 8.6** (Andreotti, Vesentini, 1965). *Let D be a domain on a complex manifold $\Omega$ such that $D=D'\setminus D''$, where $\bar D''\subset D'$, $\bar D'\subset\Omega$, the domain $D'$ is strongly $(n-q)$-convex, and the domain $D''$ is strongly q-convex. Then, for any holomorphic vector bundle E over $\Omega$, the space $H^q(D, E)$ is separable. Moreover, a form $f\in C^{(\infty)}_{0,q}(D, E)$ is $\bar\partial$-exact on D if and only if the "moment" condition:*

$$\int_D f\wedge\varphi=0$$

*is satisfied, for each $\bar\partial$-closed form $\varphi\in C^{(\infty)}_{n,q}(D,E^*)$ of finite support in D.*

The conditions of Theorems 8.5 and 8.6 are satisfied, for example, by domains in $\mathbb{CP}^n$ of the form

$$D=\{z\in\mathbb{CP}^n: |z_1|^2+\ldots+|z_{q+1}|^2-|z_{q+2}|^2-\ldots-|z_{n+1}|^2>0\} \quad (8.4)$$

where $(z_1, \ldots, z_{n+1})$ are homogeneous coordinates in $\mathbb{CP}^n$. The domain (8.4) is a strictly $q$-concave domain. According to Theorems 8.4–8.6, for this domain, the space $H^r(D, \mathcal{O}(l))$ is finite dimensional for $r\neq q$ and infinite dimensional but separable for $r=q$.

The result of Theorems 8.2 and 8.4–8.6 were first obtained by the methods of coherent analytic sheafs (see the article Vol. 10, I).

Kohn, Andreotti and Vesentini, and Hörmander obtained these results anew with boundary $L^2$-estimates using the method of the $\bar\partial$-problem of Neuman–Spencer. We formulate here a fundamental consequence of these works (see [22], [44], [55]).

For a domain $D$ on a complex manifold $\Omega$ and a vector bundle $E$ over $\Omega$, we denote by $L^{2;s}_{0,r}(D, E)$ the space of $(0, r)$-forms on $D$ whose coefficients are sections of $E$ of class $L^{2,s}(D)$. Let $H^{2;s}_{0,r}(D, E)$ denote the quotient space of the space of $\bar\partial$-closed forms in $L^{2;s}_{0,r}(D, E)$ by the space of $\bar\partial$-exact such forms.

**Theorem 8.7** (Kohn, 1963, 1975; Andreotti, Vesentini, 1965; Hörmander, 1965). *Suppose under the hypotheses of Theorems 8.4–8.6 that the domains $D^+_\alpha$, $D^-_\alpha$ and $D$ have $C^\infty$-smooth boundaries. Then, the assertions of these theorems remain valid if we replace the cohomology spaces without estimates $H^r(D, E)$ by the corresponding cohomology with $L^2$-estimates $H^{2;s}_{0,r}(D, E)$. Moreover, under the hypotheses of Theorem 8.6, for any $\bar\partial$-exact form $f\in L^{2;s}_{0,q}(D, E)$, the equation $\bar\partial u=f$ has a solution with estimate*

$$\|u\|_{L^{2,s}(D)}\leq\gamma(D)\|f\|_{L^{2,s}(D)}.$$

**8.2. Integral Representations of Differential Forms and $\bar\partial$-Cohomology with Uniform Estimates.** Integral formulas of Cauchy–Fantappié type have made it possible to give elementary proofs of the above results on the $\bar\partial$-cohomology of complex manifolds with precise uniform boundary estimates and to extend these results to domains with piecewise smooth boundaries. Thanks to these refinements, it has been possible to obtain criteria for solvability as well as formulas for the solutions of the tangential Cauchy–Riemann equations (see section 8.3), and also to write down the kernel and the image for the Radon–Penrose transform in precise form (see section 8.4).

In order to obtain these results, it turned out to be necessary to extend the fundamental formulas of Cauchy–Fantappié in §4 to the case of differential forms on $p$-convex and $q$-concave domains in $\mathbb{C}^n$. For a smooth $\mathbb{C}^n$-valued vector function $\eta=\eta(\zeta, z, t)$ of the variables $\zeta, z\in\mathbb{C}^n$ and $t\in\mathbb{R}^k$ we have the formula (see [49]):

$$\omega'(\eta)=\sum_{q=0}^{n-1}\omega'_q(\eta),$$

where $\omega'_q(\eta)$ is a form of order $q$ with respect to $d\bar{z}$ and respectively of order $(n-q-1)$ with respect to $d\bar{\zeta}$ and $dt$. Integral formulas for differential forms make use of the forms $\{\omega'_q\}$ and of the following important relation between them

$$d_t\omega'_q(\eta)+\bar{\partial}_\zeta\omega'_q(\eta)+\bar{\partial}_z\omega'_{q-1}(\eta)=0. \tag{8.5}$$

An important role in the sequel is played by the following generalization of the Martinelli–Bochner integral representation 1.1 to the case of differential forms (see [4]. (40]).

**Theorem 8.8** (Koppelman, 1967). *Let $D$ be a bounded domain in $\mathbb{C}^n$ with rectifiable boundary and $f$ a $(0,r)$-form in $D$ such that both $f$ and $\bar{\partial}f$ are continuous on $\bar{D}$. Then for any $r=0,1,\ldots,n$ and $z\in D$ we have*

$$f(z)=(-1)^q\frac{(n-1)!}{(2\pi i)^n}\left[\int_{\zeta\in\partial D} f(\zeta)\wedge\omega'_r\left(\frac{\bar{\zeta}-\bar{z}}{|\zeta-z|^2}\right)\wedge\omega(\zeta)-\right.$$

$$\left.-\int_{\zeta\in D}\bar{\partial}f(\zeta)\wedge\omega'_r\left(\frac{\bar{\zeta}-\bar{z}}{|\zeta-z|^2}\right)\wedge\omega(\zeta)-\bar{\partial}\int_{\zeta\in D} f(\zeta)\wedge\omega'_{r-1}\left(\frac{\bar{\zeta}-\bar{z}}{|\zeta-z|^2}\right)\wedge\omega(\zeta)\right].$$

The following generalization to differential forms of integral formulas in Theorem 4.6 was obtained on the basis of Theorem 8.8 together with (8.5) and the Stokes formula.

**Theorem 8.9** (Lieb, 1971; Range, Siu, 1973; R.A. Ajrapetyan, G.M. Khenkin, 1984). *Let $G$ be a domain in $\mathbb{C}^n$ having piecewise-smooth boundary given by (4.16) and let $\{\eta_J\}$ a family of vector-functions having the property (4.17). Then for each $z\in G$ and each $(0,r)$-form $f$ such that both $f$ and $\bar{\partial}f$ are continuous on $\bar{G}$, the equality*

$$f(z)=K^rf-T^{r+1}\bar{\partial}f-\bar{\partial}T^rf \tag{8.6}$$

*holds, where*

$$K^rf=(-1)^r\frac{(n-1)!}{(2\pi i)^n}\sum_{|J|=1}^{n-r}K^r_Jf,$$

$$T^rf=(-1)^r\frac{(n-1)!}{(2\pi i)^n}\sum_{|J|=1}^{n-r}T^r_Jf+T^r_0f,$$

$$K^r_Jf=\int_{(\zeta,t)\in\Gamma_J\times\Delta_J} f(\zeta)\wedge\omega'_r\left(\frac{\eta_J(\zeta,z,t)}{\langle\eta_J(\zeta,z,t),\zeta-z\rangle}\right)\wedge\omega(\zeta),$$

$$T^r_Jf=(-1)^{|J|}\int_{(\zeta,t_0,t)\in\Gamma_J\times\Delta_{0,J}} f(\zeta)\wedge$$

$$\wedge\,\omega'_{r-1}\left((1-t_0)\frac{\bar{\zeta}-\bar{z}}{|\zeta-z|^2}+t_0\frac{\eta_J(\zeta,z,t)}{\langle\eta_J(\zeta,z,t),\zeta-z\rangle}\right)\wedge\omega(\zeta)$$

$$T^r_0f=\int_G f(\zeta)\wedge\omega'_{r-1}\left(\frac{\bar{\zeta}-\bar{z}}{|\zeta-z|^2}\right)\wedge\omega(\zeta).$$

The assertion of Theorem 8.9, for domains with smooth boundary, was first obtained by Lieb (1971). For domains with piecewise-smooth boundary and under the supplementary condition 4.21, it was obtained in the work of Range and Sin (1973). In its general form, the result of Theorem 8.9 is contained in [3].

Applications to the tangential Cauchy–Riemann equations (see section 8.3) require that Theorems 8.4–8.7 be generalized to $p$-convex and $q$-concave domains with piecewise-smooth boundaries.

Let $G$ be a domain on a complex manifolds $\Omega$. A $q$-dimensional complex manifold $M^q(\zeta)$ passing through a point $\zeta \in \partial G$ is called a *local $q$-dimensional barrier* to $G$ at the point $\zeta$ if there exists a neighbourhood $U^\zeta$ of the point $\zeta$ such that $M^q(\zeta) \cap U_\zeta \cap \bar{G} = \{\zeta\}$.

Taking off from Rothstein's Theorem 8.3, we give the following definition.

A *domain $G$* given by 4.16 on a complex manifold $\Omega$ is called *$q$-convex* with *piecewise-smooth boundary* if for each multi-index $J \subset \{1, \dots, N\}$ and each point $\zeta \in \Gamma_J \subset \partial G$, there exists a family of local $q$-dimensional barriers $M^q_J(\zeta, t)$ to $G$ at the point $\zeta$, depending on the parameter $t \in \Delta_J$, such that the barriers $M^q_J(\zeta, t)$ are smooth of order Lip 1 in the variables $\zeta \in \Gamma_J$, $t \in \Delta_J$, and for

$$(\zeta, t) \in \Gamma_J \cap \Delta_{J_{\hat{\nu}}},$$

we have

$$M^q_J(\zeta, t) = M^q_{J_{\hat{\nu}}}(\zeta, t).$$

A *domain $G$* is called *$q$-concave with piecewise-smooth boundary* if the domain $\Omega \setminus \bar{G}$ is $q$-convex at each point $\zeta \in \partial G$.

The following assertion yields an important class of $p$-convex (and correspondingly $q$-concave) domains with piecewise smooth boundaries.

**Proposition 8.10** ([39], [3]). *In order that a domain $G$ in $\Omega$ be $q$-convex with piecewise-smooth boundary in the neighbourhood $U$ of a point $z^* \in \partial G$, it is sufficient that the domain $U \cap G$ be representable in one of the two following ways*

$$G \cap U = \{z \in U \colon p_1(z) < 0, \dots, p_k(z) < 0\} \tag{8.7}$$

*where $\{\rho_j\}$ are function of class $C^{(2)}(U)$ such that $\rho_j(z^*) = 0$ and for each $t \in \mathbb{R}^k \colon \Sigma t^j = 1$, $t^j \geq 0$, the function $\langle t, \rho \rangle$ is strictly $q$-convex; or*

$$G \cap U = \bigcup_{j=1}^{k} \{z \in U \colon p_j(z) < 0\}, \tag{8.8}$$

*where $\{\rho_j\}$ are functions of class $C^{(2)}(U)$ such that $\rho_j(z^*) = 0$ and for each $t \in \mathbb{R}^k \colon \Sigma t^j = 1$, $t^j \geq 0$, the function $\langle t, \rho \rangle$ is strictly $(q+k-1)$-convex.*

The construction of integral formuli for solving $\bar{\partial}$-equations on $q$-convex domains relies on the following elementary assertion, similar to Proposition 7.11.

**Proposition 8.11** ([50], [39]). *Let $G$ be a $q$-convex domain with piecewise-smooth boundary given by (4.16) on $\Omega$ and $\{M^q_J(\zeta, t)\}$ a family of local $q$-dimensional barriers at the point $\zeta \in \Gamma_J \subset \partial G$. Then, for each $J \subset \{1, \dots, N\}$,* *and each* $\zeta \in \Gamma_J$, $t \in \Delta_J$ *and* $z \in U_\zeta$, *there exists a function* $\Phi_J(\zeta, z, t) = \langle P_J(\zeta, z, t), \zeta - z\rangle$ *of class* $C^\infty$ *in* $z \in U_\zeta$ *and class* Lip 1 *in* $(\zeta, t) \in \Gamma_J \times \Delta_J$ *such that:*

*a)* $M_J^q(\zeta, t) \cap U_\zeta = \{z \in U_\zeta : \Phi_J(\zeta, z, t) = 0\}, \quad (\zeta, t) \in \Gamma_J \times \Delta_J$;

*b)* $\Phi_J(\zeta, z, t) = \Phi_{J_{\hat{v}}}(\zeta, z, t), \ (\zeta, t) \in \Gamma_J \times \Delta_{J_v}, \quad z \in U_\zeta$;

*c)* $\langle \bar{\partial}_z P_J(\zeta, z, t) \wedge dz\rangle^{n-q+1} = 0, \quad (\zeta, t) \in \Gamma_J \times \Delta_J, \quad z \in U_\zeta$.

Let $H^{(s)}_{0,r}(\bar{D}, E)$ denote the space of $(0, r)$-cohomology of the closure $\bar{D}$ of a domain $D \subset \Omega$ with $C^{(s)}(\bar{D})$ estimates, i.e. the quotient space of the $\bar{\partial}$-closed forms $f$ of class $C^{(s)}_{0,r}(\bar{D}, E)$ by the forms $f = \bar{\partial}u$, with $u \in C^{(s+1)}_{0,r-1}(\bar{D}, E)$.

Propositions 8.9–8.11 together with simple reductions to the classical theorem of Fredholm allow one to prove the following generalization of Theorems 8.4–8.7 (see [3], [49], [30], [39], [40], [66]).

**Theorem 8.12** (Grauert, 1981; G.M. Khenkin, 1981). *Let D be a domain on an n-dimensional complex manifold* $\Omega$ *of the form* $D = D' \setminus D''$, *where* $D'$ *in an* $(n-q)$-*convex domain with piecewise-smooth boundary and* $D''$ *is a q-convex domain with piecewise-smooth boundary. Then:*

*a)* $\dim_{\mathbb{C}} H^{(\infty)}_{0,r}(D', E) = \dim_{\mathbb{C}} H^{(\infty)}_{0,r}(\bar{D}', E) < \infty, \quad$ *for* $r \geq q$;

*b)* $\dim_{\mathbb{C}} H^{(\infty)}_{0,r}(\Omega \setminus \bar{D}'', E) = \dim_{\mathbb{C}} H^{(\infty)}_{0,r}(\Omega \setminus D'', E) < \infty, \quad$ *for* $r < q$;

*c) the spaces* $H^{(\infty)}_{0,r}(D, E)$ *and* $H^{(\infty)}_{0,r}(\bar{D}, E)$ *are separable;*

*d) if moreover the domain* $D'$ *locally has the form* (8.7) *and the domain* $D''$ *has the form* (8.8) *and* $d\rho_1(z) \wedge \ldots \wedge d\rho_k(z) \neq 0$, $z \in \partial D$, *then the assertions a), b), and c) are valid also for the cohomology spaces with uniform estimates*

$$H^{(o)}_{0,r}(\bar{D}', E),\ H^{(o)}_{0,r}(\Omega \setminus D'', E),\ H^{(o)}_{0,r}(\bar{D}, E).$$

The assertions a) and b) for cohomology spaces without estimates $H^{(\infty)}_{0,r}(D', E)$ and $H^{(\infty)}_{0,r}(\Omega \setminus \bar{D}'', E)$ follows from the work of Grauert [30]. The remaining assertions of the theorem follow from the works [39], [3]. We remark that assertion d), for strictly pseudoconvex domains $D'$, was first obtained in the works of Kerzman (1971), Lieb (1971), Øvrelid (1971); for piecewise strictly pseudoconvex domains $D'$ in the works of P.L. Polyakov (1972) and Range, Siu (1973); for strictly $(n-q)$-convex domains $D'$ with smooth boundary in the work of Lieb, Fischer (1974); for strictly $q$-convex domains $\Omega \setminus D''$ with smooth boundary in the works of S.A. Dautov (1972), Øvrelid (1976), G.M. Khenkin (1977).

**8.3. The Cauchy–Riemann Equations on *q*-Concave *CR*-Manifolds.** Let $M$ be a $CR$-submanifold of a complex manifold $\Omega$, i.e. a smooth, closed submanifold of $\Omega$ locally representable in the form (7.1). Let $E$ be a holomorphic vector bundle over $\Omega$. By $C^{(s)}_{p,q}(M, E)$ we denote the space of differential forms of type $(p, q)$ on $M$ whose coefficients take their values in $E$ and are smooth of class $C^{(s)}$. Here, two forms $f$ and $g$ in $C^{(s)}_{p,q}(M, E)$ are considered equal if and only if for each form $\varphi \in C^{(\infty)}_{n-p,n-k-q}(\Omega, E^*)$ of compact support, we have

$$\int_M f \wedge \varphi = \int_M g \wedge \varphi. \tag{8.9}$$

We denote by $L^{(-s)}_{p,q}(M, E)$ the space of $(p, q)$-forms on $M$ dual to the space $C^{(s)}_{n-p,\,n-k-q}(M, E^*)$.

We define the *tangential Cauchy–Riemann operator* on forms in $L^{(-s)}_{0,q}(M, E)$. If $u \in C^{(s)}_{0,q}(M, E)$, $s \geq 1$, then $u$ can be extended to a smooth form $\tilde{u} \in C^{(s)}_{0,q}(\Omega, E)$ and we may set

$$\bar{\partial}_M u = \bar{\partial} u|_M.$$

Moreover, the condition for equality of forms on $M$ shows that the given definition does not depend on the choice of the extended form $\tilde{u}$. In general, for forms $u \in L^{(-s)}_{0,q-1}(M, E)$ and $f \in L^{(-s)}_{0,q}(M, E)$, by definition,

$$\bar{\partial}_M u = f \tag{8.10}$$

will mean that for each form $\varphi \in C^{(\infty)}_{n,n-k-q}(\Omega, E^*)$ of compact support we have

$$\int_M f \wedge \varphi = (-1)^q \int_M u \wedge \bar{\partial} \varphi.$$

As a (local) necessary condition for the solvability on $M$ of equation (8.10), we must have first of all $\bar{\partial}_M f = 0$ on $M$. Forms satisfying this condition will be called *CR-forms.* In case $f$ is a function, this definition reduces to the definition of a *CR*-function given in §7.

If the manifold $M$ and the form $f$ are both real analytic, then the condition $\bar{\partial}_M f = 0$ is not only necessary but also sufficient for the local solvability of the equation (8.10). Here, the size of the domains on $M$ on which (8.10) has a solution depends not only on the manifold $M$ but also on the real-analyticity properties of the *CR*-form $f$ (see [6]).

If, however, the form $f$ is not real-analytic on $M$, then the condition $\bar{\partial}_M f = 0$ in general is no longer sufficient for the local solvability of (8.10). This unexpected effect was first observed by H. Lewy in 1957 in the following example. Let $M = \{z \in \mathbb{C}^2 : \operatorname{Im} z_2 - |z_1|^2 = 0\}$. For functions $u$ on this hypersurface, we have

$$\bar{\partial}_M u = \frac{\partial u}{\partial \bar{z}_1} d\bar{z}_1 + \frac{\partial u}{\partial \bar{z}_2} d\bar{z}_2 = \left( \frac{\partial u}{\partial \bar{z}_1} - i z_1 \frac{\partial u}{\partial \operatorname{Re} z_2} \right) d\bar{z}_1.$$

Thus equation (8.10) in this case is equivalent to the following equation in $\mathbb{R}^3$

$$Lu = \frac{1}{2}\left( \frac{\partial u}{\partial x_1} + i \frac{\partial u}{\partial x_2} \right) - i(x_1 + i x_2) \frac{\partial u}{\partial x_3} = f.$$

H. Lewy constructed a function $f$ of class $C^{(\infty)}$ such that this equation is not locally solvable even in the class of generalized functions $u \in L^{(-\infty)}$.

Treves [79] showed, however, that on any *CR*-manifold $M$ of codimension $k$ there is an approximate solvability of (8.10) for a *CR*-form $f \in C^{(s)}_{0,n-k}(M)$.

**Theorem 8.13** (Treves, 1981). *Let $M$ be an arbitrary $C^\infty$-smooth CR-manifold of codimension $k$ on a complex manifold $\Omega$. Then for each point $z \in M$ there exists a* *neighbourhood* $U_z$ *such that any CR-form* $f \in C^{(s)}_{0,n-k}(M)$, *can be approximated on* $M \cap U_z$ *with arbitrary precision, in the* $C^{(s)}$*-topology, by* $\bar\partial_M$*-exact forms of class* $C^{(\infty)}_{0,n-k}(U_z \cap M)$.

This result is a natural generalization of the approximation theorem for *CR*-forms of Range, Siu (Math. Ann., 1974, *210*, 105–122).

The clarification of conditions for actual (rather than approximate) local and global solvability of the tangential Cauchy–Riemann equations (8.10) has advanced rather far for the case when $M$ is a hypersurface (Kohn, 1965, 1975; Andreotti, Hill, 1979; Sato, 1972; Folland, 1972; Folland, Stein, 1974; A.V. Romanov, 1975; Greiner, Kohn, Stein, 1975; G.M. Khenkin, 1975; Skoda, 1975).

We formulate here only one of the criteria for the global solvability of the equation (8.10) on a compact hypersurface (see [49], [55]).

**Theorem 8.14** (Kohn, 1965, 1975, $s=\infty$; A.V. Romanov, 1975; G.M. Khenkin, 1975; Skoda, 1975, $s<\infty$). *Let* $M=\{z\in\Omega: \rho(z)=0\}$ *be a compact* $C^\infty$*-smooth hypersurface on a holomorphically convex manifold* $\Omega$; *let* $E$ *be a holomorphic vector bundle over* $\Omega$. *Suppose the Levi form of the function* $\rho$ *has on* $T^{\mathbb{C}}_z(M)$, $z\in M$, *either* max $(q, n-q)$ *eigenvalues of the same sign or* $\min(q, n-q)$ *pairs of corresponding values of opposite signs. Then, the equation* $\bar\partial_M u=f$, *where* $f\in C^{(s)}_{0,q}(M, E)$ *has a solution* $u\in C^{s+1/2}_{0,q-1}(M, E)$, *if and only if*

$$\int_M f \wedge \varphi = 0$$

for each $\bar\partial$*-closed form* $\varphi \in C^{(\infty)}_{n,n-1-q}(M, E^*)$.

In case $M$ is the boundary of a strictly convex domain in $\mathbb{C}^n$, the result of the Theorem 8.14 is a rapid consequence of the following integral representation for an arbitrary form $f\in C^{(s)}_{0,q}(M)$, $q=0, 1, \ldots, n-1$ (see [49]):

$$f=\bar\partial_M T_q f + T_{q+1}(\bar\partial_M f) + K_q f - K^*_q f, \tag{8.11}$$

where

$$T_q f(z) = (-1)^q \frac{(n-1)!}{(2\pi i)^n} \int_{(\zeta, t)\in M\times[0,1]} f(\zeta)\wedge$$

$$\wedge\, \omega'_{q-1}\left((1-t)\frac{P^*(\zeta,z)}{\Phi^*(\zeta,z)} + t\frac{P(\zeta,z)}{\Phi(\zeta,z)}\right)\wedge\omega(\zeta),$$

$$K_q f(z) = (-1)^q \frac{(n-1)!}{(2\pi i)^n} \int_{\zeta\in M} f(\zeta)\wedge\omega'_q\left(\frac{P(\zeta,z)}{\Phi(\zeta,z)}\right)\wedge\omega(\zeta),$$

$$K^*_q f(z) = (-1)^q \frac{(n-1)!}{(2\pi i)^n} \int_{\zeta\in M} f(\zeta)\wedge\omega'_q\left(\frac{P^*(\zeta,z)}{\Phi^*(\zeta,z)}\right)\wedge\omega(\zeta),$$

$$\Phi(\zeta,z)=\langle P(\zeta,z), \zeta-z\rangle, \quad \Phi^*(\zeta,z)=\langle P^*(\zeta,z), \zeta-z\rangle,$$

$$P^*(\zeta,z)=P(\zeta,z)=\frac{\partial\rho}{\partial z}(z).$$

For manifolds of arbitrary codimension, rather general results on the solvability of the tangential Cauchy–Riemann equations have been obtained for the class of $q$-concave (or $q$-convex) manifolds (see [3], [39]).

We denote by $H^{(s)}_{0,r}(M, E)$ the quotient space of the $\bar{\partial}$-closed forms $f$ in $C^{(s)}_{0,r}(M, E)$ by the subspace of forms $f = \bar{\partial}_M u$, with $u \in C^{(s)}_{0,r-1}(M, E)$.

**Theorem 8.15** (Naruki, 1972; G.M. Khenkin, 1981). *Let $\Omega$ be an n-dimensional complex manifold, E a holomorphic vector bundle over $\Omega$, and M a strongly q-concave $C^\infty$-smooth submanifold of codimension k in $\Omega$. Then, for each r, $1 \leq r \leq q$ (respectively $n-k-q < r \leq n-k$), for each strongly $(n-1)$-concave (respectively strongly $(n-1)$-convex) domain $D \Subset \Omega$, and for each CR-form $f \in C^{(s)}_{0,r}(M \cap \bar{D}, E)$, the equation* (8.10) *has a solution $u \in C^{(s+1/2-\varepsilon)}_{0,r-1}(M \cap \bar{D}, E)$, $\varepsilon > 0$, if and only if*

$$\int_{M \cap D} f \wedge \varphi = 0$$

*for each $\bar{\partial}$-closed form $\varphi \in C^{(\infty)}_{n,n-k-r}(\Omega, E^*)$ with support in the domain D. Moreover, for each $r, r < q$ or $r > n-k-q$, the space $H^{(s)}_{0,r}(M \cap \bar{D}, E)$ is finite dimensional.*

For the case $r > n-k-q$ and $s = \infty$, the result of Theorem 8.15 was obtained first by Naruki on the basis of the techniques of Kohn and Hörmander. The general case is considered in [39].

The proof of Theorem 8.15 is based on representing a $q$-concave or $q$-convex $CR$-manifold as the intersection respectively of $q$-concave or $q$-convex domains with piecewise-smooth boundaries and applying to these domains Theorem 8.12 as well as the formulas yielding this theorem (see [39], [3], and [66]).

The requirement of $q$-concavity of the $CR$-manifold $M$ in order that the assertion of Theorem 8.15 hold is essential because of the following result which sheds light on the effect of H. Lewy (see [6] and [3]).

**Theorem 8.16** (Andreotti, Nacinovich, Fredricks, 1981). *Let M be a $C^\infty$-smooth CR-manifold of codimension k in $\Omega$. If the CR-manifold M is simultaneously strongly q-concave and strongly $(n-k-q)$-convex at some point $z \in M$ in some direction $t \in S^k$, then in each neighbourhood U of the point z, there exists a CR-form $f \in C^\infty_{0,q}(M)$ which is not $\bar{\partial}_M$-exact on $M \cap U$.*

**8.4. The Radon–Penrose Transform.** A domain (or compact set), $D$ in the space $\mathbb{CP}^n$ is called *$q$-linearly concave* if for each point $z \in D$ there exists a $q$-dimensional complex linear subspace $A(z) \subset D$ containing the point $z$ and depending smoothly on $z \in D$. A *domain $D \subset \mathbb{CP}^n$* is called *$q$-linearly convex* if the compact set $\mathbb{CP}^n \backslash D$ is $q$-linearly concave.

Examples of $q$-linearly concave (or $(n-q-1)$-linearly convex) domains are given by domains in $\mathbb{CP}^n$ of the form (8.4).

The notion of $q$-linearly convex domains was introduced by Rothstein (1955) without the requirement that the subspace $A(\zeta) \subset \mathbb{CP}^n \backslash D$ depend smoothly on the point $\zeta \in \mathbb{CP}^n \setminus D$. The latter requirement was added by us in the definition of $q$-linearly concave and $q$-linearly convex domains because it figures in subsequent formulas. For $q$-linearly concave domains it turns out that one can construct a global integral representation of $(0, q)$-forms in which only integral-residues along a subspace $A(z) \subset D$ play a role. Such representations allow us to significantly clarify the theory of Radon–Penrose transforms in connection with the integration of $(0, q)$-forms on $q$-dimensional complex subspaces of $D$.

For a $q$-dimensional projective subspace $A \subset D$, we set $\tilde{A} = \pi^{-1}(A) \subset \mathbb{C}^{n+1}$. Let $a = \{a_1, \ldots, a_{n-q}\}$ be an orthonormal $(n-q)$-repère in $\mathbb{C}^{n+1}$ which gives the subspace $\tilde{A} \subset \mathbb{C}^{n+1}$, i.e.

$$\tilde{A} = \{\zeta \in \mathbb{C}^{n+1}: A_j(\zeta) = \langle a_j, \zeta \rangle = 0, j = 1, \ldots, n-q\}.$$

Let $G(q+1, n+1)$ denote the manifold of $(q+1)$-dimensional subspaces of the space $\mathbb{C}^{n+1}$. For a $q$-linearly concave domain $D \subset \mathbb{CP}^n$, we denote by $D_q^*$ the domain in $G(q+1, n+1)$ consisting of all $\tilde{A}$ such that $\pi(\tilde{A})$ is a $q$-dimensional projective subspace $A$ lying in $D$, and we denote by $F$ the fibration by $(n-q)$-repères $a = \{a_1, \ldots, a_{n-q}\}$ over $D_q^*$. Let $C(D_q^*, F)$ denote the space of continuous functions $\psi$ on $F$ satisfying the condition

$$\psi(ga) = (\det g)^{-1} \psi(a), \qquad \forall g \in GL(n-q, \mathbb{C}).$$

By $d_\zeta A$ we denote a holomorphic form of maximal degree $(q+1)$ on $\tilde{A}$ such that at the points of $\tilde{A}$ we have

$$d\zeta^1 \wedge \ldots \wedge d\zeta^{n+1} = d_\zeta A \wedge d_\zeta A_1 \wedge \ldots \wedge d_\zeta A_{n-q}.$$

For forms $f \in C_{0,q}^{(s)}(D, \mathcal{O}(l))$, we set $\tilde{f} = \pi^* f$.

Following [28] and [52] we define the *Radon-Penrose transformation*

$$R: C_{0,q}(D, \mathcal{O}(-q-1)) \to C(D_q^*, F) \tag{8.12}$$

by the formula

$$Rf(a) = \int_{\tilde{A} \cap S} \tilde{f}(\zeta)\, d_\zeta A,$$

where $f \in C_{0,q}(D, \mathcal{O}(-q-1))$ and $S$ is the unit sphere in $\mathbb{C}^{n+1}$.

Let $\partial_a^m \tilde{f}$, $m = (m_1, \ldots, m_{n-q})$ denote the mixed derivative

$$\frac{\partial^{|m|}}{(\partial a_1')^{m_1} \ldots (\partial a_{n-q}')^{m_{n-q}}} \tilde{f}$$

in the direction of an arbitrary repère $a' = \{a_1', \ldots, a_{n-q}'\} \in \mathbb{C}^{n+1}$ dual to the repère $\{a_1, \ldots, a_{n-q}\}$, i.e. $\langle a_j, a_k' \rangle = \delta_{jk}$. We complete the repère $\{a_1, \ldots, a_{n-q}\}$ to an orthonormal basis of $\mathbb{C}^{n+1}$ by adding vectors $\{b_1, \ldots, b_{q+1}\}$; set $\alpha_j = \langle a_j, d\zeta \rangle$, $\beta_k = \langle b_k, d\zeta \rangle$ and expand the $(0, q)$-form $\tilde{f}$ using these basis forms

$$\tilde{f} = \sum_{j_1 < \ldots < i_s} \left( \sum_{i_1 < \ldots < i_{q-s}} f_{i_1, \ldots, i_{q-s}}^{j_1, \ldots, j_s} \bar{\beta}_{i_1} \wedge \ldots \wedge \bar{\beta}_{i_{q-s}} \right) \wedge \bar{\alpha}_{j_1} \wedge \ldots \wedge \bar{\alpha}_{j_s}.$$

We denote the coefficient of the monomial $\bar{\alpha}_{j_1} \wedge \ldots \wedge \bar{\alpha}_{j_s}$ in this expansion by $[f]^{j_1, \ldots, j_s}$. For a repère $a = a(z)$ depending smoothly on the point $z \in D$ (i.e. if $a$ is a smooth frame) and for $r \leq q$, we set

$$\det{}^r_m(a, z, \zeta) = \det(\bar{z}, \bar{\zeta}, a_1, \ldots, a_{n-q}, \overbrace{d\bar{z}}^{r-|m|}, \overbrace{\bar{\partial}_z a_1}^{m_1}, \ldots, \overbrace{\bar{\partial}_z a_{n-q}}^{m_{n-q}}, \overbrace{d\zeta}^{q-r-1})$$

if $r < q$, and

$$\det{}^q_m(a, z) = \det(\bar{z}, a_1, \ldots, a_{n-q}, \overbrace{d\bar{z}}^{r-|m|}, \overbrace{\bar{\partial}_z a_1}^{m_1}, \ldots, \overbrace{\bar{\partial}_z a_{n-q}}^{m_{n-q}})$$

if $r = q$, where $\overbrace{d\bar{z}}^{r-|m|}$ means that column $d\bar{z}$ is repeated $r - |m|$ times.

We introduce the *inverse transform* $G^a\colon C(D_q^*, F) \to C_{0,q}(D, \mathcal{O}(-q-1))$ by the formula

$$G^a\psi(z) = \sum_{|m|=0}^{q} d(m, q) \det{}^q_m(a, z) \times$$

$$\times \left(\left\langle \bar{z}\frac{\partial}{\partial a_1}\right\rangle^{m_1} \ldots \left\langle \bar{z}\frac{\partial}{\partial a_{n-q}}\right\rangle^{m_{n-q}} \psi\right)(a_1, \ldots, a_{n-q}). \tag{8.13}$$

In addition, we introduce the operators $I_r^a$ mapping from $C_{0,r}^{(s+q-1)}(D, \mathcal{O}(-q-1))$ to $C_{0,r-1}^{(s)}(D, \mathcal{O}(-q-1))$ by the formulas

$$I_r^a f = \frac{(-1)^{r-1}}{(2\pi i)^{q+1}} \sum_{p=0}^{r-1} (-1)^p \sum_{|m| \leq r-1-p} \sum_{\substack{(i_j < \ldots < i_p) \\ (1 \leq (j_1, \ldots, j_p) \leq n+1}}$$

$$\bar{\partial} a_{i_1 j_1} \wedge \ldots \wedge \bar{\partial} \bar{a}_{i_p j_p} \wedge$$

$$\wedge \int_{\tilde{A} \cap S} \left(\prod_{l=1}^{p} \xi_{j_l}\right) (\partial_a^m [\tilde{f}]^{i_1 \ldots i_p}) \wedge \frac{\det{}^{r-p}_m(a, z, \zeta)}{(\phi^*)^{r-p-|m|} \phi^p} \wedge d_\zeta A$$

where $\Phi^*(\zeta, z) = \langle \bar{z}, \zeta - z \rangle$, $\Phi(\zeta, z) = \langle \bar{\zeta}, \zeta - z \rangle$, and $b(m)$, $c(m)$, $d(m)$ are constants depending only on $m$ and $n$.

It is possible [52] to obtain global integral formulas for forms on $\mathbb{CP}^n$ by making use of the integral representation (8.11) for forms on the sphere $S$ in $\mathbb{C}^{n+1}$. For $(0, q)$ forms on $q$-linearly concave domains in $\mathbb{CP}^n$, the formula is giving by the following.

**Theorem 8.17** (G.M. Khenkin, P.L. Polyakov [53]). *Let $D$ be a $q$-linearly concave domain in $\mathbb{CP}^n$. We fix a smooth mapping $z \mapsto a(z)$, where $z \in D$ and $a(z)$ is an orthonormal $(n-q)$-repère giving a $q$-dimensional subspace $A(z) \subset D$. Then, for any form $f \in C_{0,q}^{(s)}(D, \mathcal{O}(-q-1))$, $s \geq q-1$, we have an integral representation* *given by*

$$f = I^a_{q+1}(\bar{\partial} f) + \bar{\partial} J^a_q f + G^a \circ Rf. \tag{8.14}$$

In [52] and [53], modifications of formulas (8.12) and (8.13) for the Radon–Penrose transform and the inverse transform $G$ for the case of $(0, q)$-forms with coefficients in any vector bundle $\mathcal{O}(-l)$, $l > 0$ are also constructed. By using these modifications, the assertion of Theorem 8.17 extends to any form $f \in C^{(s)}_{0,q}(D, \mathcal{O}(-l))$, $l > 0$.

Theorem 8.17 applied to $\bar{\partial}$-closed forms shows, in particular, that the operator $G^a$ is the inverse of the Radon–Penrose transformation restricted to the $\bar{\partial}$-cohomology space $H^q(D, \mathcal{O}(-q-1))$. In form, the operator $G^a$ is similar to the inverse transformations of I.M. Gel'fand, S.G. Gindikin and M.I. Graev (1980) for the Radon transformation in real integral geometry. A different inverse for the Radon–Penrose transformation on the $\bar{\partial}$-cohomology spaces was constructed earlier in [28].

We denote by $G^0(q+1, n+1)$ the dense subdomain of $G(q+1, n+1)$ consisting of all $(q+1)$-dimensional subspaces of $\mathbb{C}^{n+1}$ given by the equations

$$z_{\beta'} = \sum_{\alpha=1}^{q+1} a_{\alpha\beta'} \cdot z_\alpha, \quad \beta' = q+2, \ldots, n+1.$$

The elements of the matrix $\{a_{\alpha\beta'}\}$ can be considered as coordinates in the domain $G^0(q+1, n+1)$. We set $D^{*,0}_q = D^*_q \cap G^0(q+1, n+1)$. The restriction to $D^{*,0}_q$ of a function in $C(D^*_q, F)$ can naturally be considered as a function of the variables $a_{\alpha,\beta'}$. An immediate verification shows that the Radon–Penrose transform $\psi = Rf$ given by (8.12) satisfies a system of "wave" equations

$$\frac{\partial^2 \psi}{\partial a_{\alpha\alpha'} \cdot \partial a_{\beta\beta'}} - \frac{\partial^2 \psi}{\partial a_{\alpha\beta'} \cdot \partial a_{\beta\alpha'}} = 0, \tag{8.15}$$

on $D^{*,0}_q$, where $\alpha, \beta = 1, \ldots, q+1$; $\alpha', \beta' = q+2, \ldots, n+1$.

As a corollary of Theorem 8.17 we obtain the following result.

**Theorem 8.18** (Penrose, 1977; S.G. Gindikin, G.M. Khenkin, 1978). *Let $D$ be a $q$-linearly concave domain in $\mathbb{CP}^n$. The Radon–Penrose transformation* (8.12) *establishes an isomorphism between the cohomology space $H^q(D, \mathcal{O}(-q-1))$ and the space of holomorphic functions in $C(D^*_q, F)$ satisfying the system of equations* (8.15) *on $D^{*,0}_q$.*

In the case $q = n - 1$ the result of Theorem 8.18 is equivalent to the theorem of Fantappié–Martineau 4.5. In this situation, the equations (8.15) disappear. For the case $q = 1$, $n = 3$, the result of Theorem 8.18 was initiated by Penrose (1969, 1977). In this case the system (8.15) consists in one "wave" equation. The general situation was considered first in [28].

Theorem 8.17 gives us a natural way of looking at the Radon–Penrose transformation not only on $\bar{\partial}$-closed $(0, q)$-forms on $D$ but also on arbitrary

(0, $q$)-forms $f \in C^{(s)}_{0,q}(D, \mathcal{O}(-q-1)$ for which the first term $I^a_{q+1}(\bar\partial f)$ on the right side of (8.14) vanishes. The class of such forms depends essentially on the choice of the mapping $z \mapsto a(z)$, $z \in D$. Let us indicate the effectiveness of such a point of view on a "physical" example. In the following let $D$ be a 1-linearly concave domain in $\mathbb{CP}^3$ given by

$$D = \{\zeta \in \mathbb{CP}^3 \colon -|\zeta_1|^2 - |\zeta_2|^2 + |\zeta_3|^2 + \zeta_4|^2 < 0\}. \tag{8.16}$$

In this case the domain $D_1^* \subset G(2,4)$ is a classical domain in $G^0(2,4) \simeq \mathbb{C}^4$ and consists of $(2 \times 2)$-matrices $\mathscr{A} = \{a_{kj}\}$ satisfying the condition $I - \mathscr{A} \cdot \mathscr{A}^* \geq 0$. Let us fix the mapping $z \mapsto a(z)$ given by $a(z) = \{\mathscr{A}(z), -I\}$, where

$$\mathscr{A}(z) = \left|\begin{matrix} \dfrac{\bar z_1 z_3 - \bar z_4 z_2}{|z_1|^2 + |z_2|^2}, & \dfrac{\bar z_2 z_3 + \bar z_4 z_1}{|z_1|^2 + |z_2|^2} \\ \dfrac{\bar z_1 z_4 + \bar z_3 z_2}{|z_1|^2 + |z_2|^2}, & \dfrac{\bar z_2 z_4 - \bar z_3 z_1}{|z_1|^2 + |z_2|^2} \end{matrix}\right|, \quad I = \begin{pmatrix} 1 & 0 \\ 0 & 1 \end{pmatrix}.$$

The family of 2-repères $a(z)$ in $\mathbb{C}^4$ gives a fibration over the domain $D$ by non-intersecting projective lines $A(z) \subset D$. We introduce on $D_1^*$ the coordinates $w = (w_1, \ldots, w_4)$ such that

$$a_{11} = w_1 + iw_2, \qquad a_{12} = w_3 - iw_4$$

$$a_{21} = w_3 + iw_4, \qquad a_{22} = -w_1 + iw_2,$$

where $w_j = x_j + iy_j$. Consider, in $\mathbb{C}^4$, the real Euclidean subspace $E = \{w \in \mathbb{C}^4 \colon y_1 = \ldots = y_4 = 0\}$. We remark that the four-dimensional ball $E \cap D_1^*$ is precisely the image of the mapping $z \mapsto w(\mathscr{A}(z))$, $z \in D$.

Let us denote by $\mathscr{L}^E_{0,1}(D, \mathcal{O}(-2))$ the space of ($\cdot$0, 1)-forms $f$ on $D$ such that the coefficients of the form $f$ and of $\bar\partial f$ are generalized sections of measure type for the vector bundle and $[\bar\partial f]^j = 0$, $j = 1, 2$.

**Proposition 8.19** ([53]). *Let $f \in \mathscr{L}^E_{0,1}(D, \mathcal{O}(-2))$. Then there exists a generalized function of measure-type $\varphi = \varphi(x)$, $x \in E \cap D_1^*$ such that*

$$\bar\partial f(z) = \varphi(x(z))\bar\alpha_1(a(z), z) \wedge \bar\alpha_2(a(z), z), \tag{8.17}$$

*where $x_j(z) = \operatorname{Re} w_j(z)$; and $w_j(z)$ are the components of the mapping $z \mapsto w(\mathscr{A}(z))$.*

Let us introduce for consideration two more spaces: $\mathscr{H}^E_{0,1}(D, \mathcal{O}(-2))$ is the quotient space of $\mathscr{L}^E_{0,1}(D, \mathcal{O}(-2))$ by the subspace of $\bar\partial$-exact forms; $\Sigma^E(D_1^*)$ is the space of all generalized functions of measure-type in the domain $D_1^*$ which satisfy the equations

$$\begin{gathered} \bar\partial\psi|_{E \cap D_1^*} = 0, \qquad \sum_j \frac{\partial^2\psi}{\partial w_j^2}(w) = 0 \\ \sum_j \frac{\partial^2\psi}{\partial \bar w_j^2}(w) + \sum_{k,j} P_{k,j}(w)\frac{\partial^2\psi}{\partial w_k \partial \bar w_j} = 0, \end{gathered} \tag{8.18}$$

where $w \in D_1^*$, $P_{ij}(w) = 2y_j^2/|y|^2 - 1$ for $k = j$; and $P_{kj}(w) = P_{jk}(w) = 2y_k \cdot y_j/|y|^2$ for $k \neq j$.

On the basis of Theorem 8.17, one obtains an unexpected strengthening of Penrose's Theorem.

**Theorem 8.20** (G.M. Khenkin, P.L. Polyakov [53]). *Let $D$ be a 1-linear convex domain in $\mathbb{CP}^3$ given by (8.16). The Radon-Penrose transformation yields an isomorphism between the spaces $\mathscr{H}^E_{0,1}(D, \mathcal{O}(-2))$ and $\Sigma^E(D_1^*)$. Moreover, the inhomogeneous Cauchy-Riemann equation (8.17) on $D$ is carried over to the inhomogeneous Laplace equation*

$$\sum_j \frac{\partial^2}{\partial x_j^2} \psi(x) = 16\pi^2 \varphi(x),$$

*where $\psi = Rf$, $x \in E \cap D_1^*$.*

Theorem 8.20 also allows us to give the following analytic reformulation of a well-known result of Atiyah (1981).

Let us fix a point $x^\circ \in E$ and a projective line $A^0 \subset \mathbb{CP}^3$ corresponding to the point $x^0$. Let $f^0$ be a (0, 1)-form from $\mathscr{L}^E_{0,1}(\mathbb{CP}^3, \mathcal{O}(-2))$ such that

$$\int_{\zeta \in S} \bar{\partial} f^0(\zeta) \wedge \tilde{\varphi}(\zeta) \wedge \omega^*(\zeta) = \int_{\zeta \in A^0 \cap S} \tilde{\varphi}(\zeta) d_\zeta A^0,$$

$\forall \varphi \in C^{(\infty)}_{0,1}(\mathbb{CP}^3, \mathcal{O}(-2))$. Then, for the form $f^0$ we have $Rf^0 = \psi^0(w) = C\langle w - x^0, w - x^0 \rangle^{-1}$, i.e. $Rf^0$ is a holomorphic extension to $D_1^*$ of the fundamental solution to the Laplace equation on $E \cap D_1^*$.

We have touched upon, here, only a few aspects of the theory of Radon-Penrose Transformations connected with integral representations of differential forms. More complete information on the Radon-Penrose transformation and its applications is contained in the papers of Vol. 9, VII, Vol. 10, II (see also [52], [51] and [64]).

# References*


1. Ajrapetyan, R.A.: Boundary properties of an integral of Cauchy-Leray type in piecewise smooth domains in $\mathbb{C}^n$, and some applications. Mat. Sb., Nov. Ser. *112*, No. 1, 3–23 (1980); English transl.: Math. USSR, Sb. *40*, 1–20 (1981). Zbl. 445.32004
2. Ajrapetyan, R.A., Khenkin, G.M. (= Henkin, G.M.): Analytic continuation of *CR*-functions through the "edge of the wedge". Dokl. Akad. Nauk SSSR *259*, No. 4, 777–781 (1981); English transl.: Sov. Math., Dokl. *24*, 129–132 (1981).


* For the convenience of the reader, references to reviews in Zentralblatt für Mathematik (Zbl.), compiled using the MATH database, have, as far as possible, been included in this bibliography.

3. Ajrapetyan, R.A., Khenkin, G.M. (= Henkin, G.M.): Integral representations of differential forms on Cauchy–Riemann manifolds and the theory of *CR*–functions I, II. Usp. Mat. Nauk *39*, No. 3, 39–106 (1984); English transl.: Russ. Math. Surv. *39*, No. 3, 41–118 (1984). Zbl. 589.32035; Mat. Sb., Nov. Ser. *127*, No. 1, 92–112 (1985); English transl.: Math. USSR, Sb. *55*, No. 1, 91–111 (1986). Zbl. 589.32036
4. Ajzenberg, L.A., Yuzhakov, A.P.: Integral representations and residues in multidimensional complex analysis. Nauka, Novosibirsk 1979 [Russian]; Engl. transl.: Am. Math. Soc., Providence, R.I. 1983. Zbl. 445.32002
5. Amar, E.: Extension de fonctions holomorphes et courants. Bull. Sci. Math., II. Sér. *107*, No. 1, 25–48 (1983). Zbl. 543.32007
6. Andreotti, A., Fredricks, G., Nacinovich, M.: On the absence of Poincaré lemma in tangential Cauchy–Riemann complexes. Ann. Sc. Norm. Super. Pisa, Cl. Sci., IV. Ser. *8*, No. 3, 365–404 (1981). Zbl. 482.35061
7. Baouendi, M.S., Chang, C.H., Treves, F.: Microlocal hypo-analycity and extension of *CR*–functions. J. Differ. Geom. *18*, No. 3, 331–391 (1983). Zbl. 575.32019
8. Baouendi, M.S., Treves, F.: A property of the functions and distributions annihilated by a locally integrable system of complex vector fields. Ann. Math., II. Ser. *113*, 387–421 (1981). Zbl. 491.35036
9. Beals, M., Fefferman, Ch., Grossman, R.: Strictly pseudoconvex domains in $\mathbb{C}^n$. Bull. Am. Math. Soc., New Ser. *8*, 125–322 (1983). Zbl. 546.32008
10. Behnke, H., Thullen, P.: Theorie der Funktionen mehrerer komplexer Veränderlichen. Ergebnisse der Mathematik, Springer-Verlag, Berlin 1934. Zbl. 8,365
11. Berndtsson, B.: A formula for interpolation and division in $\mathbb{C}^n$. Math. Ann. *263*, No. 4, 399–418 (1983). Zbl. 499.32013
12. Berndtsson, B., Andersson, M.: Henkin–Ramirez kernels with weight factors. Ann. Inst. Fourier *32*, No. 3, 91–110 (1982). Zbl. 498.32001
13. Bochner, S., Martin, W.: Several complex variables. Princeton University Press, Princeton 1948. Zbl. 41, 52
14. Boggess, A., Pitts, J.: *CR* extension near a point of higher type. C.R. Acad. Sci., Paris, Sér. I *298*, No. 1, 9–12 (1984). Zbl. 574.32030
15. Boggess, A., Polking, J.C.: Holomorphic extension of *CR*–functions. Duke Math. J. *49*, No. 4, 757–784 (1982). Zbl. 506.32003
16. Bonneau, P., Cumenge, A., Zeriahi, A.: Division dans les espaces de Lipschitz de fonctions holomorphes. C.R. Acad. Sci., Paris, Sér. I *297*, No. 9, 517–520 (1983). Zbl. 576.32025
17. Charpentier, Ph.: Résolution de l'équation $\bar{\partial}u = f$ et application aux zéros des fonctions holomorphes dans le bidisque.: Publ. Math. Orsay *85-02*, 126–147 (1985). Zbl. 568.32010
18. Chirka, E.M.: Currents and some of their applications. In.: Harvey, R.: Holomorphic chains and their boundaries (Russian translation), Mir, Moscow, 122–154 (1979)
19. Chirka, E.M., Khenkin, G.M. (= Henkin, G.M.): Boundary properties of holomorphic functions of several complex variables. [In: Current problems in mathematics, vol. *4*, 13–142, VINITI, Moscow (1975).] Itogi Nauki Techn., Ser. sovrem. Probl. Mat. *4*, 13–142 (1975). Zbl. 335.32001
20. Dautov, S.A., Khenkin, G.M. (= Henkin, G.M.): Zeros of holomorphic functions of finite order and weighted estimates for solutions of the $\bar{\partial}$–equation. Mat. Sb., Nov. Ser. *107*, 163–174 (1978); Engl. transl.: Math. USSR, Sb. *35*, 449–459 (1979). Zbl. 392.32001
21. Dzhrbashyan, M.M.: Integral transforms and the representation of functions in complex domains. Nauka, Moscow 1966 [Russian]. Zbl. 154,377
22. Folland, G.B., Kohn, J.J.: The Neumann problem for the Cauchy–Riemann complex. Princeton University Press and University of Tokyo Press, Princeton 1972. Zbl. 247.35093
23. Fuks, B.A.: Theory of analytic functions of several complex variables. Nauka, Moscow 1962; Engl. transl.: Am. Math. Soc., Providence, R.I. 1963. Zbl. 138,309
24. Fuks, B.A.: Special chapters in the theory of analytic functions of several complex variables.

Nauka, Moscow 1963; Engl. transl: Transl. Math. Monogr. *14*, Am. Math. Soc., Providence, R.I. 1965, Zbl. 146.308
25. Gamelin, T.W.: Uniform algebras. Prentice-Hall, Englewood Cliffs, N.J. 1969. Zbl. 213,404
26. Garnett, J.: Bounded analytic functions. Academic Press, New York 1981. Zbl. 469.30024
27. Gindikin, S.G.: Analysis in homogeneous domains. Usp. Mat. Nauk *19*, No. 4 (118), 3–92 (1964); Engl. transl.: Russ. Math. Surv. *19*, No. 4, 1–89 (1964). Zbl. 144,81
28. Gindikin, S.G., Khenkin, G.M. (= Henkin, G.M.): Integral geometry for $\bar{\partial}$-cohomology in $q$–linear concave domains in $\mathbb{C}P^n$. Funkts. Anal. Prilozh. *12*, No. 4, 6–23 (1978); Engl. transl.: Funct. Anal. Appl. *12*, 247–261 (1979). Zbl. 409.32020
29. Gindikin, S.G., Khenkin, G.M. (= Henkin, G.M.): The Cauchy–Fantappié formula in projective space. In: Multidimensional complex analysis, 50–63, Krasnoyarsk Inst. Fiziki, 1985 [Russian]: Zbl. 621.32008
30. Grauert, H.: Kantenkohomologie. Compos. Math. *44*, No. 1–3, 79–101 (1981). Zbl. 512.32011
31. Griffiths, Ph., Harris, J.: Principles of algebraic geometry. Wiley, New York 1978. Zbl. 408.14001
32. Gruman, L., Lelong, P.: Entire functions of several complex variables. Grundlehren der Mathematischen Wissenschaften *282*, Springer-Verlag, Berlin-Heidelberg 1986. Zbl. 583.32001
33. Gunning, R.C., Rossi, H.: Analytic functions of several complex variables. Prentice-Hall, Englewood Cliffs, N.J. 1965. Zbl. 141,86
34. Hanges, N., Treves, F.: Propagation of holomorphic extendability of $CR$–functions. Math. Ann. *236*, 157–177 (1983). Zbl. 494.32004
35. Harvey, R.: Holomorphic chains and their boundaries. Proc. Symp. Pure Math. *30*, Part 1, 309–382 (1977). Zbl. 374.32002
36. Harvey, R., Polking, J.: Fundamental solutions in complex analysis, I, II. Duke Math. J. *46*, No. 2, 253–300. Zbl. 441.35043; 301–340 (1979). Zbl. 441.35044
37. Harvey, R., Polking, J.: The $\bar{\partial}$-Neumann kernel in the ball in $\mathbb{C}^n$. Proc. Symp. Pure Math. *41*, 117–136 (1984). Zbl. 578.32030
38. Henkin, G.M. (= Khenkin, G.M.): Analytic representation for $CR$–functions on submanifolds of codimension 2 in $\mathbb{C}^n$. Lect. Notes Math. *798*, 169–191 (1980). Zbl. 431.32007
39. Henkin, G.M. (= Khenkin, G.M.): Solutions des équations de Cauchy–Riemann tangentielles sur des variétés de Cauchy–Riemann $q$-concaves. C.R. Acad. Sci., Paris, Sér. A *292*, No. 1, 27–30 (1981). Zbl. 472.32014
40. Henkin, G.M. (= Khenkin, G.M.), Leiterer, J.: Theory of functions on complex manifolds. Akademie-Verlag, Berlin 1983. Zbl. 573.32001
41. Henkin, G.M. (= Khenkin, G.M.), Polyakov, P.L.: Les zéros des fonctions holomorphes d'ordre fini dans le bidisque. C.R. Acad. Sci., Paris. Sér. I *298*, No. 2, 5–8 (1984). Zbl. 583.32006
42. Henkin, G.M. (= Khenkin, G.M.), Polyakov, P.L.: Prolongement des fonctions holomorphes bornées d'une sous-variété du polydisque. C.R. Acad. Sci., Paris, Sér. I *298*, No. 4, 221–224 (1984). Zbl. 585.32009
43. Hill, C.D., Taiani, G.: Families of analytic discs in $\mathbb{C}^n$ with boundaries on a prescribed $CR$-submanifold. Ann. Sc. Norm. Super. Pisa, Cl. Sci., IV. Ser. *5*, 327–380 (1978). Zbl. 399.32008
44. Hörmander, L.: $L^2$-estimates and existence theorems for the $\bar{\partial}$-operator. Acta Math. *113*, No. 1-2, 89–152 (1965). Zbl. 158,110
45. Hörmander, L.: An introduction to complex analysis in several variables. Princeton University Press, Princeton 1966. Zbl. 138,62
46. Hua, Loo–keng: Harmonic analysis of functions of several complex variables in the classical domains. Science Press, Peking 1958; Engl. transl.: Am. Math. Soc., Providence, R.I. 1963. Zbl. 112,74
47. Jöricke, B.: Continuity of the Cauchy projector in Hölder norms for classical domains. Math. Nachr. *112*, 227–244 (1983). Zbl. 579.32006
48. Kerzman, N., Stein, E.M.: The Szegö kernel in terms of Cauchy–Fantappié kernels. Duke Math. J. *45*, No. 2, 197–224 (1978). Zbl. 387.32009

49. Khenkin, G.M. (= Henkin, G.M.): The Lewy equation and analysis on pseudoconvex manifolds I. Usp. Mat. Nauk *32*, No. 3, 57–118 (1977); Engl. transl.: Russ. Math. Survey *32*, No. 3, 59–130 (1977). Zbl. 358.35057
50. Khenkin, G.M. (= Henkin, G.M.): The Hartogs–Bochner effect on *CR*-manifolds. Dokl. Akad. Nauk SSSR *274*, 553–558 (1984); Engl. transl.: Sov. Math., Dokl. *29*, No. 1, 78–82 (1984). Zbl. 601.32021
51. Khenkin, G.M. (= Henkin, G.M.): The tangential Cauchy–Riemann equations and the Dirac. Higgs, and Yang-Mills fields. In: Proc. Int. Congr. Math., 809–827, Warszawa 1984. Zbl. 584.58050
52. Khenkin, G.M. (= Henkin, G.M.), Polyakov, P.L.: Homotopy formulae for the $\bar{\partial}$-operator on $\mathbb{CP}^n$ and the Radon–Penrose transform. Izv. Akad. Nauk SSSR, Ser. Mat. *50*, No. 3, 566–597 (1986); Engl. transl.: Math. USSR, Izv. *28*, 555–587 (1987). Zbl. 607.32003
53. Khenkin, G.M. (= Henkin, G.M.), Polyakov, P.L.: Residue integral formulas and the Radon–Penrose transform for nonclosed forms. Dokl. Akad. Nauk SSSR *283*, 298–303 (1985); Engl. transl.: Sov. Math., Dokl. *32*, 90–95 (1985). Zbl. 599.32008
54. Khenkin, G.M. (= Henkin, G.M.), Sergeev, A.G.: Uniform estimates for solutions of the $\bar{\partial}$-equation in pseudoconvex polyhedra. Mat. Sb., Nov. Ser. *112*, No. 4, 522–567 (1980); Engl. transl.: Math. USSR, Sb. *40*, 469–507 (1981). Zbl. 452.32012
55. Kohn, J.J.: Methods of P.D.E. in complex analysis. Proc. Symp. Pure Math. *30*, No. 1, 215–237 (1977)
56. Kohn, J.J.: Subellipticity of the $\bar{\partial}$-Neumann problem on pseudoconvex domains: sufficient conditions. Acta Math. *142*, No. 1–2, 79–122 (1979). Zbl. 395.35069
57. Koranyi, A., Stein, E.M.: $H^2$-spaces of generalized half-planes. Studia Math. *44*, No. 4, 379–388 (1972). Zbl. 236.32003
58. Krantz, S.: Function theory of several complex variables. Wiley, New York 1982. Zbl. 471.32008
59. Leray, J.: Le calcul différentiel et intégral sur une variété analytique complexe. Bull. Soc. Math. Fr. *87*, 81–180 (1959). Zbl. 199,412
60. Leray, J.: Complément à l'exposé de Waelbroeck: étude spectrale des *b*-algèbres. Atti della 2ª Riunione del Groupement de Mathématiciens d'expression latine, Firenze–Bologna, 105–110 (1961). Zbl. 145,166
61. Lieb, I., Range, R.M.: Integral representations on Hermitian manifolds: the $\bar{\partial}$-Neumann solution of the Cauchy–Riemann equations. Bull. Am. Math. Soc., New Ser. *11*, 355–358 (1984). Zbl. 574.35066
62. Ligocka, E.: The Hölder continuity of the Bergman projection and proper holomorphic mappings. Stud. Math. *80*, 89–107 (1984). Zbl. 566.32017
63. Lu, Qi-keng: On the Cauchy–Fantappié formula. Acta Math. Sin. *16*, No. 3, 344–363 (1966) [Chinese]. Zbl. 173,329
64. Manin, Yu. I.: Gauge fields and complex geometry. Nauka, Moscow 1984 [Russian]; Engl. transl.: Springer-Verlag, Heidelberg 1988. Zbl. 576.53002
65. Markushevich, A.I.: Theory of analytic functions. In: Mathematics of the XIX century. Nauka, 115–269, Moscow 1981 [Russian]. Zbl. 506.01013
66. Nacinovich, M.: Poincaré lemma for tangential Cauchy–Riemann complexes. Math. Ann. *268*, No. 4, 449–471 (1984). Zbl. 574.32045
67. Passare, M.: Residues, Currents and their relation to ideals of holomorphic functions. Thesis, Uppsala University 1984
68. Pellegrino, F.: La théorie des fonctionnelles analytiques et ses applications. In: Levy, P. Problèmes concrets d'analyse fonctionnelle. Gauthier-Villars, 357–470, Paris 1951. Zbl. 43,323
69. Petrosyan, A.I.: Uniform approximation by polynomials on Weil polyhedra. Izv. Akad. Nauk SSSR, Ser. Mat. *34*, 1241–1261 (1970); Engl. transl.: Math. USSR, Izv. *34*, 1250–1271 (1970). Zbl. 217,394
70. Polyakov, P.L.: Zeros of holomorphic functions of finite order in the polydisc. Mat. Sb., Nov. Ser. *133*, No. 1, 103–111 (1987). Zbl. 624.32002

71. Piatetski-Shapiro, I.I.: Automorphic functions and the geometry of classical domains. Nauka, Moscow 1961, Engl. transl.: Mathematics and its applications, Vol. *8*, Gordon and Breach Science Publishers, New York-London-Paris 1969; French transl.: Dunod, Paris 1966. Zbl. 137,275
72. Range, R.M.: Holomorphic functions and integral representations in several complex variables. Springer–Verlag, New York 1986. Zbl. 591.32002
73. Ronkin, L.I.: Introduction to the theory of entire functions of several variables. Nauka, Moscow 1971; Engl. transl.: Amer, Math. Soc., Providence, R.I. 1974. Zbl. 286.32004
74. Roos, G.: Fonctions de plusieurs variables complexes et formules de représentation intégrale.: Lecture Notes in Mathematics *1118*, 45–182, Springer-Verlag 1986. Zbl. 594.32005
75. Rossi, H., Vergne, M.: Equations de Cauchy–Riemann tangentielles, associées à un domaine de Siegel. Ann. Sci. Ec. norm. sup., IV. Sér. *9*, No. 1, 31–80 (1976). Zbl. 398.32018
76. Rudin, W.: Function theory in the unit ball in $\mathbb{C}^n$. Springer-Verlag, Berlin 1980. Zbl. 495.32001
77. Sibony, N.: Un exemple de domaine pseudoconvexe régulier ou l'équation $\bar{\partial}u = f$ n'admet pas de solution bornée pour $f$ bornée. Invent. Math. *62*, No. 2, 235–242 (1980). Zbl. 436.32015
78. Stoll, W.: Holomorphic functions of finite order in several complex variables. Reg. Conf. Ser. Math., No. 21, Am. Math. Soc., Providence, R.I. 1974. Zbl. 292.32003
79. Treves, F.: Approximation and representation of functions and distributions annihilated by a system of complex vector fields. Ecole Polytechnique, Centre de Mathématiques 1981. Zbl. 515.580030
80. Vitushkin, A.G.: Analytic capacity of sets in problems of approximation theory. Usp. Mat. Nauk *22*, No. 6, 141–199 (1967); Engl. transl.: Russ. Math. Surv. *22*, 139–200 (1967). Zbl. 157,394
81. Vladimirov, V.S.: Methods of the theory of many complex variables. Nauka, Moscow 1964; Engl. transl.: MIT Press, Cambridge 1966. Zbl. 125,319
82. Wells, R.O.: Function theory on differentiable submanifolds. Contributions to analysis. Academic Press, 407–441, New York 1974. Zbl. 293.32001
83. Wells, R.O.: The Cauchy–Riemann equations and differential geometry. Bull. Am. Math. Soc., New Ser. *6*, No. 2, 187–199 (1982). Zbl. 496.32012

# III. Complex Analytic Sets

E.M. Chirka

Translated from the Russian
by P.M. Gauthier

## Contents

# Introduction

The theory of complex analytic sets is part of the modern geometric theory of functions of several complex variables. Traditionally, the presentation of the foundations of the theory of analytic sets is introduced in the algebraic language of ideals in Noetherian rings as, for example, in the books of Hervé [23] or Gunning–Rossi [19]. However, the modern methods of this theory, the principal directions and applications, are basically related to geometry and analysis (without regard to the traditional direction which is essentially related to algebraic geometry). Thus, at the beginning of this survey, the geometric construction of the local theory of analytic sets is presented. Its foundations are worked out in detail in the book of Gunning and Rossi [19] via the notion of analytic cover which together with analytic theorems on the removal of singularities leads to the minimum of algebraic apparatus necessary in order to get the theory started.

The study of singularities of analytic sets at the present time forms one of the most fascinating areas of analysis, geometry, and topology. However, this direction has developed to such a degree that it has practically distinguished itself as an independent branch of analysis, bordering on the theory of singularities of differentiable mappings. Without going deeply into this theory, we briefly set forth the infinitesimal properties of analytic sets, but only a first approximation thereof (sufficiently rich for applications). In this context, tangent cones, various multiplicities and intersection theory come to the fore.

In the last 10 years the direction connected with differential geometry and the theory of currents (generalized differential forms) has flourished intensively. Initiated in the work of Lelong, it was significantly pushed ahead by Griffiths and his school. This direction is reflected in §§3–4: here, we briefly set forth the basis of integration on complex analytic sets as well as some metric, functional-analytic, and differential-geometric properties.

The boundary properties of complex analytic sets are perhaps most connected with the classical theory of functions. This direction has begun to develop very recently, but already it has to its credit a series of impressive results (theorems on the removal of singularities, the theorem of Harvey–Lawson on the structure of the boundary and others). There are very interesting connections here with the boundary properties of holomorphic mappings. The basic results are presented in §5 along with several problems of this prospective direction.

## §1. Local Structure of Analytic Sets

**1.1. Zeros of Holomorphic Functions.** Let $\mathcal{O}_a$ denote functions holomorphic in a neighbourhood of a point $a$, and $\mathcal{O}(D)$-functions holomorphic in a domain $D$. A holomorphic function $f \not\equiv 0$ of one complex variable can be represented, in the neighbourhood of one of its zeros $z_0$, in the form $f(z) = (z - z_0)^k g(z)$, where $k = \operatorname{ord}_{z_0} f \in \mathbb{N}$ is the order (multiplicity) of the zero and $g \in \mathcal{O}_{z_0}$, $g(z_0) \neq 0$. From this it follows that the zeros of $f$ form a discrete set $Z_f$. Conversely, if $Z = \{z_j\}$ is an arbitrary sequence in a domain $D \subset \mathbb{C}$, having no limit points in the interior of $D$, and $\{k_j\}$ are arbitrary natural numbers, then, there exists a function $f \in \mathcal{O}(D)$ for which the set of zeros coincides with $\{z_j\}$ and $\operatorname{ord}_{z_j} f = k_j$.

The local structure of the zeros of a holomorphic function of several complex variables yields the following most important result.

**Weierstrass Preparation Theorem.** *Let $f$ be a function holomorphic in a neighbourhood of* 0 *in* $\mathbb{C}^n$, $f(0) = 0$, *but* $f(0', z_n) \not\equiv 0$ *in a neighbourhood of* 0. *Then* $f(z) = (z_n^k + c_1(z')z_n^{k-1} + \dots + c_k(z')) \cdot g(z)$, *where* $c_j$ *and* $g$ *are holomorphic in a neighbourhood of* 0, $c_j(0) = 0$, *and* $g(0) \neq 0$. (*Here* $z = (z', z_n)$.)

Thus, in the vicinity of 0, the zeros of $f$ have the same structure as the zeros of the Weierstrass polynomial $z_n^k + c_1(z')z_n^{k-1} + \dots$. The proof of this theorem is in all textbooks on multidimensional complex analysis (see, for example, [39], [19], [53]). The only algebraic theorem which is used in its proof is the well-known theorem on symmetric polynomials.

From Rado's Theorem (see [39]) follows easily the existence of a function $\Delta(z') \not\equiv 0$ holomorphic in a neighbourhood $U' \ni 0$ in $\mathbb{C}^{n-1}$ and a disc $U_n$ such that for $\Delta(z') \neq 0$, the number of geometrically different zeros of $f(z', z_n)$ in the disc $U_n$ is the maximum possible (and one and the same), while for $\Delta(z') = 0$ they are sort of "glued together". From the classical Rouché Theorem it follows easily that the projection $Z_f \cap (U' \times U_n) \to U'$ onto $U' \backslash Z_\Delta$ is locally biholomorphic and this yields a finite sheeted covering space.

**1.2. Analytic Sets. Regular Points. Dimension.** Let $\Omega$ be a complex manifold. A set $A \subset \Omega$ is called a (closed) *analytic set* in $\Omega$ if for each point $a \in \Omega$, there is a neighbourhood $U \ni a$ and functions $f_1, \ldots, f_N$ holomorphic in $U$ such that $A \cap U = Z_{f_1} \cap \ldots \cap Z_{f_N} \cap U$. Thus, locally, $A$ is the set of common zeros of a finite collection of holomorphic functions. Such a set is necessarily closed in $\Omega$. If the above representation holds only for points $a \in A$, then we obtain the broader class of *locally analytic sets.*

Several simple properties follow from the definition:

1) The intersection of a finite number of analytic sets is an analytic set. The union of a finite number of analytic sets is an analytic set. (In this, analytic sets advantageously differ from submanifolds.)

2) If $\varphi: X \to Y$ is a holomorphic mapping of complex manifolds, then the preimage of an arbitrary analytic set $A \subset Y$ is an analytic set in $X$.

3) If $\Omega$ is connected and an analytic subset $A \subset \Omega$ contains a non-empty open subset of $\Omega$, then $A = \Omega$ ($\Rightarrow$ if $A \neq \Omega$, then $A$ is nowhere dense in $\Omega$).

4) On a Riemann surface (= connected one-dimensional complex manifold), each proper analytic subset is locally finite.

5) If $\Omega$ is connected and $A \subset \Omega$ is a proper analytic subset, then $\Omega \backslash A$ is arcwise connected.

A point $a$ of an analytic set $A$ is called a *regular point* if there exists a neighbourhood $U \ni a$ in $\Omega$ such that $A \cap U$ is a complex submanifold of $U$. The complex dimension of $A \cap U$ is then called the dimension of $A$ at the point $a$ and it is denoted by $\dim_a A$. The set of all regular points of an analytic set $A$ is denoted by $\operatorname{reg} A$. Its complement, $A \backslash \operatorname{reg} A$, is denoted by $\operatorname{sng} A$. The set $\operatorname{sng} A$ is called the set of *singular points* of the set $A$.

It can be shown by induction on the dimension of the manifold $\Omega$ that $\operatorname{sng} A$ is nowhere dense (and closed) in $A$. This allows us to define the *dimension* of $A$ at any point of $A$:

$$\dim_a A = \limsup_{z \to a,\, z \in \operatorname{reg} A} \dim_z A.$$

The dimension of $A$ itself is the number $\dim A = \max_{z \in A} \dim_a A$ and the *codimension* of $A$ is equal to $\dim \Omega \backslash \dim A$. The set $A$ is called *purely p-dimensional* if $\dim_z A \equiv p$, for all $z \in A$. Along with $\operatorname{sng} A$, it is convenient to single out also the set

$$S(A) = (\operatorname{sng} A) \cup \{z \in A: \dim_z A < \dim A\}.$$

**1.3. Sets of Codimension 1.** From the Weierstrass Preparation Theorem, it follows that the zeros of a function $f \not\equiv 0$ holomorphic on a connected $n$-dimensional manifold $\Omega$ form a set $Z_f$ of codimension 1. Using the discriminant set $Z_\Delta$ (§1.1) and induction on $n$, it is not hard to show that the

Hausdorff dimension of $Z_f$ is equal to $2n-2$ while the Hausdorff dimension of the set sng $Z_f$ is at most $2n-4$. Since each analytic set is locally contained in the set of zeros of a single holomorphic function, then the following metric estimate on the size of the set of singular points is satisfied: if $A$ is a proper analytic subset of $\Omega$, then the Hausdorff dimension of $S(A)$ is at most $2n-4$. Since sets of Hausdorff dimension $\leq 2n-4$ are removable for bounded holomorphic functions in domains in $\mathbb{C}^{n-1}$, we obtain, from this and from the Weierstrass Preparation Theorem, the following theorem on the local structure of analytic sets of codimension 1. Let $\# E$ denote the number of points in a set $E$.

**Theorem 1.** *Let $A$ be an analytic $(n-1)$-dimensional subset of the polydisc $U' \times U_n$ in $\mathbb{C}^n(U' \subset \mathbb{C}^{n-1}, U_n \subset \mathbb{C})$ without limit points on $U' \times \partial U_n$. Then:*

(1) *$A$ can be represented as the union of two analytic sets $A_{(n-1)} \cup A'$, where* $A_{(n-1)} = \{z \in A: \dim_z A = n-1\}$, $A' \supset S(A)$, and $\dim A' < n-1$;

(2) *there exists a Weierstrass polynomial $f$ on $U' \times \mathbb{C}$ such that $A_{(n-1)} = Z_f$ and for each fixed $z' \in U' \backslash Z_\Delta$, the polynomial $f$ has only simple roots (the latter is equivalent to the assertion that*

$$\deg f = \max_{c \in U'} \# A \cap \{z' = c\}).$$

Thus, a purely $(n-1)$-dimensional analytic set is locally the set of zeros of a single holomorphic function. A function $f \in \mathcal{O}(\Omega)$ is called a *minimal defining function* for an analytic set $A \subset \Omega$, if $A = Z_f$ and for each domain $U \subset \Omega$, any function $g \in \mathcal{O}(U)$, which is equal to zero on $A \cap U$, is divisible by $f$ in $\mathcal{O}(U)$. The Weierstrass polynomial $f$, mentioned in the theorem, has this property, while in general, the minimality condition on $f$ is equivalent to the condition that $df \neq 0$ at each regular point of $A$. More precisely, the following holds.

**Proposition 1.** *A holomorphic function $f \not\equiv 0$ on a connected manifold is a minimal defining function for $A = Z_f$ if and only if the set $\{z \in A: (df)_z = 0\}$ is nowhere dense in $A$. If $f$ is minimal for $A$, then* sng $A$ *coincides with the analytic set* $\{f = df = 0\}$.

**1.4. Proper Projections.** The codimension of analytic sets in $\mathbb{C}^n$ is diminished by suitable projections into subspaces of smaller dimension. Analyticity, in general, is not preserved by projections (for example, the coordinate projections of the set $A$: $z_1 z_2 = 1$ in $\mathbb{C}^2$). Additional conditions are required. The following is a fundamental assertion on the lowering of codimension ("excluded variables").

**Lemma 1.** *Let $U' \subset \mathbb{C}^p$, $U'' \Subset \mathbb{C}^q$ be open subsets and $\pi: (z', z'') \to z' \in \mathbb{C}^p$. Let $A$ be an analytic subset of $U' \times U''$ without limit points on $U' \times \partial U''$. Then, $A' = \pi(A)$ is an analytic subset of $U'$ and $A \cap \pi^{-1}(z')$ is locally finite in $U'$. Moreover,*

$$\dim_{z'} A' = \max_{\pi(z) = z'} \dim_z A.$$

In particular, $\dim A' = \dim A$.

A rather simple proof of this important assertion can be found, for example, in the book of Mumford [28]. The condition on $A$ implies that $\pi|A$ is a *proper mapping*, i.e. the preimage of any compact set in $U'$ is compact in $A$. The proof of the existence of such projections (locally) is obtained rather simply by induction on the codimension. The following holds:

**Lemma 2.** *Let $A$ be an analytic set in $\mathbb{C}^n, 0 \in A$, and $\dim_0 A \leq p$. Let $L_1, \ldots, L_k$ be an arbitrary finite collection of p-dimensional subspaces of $\mathbb{C}^n$ and $\pi_j$ the orthogonal projection on $L_j$. Then, there exists a unitary transformation $l$ and arbitrarily small neighbourhoods $U_j \ni 0$ such that each restriction $\pi_j | l(A) \cap U_j \to L_j \cap U_j, j = 1, \ldots, k$, is a proper mapping. Moreover, the set of such $l$ is open and everywhere dense in the group of all unitary transformations in $\mathbb{C}^n$.*

The following assertions are easily obtained as a corollary of these two lemmas.

1) An analytic set $A \subset \mathbb{C}^n$ is compact if and only if it is finite (i.e. $\# A < \infty$).
2) An analytic subset of a complex manifold is zero-dimensional if and only if it is locally finite.
3) If $A_j$ are analytic sets and $a \in \bigcap_1^k A_j$, then $\operatorname{codim}_a \bigcap_1^k A_j \leq \sum_1^k \operatorname{codim}_a A_j$.

**1.5. Analytic Coverings.** The lowering of codimension by proper projections and the theorem on the local structure of analytic sets of codimension 1 lead to the following representation for the local structure of analytic sets of arbitrary codimension.

**Theorem 1.** *Let $A$ be an analytic set in $\mathbb{C}^n$, $\dim_a A = p$, $0 < p < n$, $U = U' \times U''$ a neighbourhood of $a$, with $\pi | A \cap U \to U' \subset \mathbb{C}^p$ a proper projection. Then, there exists an analytic subset $\sigma \subset U'$ of dimension $< p$ and a natural number $k$ such that*

*(1) $\pi | A \cap U \setminus \pi^{-1}(\sigma) \to U' \setminus \sigma$ is a locally biholomorphic k-sheeted covering,*
*(2) $\pi^{-1}(\sigma)$ is nowhere dense in $A_{(p)} \cap U$, where $A_{(p)} = \{z \in A\colon \dim_z A = p\}$.*

For such a structure it is convenient to introduce a particular name (see [19]).

**Definition 1.** Let $A$ be a locally closed set on a complex manifold $X$ and $f\colon A \to Y$ a continuous proper mapping into another complex manifold $Y$ such that the preimage of each point is finite. The triple $(A, f, Y)$ is called an *analytic covering* (over $Y$) if there exists an analytic subset $\sigma \subset Y$ (perhaps empty) of dimension $< \dim Y$ and a natural number $k$ such that

(1) $f | A \setminus f^{-1}(\sigma)$ is a locally biholomorphic $k$-sheeted covering (over $Y \setminus \sigma$),
(2) $f^{-1}(\sigma)$ is nowhere dense in $A$.

If we drop the hypothesis that $\# f^{-1}(w) < \infty$ and replace the condition of analyticity of $\sigma$ by that of local removability, then we obtain an object $(A, f, Y)$ called a *generalized analytic covering*. (A set $\sigma$ is said to be *locally removable* if it is closed and for each domain $U \subset Y$, each function bounded and holomorphic in $U \backslash \sigma$ extends holomorphically to $U$; an example of such a set is any closed subset of Hausdorff dimension $< 2\dim_{\mathbb{C}} Y - 1$.)

Theorem 1 asserts that each purely $p$-dimensional analytic set is locally representable as an analytic covering over a domain $\mathbb{C}^p$. The converse is valid even for generalized analytic coverings.

**Theorem 2.** *Let $A$ be a closed subset of a domain $U = U' \times U'' \subset \mathbb{C}^n$ and $\pi\colon (z', z'') \mapsto z' \in \mathbb{C}^p$. Suppose $\pi|A \to U'$ is a generalized analytic covering. Then, $A$ is a purely $p$-dimensional analytic subset of $U$ and $\pi|A \to U'$ is an analytic covering.*

A convenient analytic tool for the study and the application of analytic coverings is given by canonically defined functions representing multidimensional analogues of the polynomials $\Pi(z - \alpha_j)$ on the plane (see Whitney [53]). They are defined in the following way. Let $\alpha^1, \ldots, \alpha^k$ be points of $\mathbb{C}^m$, not necessarily distinct, and set $P_\alpha(z, w) = \langle w, z - \alpha^1 \rangle \ldots \langle w, z - \alpha^k \rangle$, where $\langle a, b \rangle = a_1 b_1 + \ldots + a_m b_m$. The condition $P_\alpha(z, w) \equiv 0$ for a fixed $z \in \mathbb{C}^m$, as is easily seen, is equivalent to the assertion that $z$ belongs to the (unordered) collection $\alpha = \{\alpha^1, \ldots, \alpha^k\}$. Expanding $P_\alpha(z, w)$ in powers of $w$,

$$P_\alpha(z, w) = \sum_{|I| = k} \Phi_I(z; \alpha) w^I,$$

we obtain that this condition is equivalent to the system $\Phi_I(z; \alpha) = 0$, $|I| = k$, of

$$\binom{k + m - 1}{m - 1}$$

equations of degree $\leq k$ for $z$. Thus, the set of solutions of this system is precisely the set $\alpha$. The functions $\Phi_I$ depend nicely on $\alpha$:

$$\Phi_I(z; \alpha) = \sum_{|J| \leq k} \psi_{IJ}(\alpha) z^J,$$

where $\psi_{IJ}$ are polynomials of $k \cdot m$ variables $\alpha_v^j$ of degree no greater than $k$. In the situation, when $\alpha = \alpha(z')$ are the fibres $A \cap \pi^{-1}(z')$ of the generalized analytic covering $\pi|A \to U'$, $z' \notin \sigma$ (as in Theorem 2), $\psi_{IJ}(\alpha(z'))$ are bounded holomorphic functions on $U' \backslash \sigma$ (since $\psi_{IJ}(\alpha)$ is independent of the enumeration of the collection $\alpha$). Since $\sigma$ is removable, we obtain holomorphic functions

$$\Phi_I(z', z'') = \sum_{|J| \leq k} \varphi_{IJ}(z')(z'')^J, \qquad |I| = k,$$

where $\varphi_{IJ}(z')$ is the holomorphic continuation to $U'$ of the function $\psi_{IJ}(\alpha(z'))$. These $\Phi_I$ are called the *canonical defining functions* of the (generalized) analytic covering $\pi|A \to U'$. The set of common zeros of the $\Phi_I$ in $U' \times \mathbb{C}^{n-p}$ coincides with $A$ (and from this, of course, follows the analyticity of $A$).

The analytic set in $A$, defined by the condition

$$\operatorname{rank} \frac{\partial \Phi(z', z'')}{\partial z''} < n - p,$$

is precisely the set of critical points $\operatorname{br} \pi|A$ of the projection $\pi|A$, i.e. the singular points as well as those regular points $z$, for which $\operatorname{rank}_z \pi|A < p$. From this, easily follows, for example, the analyticity of $\operatorname{sng} A$ for purely $p$-dimensional $A$.

**1.6. Irreducible Components.** An analytic set $A$ is called *irreducible* if it is not possible to represent $A$ in the form $A_1 \cup A_2$, where $A_1$ and $A_2$ are also analytic sets, closed in $A$ and distinct from $A$ (in the opposite case, $A$ is said to be *reducible*). Irreducible sets have a simple characterization in terms of regularity.

**Proposition 1.** *An analytic set $A$ is irreducible if and only if the set $\operatorname{reg} A$ is connected.*

An analytic set $A$ is said to be *irreducible at a point $a \in A$* if there exists a fundamental system of neighbourhoods $U_j \ni a$ such that each $A \cap U_j$ is irreducible. We say that $A$ is locally irreducible if it is irreducible at each of its points. The simplest example of a locally-irreducible but reducible analytic set is a disconnected complex submanifold. The set $A$: $z_2^2 = z_1^3 + z_1^2$ in $\mathbb{C}^2$ (the image of $\mathbb{C}$ by the mapping $\lambda \to (\lambda^2 - 1, \lambda(\lambda^2 - 1))$ is globally irreducible but reducible at the point $z = 0$. The set $A$: $z_1^2 = z_3 z_2^2$ in $\mathbb{C}^3$ is irreducible globally and at 0 but is reducible at each point of the form $(0, 0, z_3)$, $z_3 \neq 0$.

An analytic set $A' \subset A$ is said to be an *irreducible component* of $A$, if $A'$ is irreducible and maximal, i.e. is contained in no other irreducible analytic set $A'' \subset A$.

**Theorem 1.** *Let $A$ be an analytic subset of a complex manifold $\Omega$. Then:*

(1) *each irreducible component of $A$ has the form $\bar{S}$, where $S$ is a connected component of $\operatorname{reg} A$;*

(2) *if $\operatorname{reg} A = \bigcup_{j \in J} S_j$ is the decomposition into connected components ($J$ finite or countable, $S_i \neq S_j$, for $i \neq j$), then $A = \bigcup_{j \in J} \bar{S}_j$, and this is the decomposition of $A$ into irreducible components;*

(3) *the decomposition into irreducible components is locally finite, i.e., any compact set $K \subset \Omega$ meets at most finitely many $\bar{S}_j$.*

From this and from §1.5, we have the following properties:

1) Let $A$ be an analytic subset of $\Omega$ of dimension $p$ and $A_{(k)} = \{z \in A: \dim_z A = k\}$. Then $\bar{A}_{(k)}$ is a purely $k$-dimensional or empty analytic subset of $\Omega$ and $A = \bigcup_0^p \bar{A}_{(k)}$.

2) For any analytic set $A$, the sets $\operatorname{sng} A$ and $S(A)$ are also analytic. Moreover, $\dim S(A) < \dim A$ and $\dim_z \operatorname{sng} A < \dim_z A$, for each $z \in \operatorname{sng} A$.

3) For each point $a$ of an analytic set $A$, there exists a neighbourhood $U \ni a$ such that each irreducible component of the set $A \cap U$ is also irreducible at $a$.

4) If $A_1$ and $A_2$ are analytic subsets, $A_1$ is irreducible and $\dim A_1 \cap A_2 = \dim A_1$, then $A_1 \subset A_2$. Moreover, $A_1$ belongs to some irreducible component of the set $A_2$.

From these properties and from the decomposition into irreducible components, we easily obtain the following:

**Theorem 2.** *Let $\{A_\alpha\}_{\alpha \in I}$ be an arbitrary family of analytic subsets of a complex manifold $\Omega$. Then, $A = \bigcap_{\alpha \in I} A_\alpha$ is also an analytic subset of $\Omega$. Moreover, for each compact set $K \subset \Omega$, there is a finite subset $J \subset I$ such that $A \cap K = \left(\bigcap_{\alpha \in J} A_\alpha\right) \cap K$.*

**1.7. One-Dimensional Analytic Sets.** The local structure of one-dimensional analytic sets is the simplest. If $\dim_a A = 1$, $U \ni a$, and $\pi | A \cap U \to U_1$ is an analytic covering over a disc $U_1 \subset \mathbb{C}$, then the critical set $\sigma \subset U_1$ of this covering is zero-dimensional, and hence, either empty or consists of the single point $a_1 = \pi(a)$, if $U_1$ is sufficiently small. Since the fundamental group of the punctured disc is $\mathbb{Z}$, from this one easily obtains, the following.

**Proposition 1.** *If an analytic set $A$ is one-dimensional and irreducible at a point $a$, then there exists a neighbourhood $U \ni a$ and a holomorphic mapping of the disc $\Delta \subset \mathbb{C}$ into $U$ which covers $A \cap U$ in a one-to-one manner.*

In the situation described above, when $A \cap U \subset \mathbb{C}^n$ is irreducible, $a = \pi^{-1}(0) \cap A = 0$, $\sigma = \{0\}$, and the number of sheets of $A \cap U \to U_1$ is $k$, such a mapping (local parametrization) can be written in the form $z_1 = \xi^k$, $z_j = \varphi_j(\xi)$, $j = 2, \ldots, n$, where $\varphi_j$ are functions holomorphic in the disc $\Delta \subset \mathbb{C}$. In other words it can be written in the form $z_j = \varphi_j(z_1^{1/k})$, $j = 2, \ldots, n$ (representation of Puiseux). Since the indicated mapping $\Delta \to A \cap U$ is one-to-one, the inverse mapping $\xi = \xi(z)$ is a function continuous on $A \cap U$ and holomorphic in $A \cap U \setminus \{0\}$. It is called a *local normalizing parameter* on $A$ in a neighbourhood of the point 0.

It is not possible, in principle, to parametrize in this way analytic sets of dimension greater than 1, because even in the neighbourhood of an isolated singular point, an analytic set which is irreducible at the point need not be a topological manifold. For example, the cone $A$: $z_3^2 = z_2 z_1$ in a neighbourhood of 0 in $\mathbb{C}^3$; $A \setminus \{0\}$ is the image of $\mathbb{C}^2 \setminus \{0\}$ by the two-sheeted covering $(\lambda_1, \lambda_2) \mapsto (\lambda_1^2, \lambda_2^2, \lambda_1 \lambda_2)$, and hence $A \cap U \setminus \{0\}$ is not simply connected for any $U \ni 0$ and hence cannot be homeomorphic to the punctured ball.

The global structure of one-dimensional analytic sets is also comparatively simple. Let $\operatorname{sng} A = \{a_j\}$ and $U_j \ni a_j$ be pairwise disjoint neighbourhoods such that $A \cap U_j$ consists of a finite number of irreducible components $S_{j\nu}$ each of which admits a normalizing parameter $\xi_{j\nu}\colon S_{j\nu} \to \Delta \subset \mathbb{C}$. Let $V_j \Subset U_j$ be smaller neighbourhoods of the $a_j$. Let the abstract set $A^*$ be defined by

$$A^* = \left(A \setminus \bigcup_i \bar{V}_i\right) \cup \left(\bigsqcup_{j,\nu} S_{j\nu}\right),$$

where $\bigsqcup$ denotes the disjoint union (points from different $S_{j\nu}$ are considered different). The set $A^*$ is naturally endowed with the structure of a complex manifold, since the usual charts on $A \setminus \bigcup \bar{V}_i \subset \operatorname{reg} A$ are holomorphically compatible with each of the charts $(S_{j\nu}, \xi_{j\nu})$. Here, the natural projection $A^* \to A$ is obviously locally biholomorphic over $\operatorname{reg} A$. The manifold $A^*$, which we have constructed, is called a *normalization* of the one-dimensional set $A$. If $A$ is irreducible, then $A^*$ is connected, i.e., is a Riemann surface. Every Riemann surface has a universal covering surface which, by Riemann's theorem, is either the disc, the plane, or the Riemann sphere. Hence, for example, each bounded one-dimensional irreducible analytic set $A$ in $\mathbb{C}^n$ is the image of the unit disc $\Delta$ by some holomorphic mapping $\Delta \to \mathbb{C}^n$ which over $\operatorname{reg} A$ is a locally biholomorphic covering.

Any analytic set of positive dimension is locally representable as a union of one-dimensional analytic sets containing the given point (this follows easily from the local representation in terms of an analytic covering). From this, we obtain, for example, the following important property.

**Theorem** (Maximum principle). *Let $u$ be a plurisubharmonic function in the neighbourhood of a connected analytic set $A$. If $u(a) = \sup_A u$ at some point $a \in A$, then $u \equiv u(a)$ on $A$.*

**Remark.** The results of this section are basically well known. The proofs can be found in the books [19], [23], [53] and the author's book [10].

## §2. Tangent Cones, Multiplicity and Intersection Theory

**2.1. The Tangent Cone.** Let $E$ be an arbitrary set in $\mathbb{R}^n$. A vector $v \in \mathbb{R}^n$, is called a *tangent vector* to the set $E$ at the point $a$ if there exist a sequence of points $a^j \in E$ and numbers $t_j > 0$ such that $a^j \to a$ and $t_j(a^j - a) \to v$, as $j \to \infty$. The collection of all such vectors $v$ is denoted by $C(E, a)$ and is called the *tangent cone* to the set $E$ at the point $a$. If $E$ is a manifold of class $C^1$, then $C(E, a)$ clearly coincides with the usual tangent space $T_a E$ of the manifold $E$.

Let us examine some properties of tangent cones for analytic sets in $\mathbb{C}^n \approx \mathbb{R}^{2n}$. If $A$ is a purely one-dimensional analytic set in $\mathbb{C}^n$, irreducible at a point $a \in A$, then, from the local paramatrization in §1.7, it follows easily that $C(A, a)$ is a complex line. Hence, the tangent cone to an arbitrary one-dimensional analytic set at any of its points is a finite union of complex lines. The case of higher dimension is considerably more complicated (for example, the tangent cone at 0 for a set $A$ given by $\rho_j(z) = 0$, $j = 1, \ldots, k$, where the $\rho_j$ are homogeneous polynomials, simply coincides with $A$). As regards an analytic description, the simplest case is that of a hypersurface.

**Proposition 1.** *Let $f \not\equiv 0$ be a function holomorphic in the neighbourhood of 0 in $\mathbb{C}^n$, and $\mu \geq 1$, the multiplicity of the zero of $f$ at 0, i.e., in the expansion $f(z) = \sum_{\mu}^{\infty} f_j(z)$ in homogeneous polynomials, $f_\mu \not\equiv 0$. Then, $C(Z_f, 0) = \{z: f_\mu(z) = 0\}$.*

In the general case, a good description is obtained in terms of the canonical defining functions.

**Proposition 2.** *Let $A$ be a purely p-dimensional analytic set in $\mathbb{C}^n$, $0 \in A$, and $C(A, 0) \cap \{z_1 = \ldots = z_p = 0\} = 0$. Let $U = U' \times U''$ be a neighbourhood of 0 such that the projection $\pi|A \cap U \to U' \subset \mathbb{C}^p$ is an analytic covering, and let $\Phi_I$, $|I| = k$, be canonical defining functions of this covering. Then $C(A, 0) = \{z: \Phi_I^*(z) = 0, |I| = k\}$, where $\Phi_I^*$ is the initial homogeneous polynomial in the expansion of the function $\Phi_I$ near 0.*

**Corollary.** *The tangent cone to an analytic set at any of its points is a homogeneous algebraic set.*

For many questions, the tangent cone successfully replaces the tangent space and this is connected, most of all, to the following proximity property.

**Proposition 3.** *Let $A$ be a purely p-dimensional analytic set in $\mathbb{C}^n$ and $0 \in A$. Then, there exist positive constants $\varepsilon$, $C$, and $r_0$ such that* $\operatorname{dist}(A_r, C(A, 0)_r) \leq Cr^{1+\varepsilon}$ *for each $r < r_0$, where* dist *denotes the standard distance and $E_r = E \cap \{|z| \leq r\}$.*

This property follows from Proposition 2; indeed, from there it follows that, as $\varepsilon$ one may choose $1/k$, where $k$ is the multiplicity of the covering in Proposition 2.

**2.2. Whitney Cones.** The tangent cone introduced above is not the only possible analogue of the tangent space in the nonregular case. Other natural generalizations of this notion, useful in the study of analytic sets, were investigated by Whitney [51], [52]. We introduce the definition of Whitney cones $C_i(A, a)$, $i = 1, \ldots, 6$, via their constituent vectors. Let $A$ be an analytic set in $\mathbb{C}^n$ and $a \in A$.

1) A vector $v \in \mathbb{C}^n$ belongs to $C_1(A, a)$ if there exists a neighbourhood $U \ni a$ in $\mathbb{C}^n$ and holomorphic vector field $v(z)$ in $U$ (i.e. a holomorphic mapping $U \to \mathbb{C}^n$) such that $v(a) = v$ and $v(z) \in T_z A$, for each $z \in \operatorname{reg} A \cap U$.

2) $v \in C_2(A, a)$ if for each $\varepsilon > 0$, there exists a $\delta > 0$ such that if $z \in \operatorname{reg} A$ and $|z - a| < \delta$, then $|v - v'| < \varepsilon$ for some $v' \in T_z A$.

3) $C_3(A, a) = C(A, a)$, the tangent cone from §2.1.

4) $v \in C_4(A, a)$ if there exists a sequence of point $z^j \in \operatorname{reg} A$ and vectors $v^j \in T_{z^j}A$ such that $z^j \to a$ and $v^j \to v$ as $j \to \infty$.

5) $v \in C_5(A, a)$ if there exist sequences of points $z^j, w^j \in A$ and numbers $\lambda_j \in \mathbb{C}$, such that $z^j \to a$, $w^j \to a$, and $\lambda_j(z^j - w^j) \to v$ as $j \to \infty$.

6) $v \in C_6(A, a)$ if $(df)_a(v) = 0$ for each function $f$, holomorphic in the vicinity of $a$ and equal to zero on $A$ (the set $C_6(A, a)$ is denoted by $T(A, a)$ and is called the *tangent space* to $A$ at the point $a$).

**Proposition 1** (Whitney [51]). *At each point of an analytic set we have the inclusions* $C_1 \subset C_2 \subset C_3 \subset C_4 \subset C_5 \subset C_6$.

The cones $C_1$, $C_2$, and $C_6$ are clearly linear subspaces of $\mathbb{C}^n$. The cone $C_3$, as shown above, is an algebraic subset of $\mathbb{C}^n$. For $C_4$ and $C_5$, this property issues from the following.

**Proposition 2** (Whitney [51]). *Let $A$ be an analytic subset of a domain $D \subset \mathbb{C}^n$ and $C_i(A) = \{(z, v): z \in A, v \in C_i(A, z)\}$, $i = 4, 5$. Then, $C_i(A)$ are analytic subsets of $D \times \mathbb{C}^n$ of dimension $2 \dim A$, where $C_4(A)$ is the closure in $D \times \mathbb{C}^n$ of the tangent bundle $\{(z, v): z \in \operatorname{reg} A, v \in T_z A\}$.*

As a corollary we also have that the cones $C_i(A, z)$, $i = 4, 5$, are "upper-semicontinuous" with respect to $z$, i.e. if $A \ni z^j \to a \in A$, then the limit set of the family of cones $C_i(A, z^j)$, as $j \to \infty$, is contained in $C_i(A, a)$.

The cone $C_6$, the tangent space, reflects the "imbedding dimension" of the analytic set. The following is easily proved.

**Proposition 3.** *Let $A$ be an analytic set in $\mathbb{C}^n$, $a \in A$, and $\dim T(A, a) = d$. Then, there exists a neighbourhood $U \ni a$ in $\mathbb{C}^n$ and a complex $d$-dimensional submanifold $M \subset U$ such that $A \cap U \subset M$. Manifolds of lesser dimension with this property do not exist.*

The cones $C_4$ and $C_5$ reflect important local properties of analytic sets (see, for example, [49] and [53]). If, for example, $\dim C_5(A, a) = \dim_z A \equiv p$, then $a$ is a regular point of $A$; if $\dim C_4(A, a) = p$ and $a \in \operatorname{sng} A$, then $\dim_a(\operatorname{sng} A) = p - 1$; etc. (see [49]). The cone $C_4$ also "assists" in normalizing certain singularities of codimension 1 (Stutz [49]):

**Proposition 4.** *Let $A$ be a purely $p$-dimensional analytic set in $\mathbb{C}^n$ irreducible at the point $0 \in \operatorname{reg}(\operatorname{sng} A)$ and let $\dim C_4(A, 0) = p$. Then, there exists a neighbourhood $U \ni 0$ and a one-to-one holomorphic mapping $z = z(\xi)$ of the polydisc $\Delta^p$: $|\xi_j| < 1$, $j = 1, \ldots, p$, onto the set $A \cap U$ such that the set $\operatorname{sng} A \cap U = z(\Delta^p \cap \{\xi_1 = 0\})$ and the restriction $\Delta^p \setminus \{\xi_1 = 0\} \to \operatorname{reg} A \cap U$ is biholomorphic. If the local coordinates are chosen such that $\operatorname{sng} A \cap U = U \cap \{z_1 = z_{p+1} = \ldots = z_n = 0\}$ and $C_4(A, 0) \cap \{z_1 = \ldots = z_p = 0\} = 0$, then one such parametrization has the form*

$$z(\xi) = (r\xi_1^k, r\xi_2, \ldots, r\xi_p, z_{p+1}(\xi), \ldots, z_n(\xi)),$$

*where* $r > 0$ *is constant, and* $k$ *is the multiplicity of the covering* $A$ *over* $\mathbb{C}^p_{z_1,\ldots,z_p}$ *in the neighbourhood of* 0, *and* $z_j(\xi) = \xi_1^k \varphi_j(\xi)$, *where* $\varphi_j$ *are holomorphic in the neighbourhood of* 0 *in* $\mathbb{C}^p$.

This parametrization can be seen as a multi-dimensional generalization of the Puiseux representation and as a first step towards the "normalization" of an arbitrary analytic set. Since the set $\{z \in A: \dim C_4(A, z) > p\}$ has dimension $\leq p-2$ (see [49]), then after such "unwindings" and "ungluings" at reducible points, we obtain a set with singularities of codimension $\geq 2$; it turns out that this can be done while preserving analyticity (see [14] and [53]).

**2.3. Multiplicity of Holomorphic Mappings.** Let $A$ be a purely $p$-dimensional analytic set in a complex manifold and $f: A \to Y$ a holomorphic mapping into another complex manifold. Suppose $a \in A$ is an isolated point of the fibre $f^{-1}(f(a))$ and that $\dim Y = p$. Then, clearly, there exist neighbourhoods $U \ni a$ and $V \ni f(a)$ such that $f|A \cap U \to V$ is a proper mapping and $f^{-1}(f(a)) \cap A \cap \bar{U} = a$. According to §1.5, $f|A \cap U$ is an analytic covering over $V$. Its multiplicity (number of sheets) is independent of the choice of $U$ and $V$ having the given properties; it is called the *multiplicity of the mapping* $f$ at the point $a$ and is denoted by $\mu_a(f|A)$ or simply $\mu_a(f)$. For each $w \in V \backslash \sigma$, where $\sigma$ is the critical set of the indicated covering, the number of preimages in $A \cap U$ is precisely $\mu_a(f)$. Each of these points are regular on $A$ and in the neighbourhood of such a point, the mapping $f$ is locally biholomorphic. For $w \in \sigma$, the number of preimages on $A \cap U$ is strictly less than $\mu_a(f)$. Since $\sigma$ is nowhere dense in $V$,

$$\mu_a(f) = \overline{\lim_{w \to f(a)}} \# f^{-1}(w) \cap A \cap U.$$

This equation may be taken as the definition of the multiplicity, even if $\dim Y > p$; in this case, the only requirement on $U$ is that $f^{-1}(f(a)) \cap A \cap \bar{U} = a$.

Each mapping reduces to the projection of its own graph, and hence, the local properties of multiplicity can be studied in the standard situation of the analytic covering, namely, projection $\pi|A \cap U \to U'$ in $\mathbb{C}^n$.

The following properties of multiplicity follow immediately from the definition.

1) The additivity of multiplicity: If $A = \bigcup_1^k A_j$, where each $A_j$ is purely $p$-dimensional and $\dim A_i \cap A_j < p$ for $i \neq j$, then $\mu_a(f|A) = \sum_1^k \mu_a(f|A_j)$.

2) Multiplicativity of multiplicity: If $A \xrightarrow{f} Y \xrightarrow{g} Z$ are holomorphic mappings with discrete fibres and $f$ is proper, then

$$\mu_a(g \circ f|A) = \mu_a(f|A) \cdot \mu_{f(a)}(g|f(A)).$$

For the calculation of the multiplicity the following generalization of Rouché's Theorem is very convenient (see, for example, [1]).

**Theorem 1.** *Let $D$ be a domain with compact closure in an $n$-dimensional complex manifold. Let $f$ and $g$ be holomorphic mappings $D \to \mathbb{C}^n$ continuous up to the boundary and such that $|f - g| < |f|$ on $\partial D$. Then,*

$$\sum_{f(a)=0} \mu_a(f|D) = \sum_{g(b)=0} \mu_b(g|D).$$

The calculation of the multiplicity of a mapping at a point begins with the study of the first non-zero terms of the Taylor expansion and the use of Rouché's Theorem for the remaining terms. In this direction, the following theorem of A.K. Tsikh and A.P. Yuzhakov is very useful (see [1]).

**Theorem 2.** *Let $f = (f_1, \ldots, f_n)$ be a holomorphic mapping in a neighbourhood of $0 \in \mathbb{C}^n$ such that $f(0) = 0$ and $f^{-1}(0) = 0$. Let $f_j^*$ be the initial homogeneous polynomial for $f_j$ at 0 and $f^* = (f_1^*, \ldots, f_n^*)$. If $(f^*)^{-1}(0) = 0$, then*

$$\mu_0(f) = \prod_1^n \operatorname{ord}_0 f_j = \prod_1^n \deg f_j^*.$$

*In the opposite situation, $\mu_0(f) > \prod_1^n \operatorname{ord}_0 f_j$.*

In place of homogeneous polynomials here, one may take quasi-homogeneous ones (assigning to each variable its own weight). The proof easily reduces to the homogeneous case (see, for example, [1]).

**2.4. Multiplicity of an Analytic Set at a Point.** Let $A$ be a purely $p$-dimensional analytic set in $\mathbb{C}^n$ and $a \in A$. For each complex $(n - p)$-plane $L \ni 0$ such that $a$ is an isolated point of $A \cap (a + L)$, the multiplicity $\mu_a(\pi_L|A)$ of the orthogonal projection to the plane $L^\perp$ is defined (§2.3). The minimum of these integers over all $(n - p)$-planes $L$ (i.e. elements of the Grassmannian $G(n - p, n)$ is called the *multiplicity of $A$ at the point $a$* and is denoted by $\mu_a(A)$.

At regular points, the multiplicity of an analytic set is clearly 1. The converse is also true: if $\mu_a(A) = 1$, then $a \in \operatorname{reg} A$. This follows easily from the local representation of $A$ as a covering with a finite number $\mu_a(A)$ of sheets (see §1.5).

The following property of biholomorphic invariance allows one to define the multiplicity of an analytic set on an arbitrary complex manifold (with the help of local coordinates).

**Proposition 1.** *Let $A$ be a purely $p$-dimensional analytic subset of a domain $D$ in $\mathbb{C}^n$ and $\varphi: D \to G \subset \mathbb{C}^n$ a biholomorphic mapping. Then,*

$$\mu_a(A) = \mu_{\varphi(a)}(\varphi(A)),$$

*for each $a \in A$.*

The proof can be obtained, for example, from Rouché's Theorem and the fact that $\varphi$ is homotopic to the identity mapping $z \mapsto z$.

The tangent cone plays an important role in the study of the multiplicity of analytic sets.

**Theorem 1.** *Let $A$ be a purely p-dimensional analytic set in the neighbourhood of 0 in $\mathbb{C}^n$, $0 \in A$, and $L \in G(n-p, n)$. The equality $\mu_0(\pi_L|A) = \mu_0(A)$ holds if and only if the subspace $L$ is transversal to $A$ at 0, i.e. $L \cap C(A, 0) = 0$.*

The proof is fairly simple in case $p = n - 1$. Indeed, if $f$ is a minimal defining function of such an analytic set in the neighbourhood of 0 (see §1.3), then $\mu_0(\pi_L|A) = \mathrm{ord}_0 f|L$, and the tangent cone is precisely the set of zeros of the initial homogeneous polynomial of the function $f$ (see §2.1). In the general case, the proof reduces to the hypersurface case by suitable projections in $\mathbb{C}^{p+1}$ and is technically rather bulky (see [10]).

The function $\mu_z(A)$, $z \in A$, is clearly upper semi-continuous. As mentioned already, $\{z: \mu_z(A) \geq 2\} = \mathrm{sng}\, A$ is an analytic set. This is also true for other level sets of $\mu_z(A)$ (Whitney [53]):

**Theorem 2.** *Let $A$ be a purely p-dimensional analytic set in $\mathbb{C}^n$ and $A^{(m)} = \{z \in A: \mu_z(A) \geq m\}$. Then, $A^{(m)}$, $m = 1, 2, \ldots$, are analytic subsets in $A$.*

A simple proof is obtained with the help of Theorem 1.

This theorem allows us to introduce the notion of multiplicity along a subset. If $A'$ is an irreducible analytic set and $A' \subset A$, then by Theorem 2, $\mu_z(A) = m$ for each $z \in A'$ with the possible exception of an analytic set $A'' \subset A'$ of lower dimension (on which $\mu_z(A) > m$). This generic value $m$ is naturally called the multiplicity of $A$ along $A'$.

**2.5. Indices of Intersection. Complementary Codimensions.** Let $A_1, \ldots, A_k$ be analytic subsets of a domain $D \subset \mathbb{C}^n$ of pure dimensions $p_1, \ldots, p_k$ respectively such that the sum of codimensions of the $A_j$'s is $n$ and $\bigcap_1^k A_j$ is zero-dimensional. Then the set $A_1 \times \ldots \times A_k = \Pi A_j$ in $C^{kn} = C^n_{z^1} \times \ldots \times \mathbb{C}^n_{z^k}$ has dimension $\Sigma p_j = (k-1)n$ and the diagonal $\Delta$: $z^1 = \ldots = z^k$ has complementary dimension $n$. If $a$ is an isolated point of $\bigcap A_j$ then $(a)^k = (a, \ldots, a) \in \mathbb{C}^{kn}$ is an isolated point of the set $(\Pi A_j) \cap \Delta$, and consequently, the multiplicity of the projection $\pi_\Delta|\Pi A_j$ to the space $\Delta^\perp$ is defined at this point. This number is called the *intersection index* of the sets $A_1, \ldots, A_k$ at the point $a$ and is denoted by $i_a(A_1, \ldots, A_k)$. Thus

$$i_a(A_1, \ldots, A_k) = \mu_{(a)^k}(\pi_\Delta|\Pi A_j).$$

For $a \notin \bigcap A_j$, we set $i_a(A_1, \ldots, A_k) = 0$. This definition, at first glance, seems somewhat artificial, however, technically, it is very useful for studying the multiplicity properties of projections (§2.3). The geometric meaning of the index of intersection (justifying such a definition) is clarified by the following assertion whose proof follows easily from the definition.

**Proposition 1.** *Let $A_1, \ldots, A_k$ be analytic subsets of $\mathbb{C}^n$ of pure complementary codimensions and a an isolated point of $\bigcap_1^k A_j$. Then there exists a neighbourhood $U \ni a$ and a number $\varepsilon > 0$ such that for almost all $c = (c^1, \ldots, c^k) \in \mathbb{C}^{kn}$ with $|c^j| < \varepsilon$, the number of points in the set $\bigcap (c^j + A_j) \cap U$ is equal to $i_a(A_1, \ldots, A_k)$. For such c, all points of the set $\bigcap (c^j + A_j) \cap U$ are regular on each $c^j + A_j$ and the tangent spaces to $c^j + A_j$ at these points intersect transversally. The exceptional values of c (for which these properties do not hold) form a proper analytic set in the domain $\{|c^j| < \varepsilon, j = 1, \ldots, k\}$, and for such c, the number of point in $\bigcap (c^j + A_j) \cap U$ is strictly less than $i_a(A_1, \ldots, A_k)$.*

The intersection index can also be defined for *holomorphic chains*, i.e. formal locally finite sums $\Sigma m_j A_j$, where $m_j$ are integers and $A_j$ are irreducible pairwise distinct analytic subsets of some fixed domain $D$. A chain $T$ is called a $p$-chain if each $A_j$ is purely $p$-dimensional and positive if each coefficient $m_j \geq 0$. The support of a chain $T$ is the set

$$\bigcup_{m_j \neq 0} A_j = |T|.$$

The intersection index of chains $T_j = \Sigma m_{jl} A_{jl}, 1 \leq j \leq k$, with complementary pure codimensions (of their supports) at a point $a \in \bigcap |T_j|$ is defined algebraically:

$$i_a(T_1, \ldots, T_k) = \Sigma m_{1l_1} \ldots m_{kl_k} \cdot i_a(A_{1l_1}, \ldots, A_{kl_k}).$$

The number

$$\sum_{a \in \cap |T_j|} i_a(T_1, \ldots, T_k) \qquad (\text{for } \# \bigcap |T_j| < \infty)$$

is called the *complete index of intersection* of the chains $T_1, \ldots, T_k$ in the domain $D$ and is denoted by $i_D(T_1, \ldots, T_k)$.

From the definitions and the multiplicity properties of projections, it follows easily that intersection index for analytic sets and holomorphic chains is invariant with respect to biholomorphic transformations; the complete intersection index is invariant under "small motions" of the chains (analogue of Rouché Theorem); etc.

**2.6. Indices of Intersection. General Case.** Let $T_1, \ldots, T_k$ be holomorphic chains in a domain $D \subset \mathbb{C}^n$ with supports $|T_j|$ of pure dimension $p_j$ respectively. We shall say that these chains *intersect properly* at a point $a \in D$ if $\dim_a \cap |T_j| = (\Sigma p_j) - (k-1)n$ (the smallest possible). If this property is satisfied at all points $a \in \cap |T_j|$, then $T_1, \ldots, T_k$ intersect properly in the domain $D$. For an affine plane $L \ni a$ of dimension $n - \dim_a \cap |T_j|$ and transversal to $A = \cap |T_j|$ (i.e. $(L - a) \cap C(A, a) = 0$), the number $i_a(T_1, \ldots, T_k, L)$ is defined in §2.5, and it is not difficult to show that it is independent of the choice of such an $L$ (see [11]). This number is called the index of intersection of the chains

$T_1, \ldots, T_k$ at the point $a$ and is denoted by $i_a(T_1, \ldots, T_k)$. As above, this notion is biholomorphically invariant and hence carries over to arbitrary complex manifolds. From the definition it follows easily that the number $i_z(T_1, \ldots, T_k)$ is locally constant on $\operatorname{reg} A$, and hence, for each irreducible component $S \subset A = \bigcap |T_j|$, the number is constant on $S \cap \operatorname{reg} A$. It is called the *intersection index* of $T_1, \ldots, T_k$ on $S$ and is denoted by $i_S(T_1, \ldots, T_k)$. If $A = \bigcup S_\nu$ is the decomposition into irreducible components, then (in the case of a proper intersection of the $T_j$'s in $D$) we define a holomorphic chain

$$T_1 \wedge \ldots \wedge T_k = \Sigma i_{S_\nu}(T_1, \ldots, T_k) \cdot S_\nu.$$

which is called the *intersection chain* of the chains $T_1, \ldots, T_k$ in $D$. The properties of the intersection chain (following from the definition and the properties of the index) are summarized in the following theorem (see [11]).

**Theorem 1.** *Let $T_1, \ldots, T_k$ be holomorphic chains of pure dimension and intersecting properly in $D$. Then the holomorphic chain $T_1 \wedge \ldots \wedge T_k$ is defined in $D$, and has dimension $(\Sigma \dim |T_j|) - (k-1)n$ and support in $\bigcap |T_j|$. Its multiplicity on each irreducible component $S$ of the set $\bigcap |T_j|$ is $i_S(T_1, \ldots, T_k)$. At each point $a \in \bigcap |T_j|$, the multiplicity of the chain $T_1 \wedge \ldots \wedge T_k$ is $i_a(T_1, \ldots, T_k)$. The operation $(T_1, \ldots, T_k) \mapsto T_1 \wedge \ldots \wedge T_k$ (under the condition of proper intersection) is multilinear over the ring $\mathbb{Z}$, commutative, associative, invariant with respect to biholomorphic mappings of the domain $D$, and continuous.*

The multiplicity of the chain $\Sigma m_j A_j$ at the point $a$ is by definition, $\Sigma m_j \mu_a(A_j)$. We shall say that $T_1, \ldots, T_k$ *intersect transversally* at $a$, if $\dim_0 \bigcap_1^k C(|T_j|, a) = \left(\sum_1^k \dim |T_j|\right) - (k-1)n$. The importance of this concept is seen, for example, by the following.

**Proposition 1.** *Let $T_1, \ldots, T_k$ be positive holomorphic chains of constant dimension in a domain $D \subset \mathbb{C}^n$ and intersecting properly at a point $a \in \bigcap |T_j|$. Then,*

$$i_a(T_1, \ldots, T_k) = \mu_a(T_1 \wedge \ldots \wedge T_k) \geq \mu_a(T_1) \ldots \mu_a(T_k),$$

*where equality holds if and only if these chains intersect transversally at the point $a$.*

In the final analysis, this follows from Theorem 1 of §2.4.

**2.7. Algebraic Sets.** An affine algebraic set is the set of common zeros in $\mathbb{C}^n$ of a finite system of polynomials. A projective algebraic set is the set of common zeros in $\mathbb{P}_n$ of a finite system of homogeneous polynomials of homogeneous coordinates. The closure in $\mathbb{P}_n \supset \mathbb{C}^n$ of an arbitrary affine algebraic set is a projective algebraic set. This follows, for example, from the results of §1.5. The famous theorem of Chow (see, for example, [19] or [28]) asserts that each analytic subset of $\mathbb{P}_n$ is a projective algebraic set. For an analytic subset of $\mathbb{C}^n$, there are many different criteria, in terms of the behaviour at infinity, in order for it to be algebraic. Here are several of them.

**Proposition 1.** *A purely p-dimensional analytic subset $A \subset \mathbb{C}^n$ is algebraic if and only if its closure in $\mathbb{P}_n \supset \mathbb{C}^n$ does not meet some complex $(n-p-1)$-plane in $\mathbb{P}_n \setminus \mathbb{C}^n$.*

The proof is, for example, in [17]. In terms of $\mathbb{C}^n$, this criterion can be formulated as follows:

**Corollary.** *A is algebraic if and only if after an appropriate unitary change of coordinates, A is contained in the union of some sphere $|z| < R$ and some cone $|z''| < C|z'|$, where $z = (z', z'')$, $z' \in \mathbb{C}^p$.*

The following criterion of Rudin [36] is analogous in form to the theorem of Liouville.

**Proposition 2.** *A purely p-dimensional analytic subset $A \subset \mathbb{C}^n$ is algebraic if and only if, after a suitable unitary change of coordinates, A belongs to a domain $|z''| < C(1 + |z'+|)^s$, for some constants C and $s(z = (z', z''), z' \in \mathbb{C}^p)$.*

The proof is easily obtained by establishing that the canonical defining functions of an analytic covering $A \to \mathbb{C}^p$ are polynomials.

If $A$ is a purely $p$-dimensional analytic subset of $\mathbb{P}_n$ and $\pi: \mathbb{C}^{n+1} \setminus \{0\} \to \mathbb{P}_n$ is the canonical mapping, then $\tilde{A} = \pi^{-1}(A) \cup \{0\}$ is a purely $(p+1)$-dimensional analytic subset of $\mathbb{C}^{n+1}$. Its multiplicity at 0, as is easily understood, is precisely $\deg A$, the *degree of the algebraic set* $A$, i.e., the maximal number of points of proper intersection of $A$ with complex $(n-p)$-planes in $\mathbb{P}_n$. Such an equivalence allows us easily to obtain various properties of the degree of an algebraic set.

From Rouché's Theorem (§2.3), one easily obtains the well-known Theorem of Bezout on the number of solutions of polynomial equations (see [1], [28]) which states that if $p_1, \ldots, p_n$ are homogeneous polynomials in $\mathbb{P}_n$ having only finitely many common zeros, then this number, taking multiplicities into account, is equal to the product of the degrees of the polynomials. The multiplicity $\mu_a(p)$ at a point $a = [a_0, \ldots, a_n]$ for which, say, $a_0 \neq 0$, is defined as the multiplicity of the mapping $p(1, z_1, \ldots, z_n)$ at the point $(a_1/a_0, \ldots, a_n/a_0)$ of $\mathbb{C}^n$. From §2.5 we obtain the following geometric analogue of this theorem:

**Proposition 3.** *Let $T_1, \ldots, T_k$ be positive holomorphic chains of complementary pure codimensions in $\mathbb{P}_n$ with algebraic supports and zero-dimensional intersection. Then, the complete index of intersection of the chains $T_1, \ldots, T_k$ in $\mathbb{P}_n$ is equal to the product of their degrees, where* $\deg \Sigma m_j A_j := \Sigma m_j \deg A_j$.

This is a particular case of the following theorem which is a consequence of §2.6.

**Proposition 4.** *Let $T_1, \dots, T_k$ be positive chains with algebraic supports and proper intersection in $\mathbb{P}_n$. Then,*

$$\deg(T_1 \wedge \dots \wedge T_k) = \Pi_1^k \deg T_j.$$

**Remarks.** Tangent cones and Whitney cones are well-studied in the works of Whitney [51], [52], and [53]. For the multiplicity of mappings of complex manifolds, see, for example, [1], [29], and [47]. For the multiplicity of analytic sets, see the book of Whitney [53]. The intersection theory for analytic sets is exposed in the paper of Draper [11], containing proofs for most of the properties of intersections introduced above. Additional proofs of results in this section appear also in the author's book [10].

## §3. Metric Properties of Analytic Sets

**3.1. Fundamental Form and Theorem of Wirtinger.** Let $\Omega$ be a Hermitian complex manifold, i.e. in each fibre $T_\zeta\Omega$ of the tangent bundle to $\Omega$, there is given a positive definite quadratic Hermitian form $H_\zeta(v, w)$ of class $C^\infty$ with respect to $\zeta$. In a local coordinate $z$, it is represented in the form $\Sigma h_{jk}\, dz_j \otimes d\bar{z}_k$, i.e. $H_\zeta(v, w) = \Sigma h_{jk} v_j \bar{w}_k$ ($T_\zeta\Omega$ has the standard complex structure). The form $S = \operatorname{Re} H$ is symmetric ($S(v, w) = S(w, v)$), $\mathbb{R}$-bilinear, and also positive definite, i.e. it is a Riemannian metric (usually denoted by $ds^2$). The form $\omega = -\operatorname{Im} H$ is clearly skew-symmetric, i.e. it is a differential form on $\Omega$ of degree 2. It is called the *fundamental form* of the Hermitian manifold $(\Omega, H)$. In local coordinates, where $H = \Sigma h_{jk}\, dz_j \otimes d\bar{z}_k$, the form $\omega$ is given by $(i/2) \Sigma h_{jk}\, dz_j \wedge d\bar{z}_k$ (this easily follows from the definition).

In the space $\mathbb{C}^n$ with the natural Hermitian structure $H = \sum_1^n dz_j \otimes d\bar{z}_j$, the fundamental form is $\omega = (i/2)\Sigma\, dz_j \wedge d\bar{z}_j = (i/2)\partial\bar{\partial}|z|^2 = (1/4)dd^c|z|^2$, where $d = \partial + \bar{\partial}$ and $d^c = i(\bar{\partial} - \partial)$. If the local coordinate system $z$ on $\Omega$ is orthonormal with respect to $H$ at the point $a$, then $(\omega)_a = (i/2)\Sigma(dz_j \wedge d\bar{z}_j)_a$, i.e. $\omega$ is locally Euclidean in "infinitely small neighbourhoods of $a$".

The form $\tilde{\omega}_0 = (1/4)\, dd^c \log|z|^2$ in $\mathbb{C}^{n+1} \setminus \{0\}$ is invariant under dilatations Thus, there is a corresponding form $\omega_0$ in $\mathbb{P}_n$ such that $\tilde{\omega}_0$ is the pull-back of $\omega_0$ by the canonical mapping. In a coordinate neighbourhood $U_j$: $z_j \neq 0$ in $\mathbb{P}_n$, we have the representation

$$\omega_0 = \frac{1}{4}\, dd^c \log\left(1 + \sum_{k \neq j} |z_k/z_j|^2\right).$$

The form $\omega_0$ is called the *Fubini-Study form* on $\mathbb{P}_n$. To $\omega_0$, these corresponds a unique Hermitian structure $H_0$ on $\mathbb{P}_n$ such that $\omega_0 = -\operatorname{Im} H_0$, and the

Riemannian metric $S_0 = \operatorname{Re} H_0$ ($S_0(v, w) = \omega_0(v, iw)$) is called the Fubini-Study metric on $\mathbb{P}_n$. In the sequel, both $\omega_0$ and $\tilde{\omega}_0$ will be denoted by $\omega_0$.

The Euclidean volume form in $\mathbb{C}^n$ is

$$\Pi dx_j \wedge dy_j = \Pi \frac{i}{2} dz_j \wedge d\bar{z}_j = \frac{1}{n!}\omega^n.$$

If $L$ is a linear subspace of $\mathbb{C}^n$ of dimension $p$, then, by a unitary transformation, $L$ is transformed to $\mathbb{C}^p = \mathbb{C}^p_{z_1, \ldots, z_p} \subset \mathbb{C}^n$. Since $\omega = (1/4) dd^c |z|^2$ is invariant by such transformations and the volume form on $\mathbb{C}^p$ is $(1/p!)\omega^p | \mathbb{C}^p$, then the Euclidean volume form on $L$ is $(1/p!)\omega^p | L$. From this, it follows easily that on any complex manifold $M$ of dimension $p$, imbedded in $\mathbb{C}^n$, the form $(1/p!)\omega^p | M$ is the volume form on $M$ with respect to the metric induced by $\mathbb{C}^n$. Since the fundamental form of an arbitrary Hermitian manifold is Euclidean in appropriate local coordinates (at a given point), we have the following.

**Theorem of Wirtinger.** *Let $\Omega$ be a complex Hermitian manifold with fundamental form $\omega$. Then for any p-dimensional complex manifold $M$, imbedded in $\Omega$, the restriction $(1/p!)\omega^p | M$ coincides with the volume form on $M$ in the induced metric. In particular,*

$$\operatorname{vol}_{2p} M = \frac{1}{p!} \int_M \omega^p.$$

From finer linear properties (the Wirtinger inequality, see [13] and [20]), there follows the property of minimality: if $M$ is an arbitrary oriented $2p$-dimensional manifold of class $C^1$, imbedded in $\Omega$, then

$$\frac{1}{p!} \int_M \omega^p \leq \operatorname{vol}_{2p} M.$$

**3.2. Integration on Analytic Sets.** From Wirtinger's Theorem, Lemma 2 of § 1.4, and the analyticity of sng $A$, we obtain the following theorem of Lelong (see [25]).

**Theorem 1.** *Each purely p-dimensional analytic subset $A$ of a complex manifold $\Omega$ has locally-finite Hausdorff 2p-measure, i.e., $\mathscr{H}_{2p}(A \cap K) < \infty$, for any compact set $K \subset \Omega$.*

In $\mathbb{C}^n$, we have the formula

$$\frac{1}{p!}\omega^p = \sum_{\#I = p} dV_I, \qquad \text{where } dV_I = \prod_1^p dx_{i_\nu} \wedge dy_{i_\nu}$$

is the volume on a coordinate $p$-plane in the variables $z_{i_1}, \ldots, z_{i_p}$ (the sum is over ordered multi-indices $I$). Hence, Wirtinger's Theorem can be interpreted as follows: the volume of a $p$-dimensional analytic set $A \subset \mathbb{C}^n$ (i.e., $\mathscr{H}_{2p}(A)$) is equal to the sum of its projections on $p$-dimensional coordinate planes counting multiplicities of these projections (the volume of $A$ and of reg $A$ coincide).

If $A$ is a purely $p$-dimensional analytic subset of a complex manifold $\Omega$, then by Lelong's Theorem, we can define a *current* $[A]$ of integration on the manifold reg $A$, a current of measure type and dimension $2p$:

$$\langle [A], \varphi \rangle = \int_{\operatorname{reg} A} \varphi,$$

$\varphi \in \mathscr{D}^{2p}(\Omega)$, where the symbol $\mathscr{D}^s(\Omega)$ denotes the space of infinitely differentiable forms of degree $s$ and of compact support in $\Omega$ (concerning currents, see, for example, [13], [20]). In the sequel, we write

$$\int_A \equiv \int_{\operatorname{reg} A} .$$

We introduce some characteristic properties of integrals on analytic sets.

**Proposition.** *A current of integration $[A]$ on a purely $p$-dimensional analytic subset of a complex manifold $\Omega$ has bidimension $(p, p)$, i.e.*

$$\int_A \varphi = 0$$

*for any form $\varphi \in \mathscr{D}^{2p}(\Omega)$ of bidegree $(r, s) \neq (p, p)$. Such a current is positive. That is,*

$$\int_A \varphi \geq 0,$$

*for each positive $\varphi \in \mathscr{D}^{2p}(\Omega)$. Also, $[A]$ is closed, i.e.,*

$$\int_A d\varphi = 0,$$

*for each $\varphi \in \mathscr{D}^{2p-1}(\Omega)$.*

A form $\varphi$ is called positive if at each point it is represented as a linear combination, with non-negative coefficients, of forms

$$\prod_1^p \frac{i}{2} dl_j \wedge d\bar{l}_j,$$

where each $l_j$ is a linear function of the local coordinates; for example, the fundamental form and its powers.

The last property of $[A]$ can be strengthened. We shall say that $A$ is an *analytic set with border bA*, if $A$ is purely $p$-dimensional, $\bar{A}$ lies in some other purely $p$-dimensional analytic set $\tilde{A}$, $bA = \bar{A} \setminus A$ has locally finite $(2p-1)$-dimensional Hausdorff measure, and there exists a closed set $\Sigma \subset \bar{A}$ such that $\mathscr{H}_{2p-1}(\Sigma) = 0$ and $\bar{A} \setminus \Sigma$ is a $C^1$-submanifold with border $(bA) \setminus \Sigma$ in $\Omega \setminus \Sigma$.

**Theorem 2** (Stokes formula). *Let $A$ be a purely $p$-dimensional analytic set with border on a complex manifold $\Omega$. Then, for any $(2p-1)$-form of class $C^1(\Omega)$ and of* *compact support, we have*

$$\int_A d\varphi = \int_{bA} \varphi,$$

*where the integral on the right is also to be understood as an improper integral along* $(bA)\setminus\Sigma$.

In terms of currents, Stokes' formula has the form $\langle [A], d\varphi \rangle = \langle [bA], \varphi \rangle$ or, more briefly, $d[A] = -[bA]$. We remark that if $A$ is a purely $p$-dimensional analytic subset of $\Omega$ and $\rho$ is a real function of class $C^{2p}(\Omega)$, then, by Sard's theorem (see [26]), almost every set $A_t = A \cap \{\rho < t\}$ is an analytic set with border $bA_t = A \cap \{\rho = t\}$.

One of the frequently applied equivalent forms of Stokes' theorem is the following complex analogue of Green's formula.

**Corollary.** *For any functions $u$ and $v$ of class $C^2(\Omega)$ and any closed form $\varphi$ of bidegree $(p-1, p-1)$, we have the following equality*

$$\int_{bA} (ud^c v - vd^c u) \wedge \varphi = \int_A (udd^c v - vdd^c u) \wedge \varphi.$$

**3.3. Crofton Formulas.** The volume of an arbitrary purely $p$-dimensional analytic set $A$ in complex projective space $\mathbb{P}_n$ (with respect to the Fubini-Study metric) can be calculated in terms of the volumes of its traces along complex $k$-dimensional planes, $k \geq n - p$, via the well-known "Crofton formula"

$$\mathcal{H}_{2p}(A) = c(p, k, n) \int_{\tilde{G}(k,n)} \mathcal{H}_{2(p+k-n)}(A \cap L)\, dL,$$

where $\tilde{G}(k, n)$ is the Grassmannian or all $k$-dimensional complex planes in $\mathbb{P}_n$, $dL$ is the standard normalized volume form in $\tilde{G}(k, n)$, invariant with respect to unitary transformations, and $c(p, k, n) = \pi^{n-k}(p + k - n)!/p!$ (see [3], [15], and [38]). Essentially, this is Fubini's theorem on the incidence manifold, more precisely, on the manifold $\{(z, L) \in \mathbb{P}_n \times \tilde{G}(k, n) : z \in (\operatorname{reg} A) \cap L\}$.

If $A$ is a purely $p$-dimensional analytic set in $\mathbb{C}^n$, $0 \notin A$, and the rank of the restriction of the canonical mapping $\pi\colon \mathbb{C}^n \setminus \{0\} \to \mathbb{P}_{n-1}$ to $\operatorname{reg} A$ is almost everywhere equal to $p$, then from the formula in $\mathbb{P}_n$, we have

$$\int_A \omega_0^p = \pi^{n-k} \int_{G(k,n)} \left( \int_{A \cap L} \omega_0^{p+k-n} \right) dL,$$

where $\omega_0 = (1/4) dd^c \log |z|^2$ and $G(k, n)$ is the Grassmannian of all $k$-dimensional subspaces of $\mathbb{C}^n$ $(L \ni 0)$. The transition from "projective" volume to Euclidean volume is further realized with the help of the relation

$$\omega_0^p | T_z A = \frac{\sin^2(z, T_z A)}{|z|^{2p}} \omega^p | T_z A, \qquad z \in \operatorname{reg} A,$$

which is easily verified in appropriate local coordinates. For an arbitrary analytic set $A \subset \mathbb{C}^n$, we obtain, in this way, a rather unwidely formula, which in case $A$ is a subset of the ball, simplifies in the following way as a consequence of Stokes' formula (§3.2).

**Proposition 1.** *Let $A$ be a purely $p$-dimensional analytic subset of the ball $|z| < R$ and $0 \notin A$. Then*

$$\int_A \omega^p = R^{2p} \int_A \omega_0^p.$$

From this and from the formula introduced earlier with $\omega_0$, we obtain the following theorem of Alexander [3]:

**Proposition 2.** *Let $A$ be a purely $p$-dimensional analytic subset of the ball $|z| < R$ in $\mathbb{C}^n$ and $0 \notin A$. Then, for any $k \geq n - p$, we have the equality*

$$\mathscr{H}_{2p}(A) = c(p, k, n) R^{2(n-k)} \int_{G(k,n)} \mathscr{H}_{2(p+k-n)}(A \cap L)\, dL.$$

**3.4. The Lelong Number.** Let $A$ be a purely $p$-dimensional analytic set in $\mathbb{C}^n$ containing 0 and let $A_r = A \cap \{|z| < r\}$. From Stokes' Theorem, one easily obtains

$$\frac{1}{R^{2p}} \int_{A_R} \omega^p - \frac{1}{r^{2p}} \int_{A_r} \omega^p = \int_{A_R \setminus A_r} \omega_0^p.$$

Since the form $\omega_0^p$ is positive, it follows that the function $\mathscr{H}_{2p}(A_r)/(c(p)r^{2p})$ decreases as $r \to 0$, and hence, has a limit which we denote by $n(A, 0)$ and which is called the *Lelong number of the set A* at 0. Here $c(p) = \pi^p/p!$ is the volume of the unit ball in $\mathbb{C}^p$. At an arbitrary point $a \in A$, the Lelong number $n(A, a)$ is defined analogously. Thus,

$$\mathscr{H}_{2p}(A_r) = n(A, 0)c(p)r^{2p} + r^{2p} \int_{A_r} \omega_0^p/p!.$$

If $A$ is a homogeneous algebraic set, then $\omega_0^p|A = 0$. Since $\omega^{p-1}|bA_1 = \omega_0^{p-1}|bA_1$, we have

$$\int_{A_1} \omega^p = \pi \int_{A^*} \omega_0^{p-1},$$

where $A^*$ is the corresponding algebraic subset of $\mathbb{P}_{n-1}$. The volume of such an $A^*$ in the Fubini-Study metric is equal to $\pi^{p-1} \cdot \deg A^*$. This follows from the fact that $A^*$ can homotopically be deformed into a $k$-fold plane of dimension $p - 1$, where $k = \deg A^*$, and from the fact that the integral of $\omega_0^{p-1}$ is invariant under such a homotopy since $\omega_0$ is closed (for more details see, for example, [28]). Thus, for cones, the Lelong number $n(A, 0)$ coincides with the degree of $A^*$, i.e. the multiplicity $\mu_0(A)$. From the homotopy of $A$ to the tangent cone (with the covering multiplicity) and with the help of the dilation $z \mapsto z/r$, we obtain analogously (see [11] and [20]):

**Proposition 1.** *The Lelong number of a purely p-dimensional analytic set $A \subset \mathbb{C}^n$ at an arbitrary point $a \in A$ agrees with the multiplicity of $A$ at this point and, in particular, is an integer.*

Thus, for analytic sets, this concept yields nothing new, however, it is more flexible than the notion of multiplicity and easily extends to essentially more general objects (see, for example, [20]):

**Proposition 2.** *Let $T$ be a positive d-closed current of bidimension $(p, p)$ in a domain $D \subset \mathbb{C}^n$ and let $n(T, a, r) = (\pi r^2)^{-p} \langle T, \chi_{a,r}\omega^p \rangle$, where $\chi_{a,r}$ is the characteristic function of the ball $|z - a| < r$. Then $n(T, a, r)$ decreases monotonically as $r \to 0$ at each point $a \in D$ and there exists*

$$\lim_{r \to 0} n(T, a, r) =: n(T, a).$$

**3.5. Lower Estimates for the Volume of Analytic Sets.** From the positivity of the form $\omega_0$ and from the previous section we easily obtain the following well-known estimate (Rutishauser, Lelong (1950), see [6]).

**Proposition.** *Let $A$ be a purely p-dimensional analytic subset of the ball $|z| < r$ in $\mathbb{C}^n$, containing 0. Then, $\mathcal{H}_{2p}(A) \geq n(A, 0) \cdot c(p) r^{2p}$.*

Beautiful applications of this estimate to the question of removing singularities of analytic sets were given by Bishop [6]. It is also interesting to obtain analogous lower estimates for arbitrary convex domains. Instead of $c(p)r^{2p}$, it is necessary, of course, to take the minimal volume of traces on complex $p$-planes. Such estimates are valid for cubes in $\mathbb{C}^2$, for certain tubular domains, etc. However, not every convex symmetric domain has this property. For example, in the domain $D$: $|x| < 1$, $|y| < R$ in $\mathbb{C}^3 (z = x + iy)$, the analytic set $D \cap \{z_3 = z_1^2 + z_2^2\}$ has 4-volume $< 20$, while any trace of $D$ on a complex 2-plane $L \ni 0$ has 4-volume $\geq R \cdot 4^{-4} > 20$ for large $R$.

By Wirtinger's Theorem, $\mathcal{H}_{2p}(A)$ is the sum of the projections of $A$ on respective coordinate planes counting multiplicities. Hence, the following estimate (Alexander–B. Taylor–Ullman [4]) is significantly stronger than the above one.

**Theorem.** *Let $A$ be a purely p-dimensional analytic subset of the ball $|z| < r$ in $\mathbb{C}^n$ and $0 \in A$. Then,*

$$\sum_{\# I = p}' \mathcal{H}_{2p}(\pi_I(A)) \geq c_p r^{2p},$$

*where $c_p > 0$ is a constant depending only on $p$ (for $p = 1$, we may take $c_1 = \pi$).*

Here $\Sigma'$ denotes the sum over ordered multi-indices $I$, i.e. $i_1 < \ldots < i_p$, and $\pi_I$ is the orthogonal projection of $\mathbb{C}^n$ onto the plane of the variables $z_{i_1}, \ldots, z_{i_p}$.

For $p = 1$ the theorem follows easily from the estimate

$$\mathscr{H}_2(f(\Delta)) \geq \frac{1}{2\pi}\int_0^{2\pi} |f(e^{i\theta})|^2\,d\theta$$

for a bounded holomorphic function $f$ in the unit disc $\Delta$ with $f(0) = 0$. One then obtains the case of arbitrary $p$ by Fubini's Theorem (with constants $c_p = \pi p^{-4p+1} c_{p-1}$ not best possible). See [4].

**3.6. Volumes of Tubes.** Let $M$ be a $p$-dimensional manifold in $\mathbb{C}^n$. The union of all balls $B(\xi, r) \cap N_\xi M$ (where $N$ denotes the affine normal space) for all $\xi \in M$ is called a tubular neighbourhood of radius $r$ around $M$ and is denoted by $\tau_r(M)$. If the affine normal spaces $N_\xi M \ni \xi$ at various points of $M$ do not intersect in some $r_0$-neighbourhood of $M (r_0 > 0)$, then for $r \leq r_0$, the tube $\tau_r(M)$ is naturally diffeomorphic to $M \times \{|t| < r\}, t \in \mathbb{C}^{n-p}$. Thus, from the formula for the change of variables in an integral, it follows that the volume of $\tau_r(M)$ is a polynomial in $r$ of the form

$$\sum_{n-p}^{n} c_j r^{2j}.$$

From the local approximation by tangent spaces, it is clear that the principal part of this polynomial (as $r \to 0$), just as for the plane, is equal to $(\operatorname{vol} M)\cdot c(n-p) r^{2(n-p)}$. The remaining coefficients can be expressed by the curvature of the manifold $M$ in the following way. Let $\tau^1, \ldots, \tau^p$ be holomorphic vector fields on $M$, forming a basis at each $T_\xi M$ (everything local). Let $H = (h_{jk})$, where $(h_{jk}) = (\tau^j, \tau^k)$, be Hermitian scalar products in $\mathbb{C}^n$. Then $\theta = (\partial H)\cdot H^{-1}$ is the canonical Euclidean connection matrix on $M$ (here $\partial H = (\partial h_{jk})$), and $\Omega_M = \bar{\partial}\theta_M$ is the curvature matrix (of the tangent bundle) of the manifold $M$; its elements are differential forms of bidegree (1, 1). The differential form $c(M) = \det(E + i(2\pi)^{-1}\Omega_M)$, where $E$ is the unit $(p \times p)$-matrix, is called the *total Chern class* of the manifold $M$. Expanding $c(M)$ according to degrees, we obtain $c(M) = 1 + c_1(M) + \ldots + c_p(M)$, where $c_q(M)$ is called the *q-th Chern class* of the manifold $M$. A rather tedious calculation of the volume of $\tau_r(M)$ according to the above scheme of H. Weyl (with essential use of the complex coordinate $\xi, t$) leads to the following formula of Griffiths [16]:

**Theorem 1.**

$$\operatorname{vol}\tau_r(M) = \sum_{m=n-p}^{n}\left(\int_M \frac{\omega^{n-m}}{(n-m)!} \wedge c_{m-n+p}(M)\right) c(m) r^{2m},$$

*for* $r \leq r_0$.

From the definitions and by calculating in local coordinates, it follows that the forms $(-1)^q c_q(M)$ are positive, and hence the signs in the Griffiths Formula are alternating.

If $M = \operatorname{reg} A$, where $A$ is an analytic set, then, in the vicinity of a singularity, the normal bundle has a very bad structure (it is not holomorphic, unlike the tangent bundle), and the Griffiths Formula already no longer has such a geometric meaning as for manifolds. From the local representation in terms of analytic covers, it is not difficult to deduce that the volume of an $r$-tube about $A(= \operatorname{vol} \tau_r(\operatorname{reg} A))$ is equal to

$$\mathscr{H}_{2p}(A) \cdot c(n-p) r^{2(n-p)}(1 + o(1)),$$

as $r \to 0$. That is, asymptotically, the singularity does not influence the main term of $\operatorname{vol} \tau_r(A)$.

**3.7. Integrability of Chern Classes.** The Chern forms $c_q(M)$ of a complex $p$-dimensional manifold $M$ (imbedded) in $\mathbb{C}^n$ are the preimages of the standard forms on the Grassmannian $G(p, n)$ under the *Gauss mapping* $\gamma_M: z \to T_z M$, $z \in M$. More precisely, over $G(p, n)$ there is a so-called universal bundle $U$ which is no other than the *incidence manifold* $\{(L, z): L \in G(p, n), z \in L \subset \mathbb{C}^n\}$. With the help of the local bases in the fibres of this bundle we define, just as before, the curvature matrix $\Omega_U$ and the complete Chern class $c = \det(E + i(2\pi)^{-1}\Omega_U) = 1 + c_1 + \ldots + c_p$. Comparing this with the already defined classes $c_q(M)$, it is not difficult to see that $c_q(M) = (\gamma_M)^*(c_q)$, $q = 0, \ldots, p$. The forms $c_q$ on $G(p, n)$ are really universal; they depend only on $p$, $n$, and $q$. Hence, the investigation of the forms $c_q(M)$ reduces to the study of the corresponding properties of the Gauss mapping. We introduce some examples.

**Lemma 1.** *Let $A$ be a purely $p$-dimensional analytic subset of a domain $D \subset \mathbb{C}^n$, and $A^*$ the closure of the graph $\{(z, T_z A): z \in \operatorname{reg} A\}$ of the Gauss mapping in $D \times G(p, n)$. Then $A^*$ is a purely $p$-dimensional analytic subset of $D \times G(p, n)$.*

From this and from the theorem of Lelong (§3.2), clearly follows

**Proposition 1.** *Let $A$ be a purely $p$-dimensional analytic set in $\mathbb{C}^n$ and let $c_q(A)$, $q = 0, \ldots, p$, be the Chern forms of the manifold* $\operatorname{reg} A$. *Then the forms $\omega^{p-q} \wedge c_q(A)$ are locally integrable on $A$ i.e.*

$$\left| \int_{K \cap \operatorname{reg} A} \omega^{p-q} \wedge c_q(A) \right| < \infty,$$

*for any compact $K \subset A$.*

Recall that the forms $(-1)^q \cdot c_q(A)$ are positive on $\operatorname{reg} A$. Using this fact, Stokes' Theorem, and the Lemma, we may more precisely estimate the local behaviour of such integrals.

**Proposition 2.** *Let $A$ be a purely p-dimensional analytic set in $\mathbb{C}^n$ and $0 \in A$. Then, the forms $\omega_0^{p-q} \wedge c_q(A)$ are locally integrable on $A$ and the functions*

$$\frac{(-1)^q}{r^{2(p-q)}} \int_{A_r} \omega^{p-q} \wedge c_q(A), \qquad q = 0, \ldots, p$$

*monotonically decrease as $r \to 0$.*

The limiting values of these functions, as $r \to 0$, have, till now, been practically uninvestigated except for the case $q = 0$, when we obtain the Lelong number.

**Remarks.** The results of the first four sections are developed in detail by Harvey in [20]. For lower estimates on volume, see Bishop [6] and [4]. Volumes of tubes and integrability properties of Chern classes are studied in detail in the paper by Griffiths [15].

## §4. Holomorphic Chains

**4.1. Characterization of Holomorphic Chains.** Formal holomorphic chains, which appear because of the necessity of considering various multiplicities in dealing with analytic sets (see §2), are most easily treated in terms of currents. In the sequel a holomorphic $p$-chain in $\Omega$ will mean a current in $\Omega$ of the form $\Sigma m_j[A_j]$, where $A_j$ are the irreducible components of a purely $p$-dimensional analytic subset $A = \cup A_j$ of $\Omega$, $[A_j]$ is the current of integration on $A_j$, and the multiplicity $m_j$ is an integer (see §3.2). Of particular importance for applications are the positive holomorphic chains (for which $m_j \geq 0$).

From the properties of integrals on analytic sets (see §3.2), it easily follows that a holomorphic $p$-chain $T$ is a current of bidimension $(p, p)$, is closed ($dT = 0$), and is of measure-type (i.e. extends to a functional on continuous $2p$-forms of compact support in $\Omega$). If $\Omega$ is a domain in $\mathbb{C}^n$, then at each point $a \in \Omega$, each such chain $T$ has a density which is an integer as in §3.4. The density of a current $T$ of measure-type at a point $a$ is defined as

$$\lim_{r \to 0} M_{B(a,r)}(T)/(c(p)r^{2p}),$$

where $B(a, r) = \{|z - a| \leq r\}$ and $M_K(T) = \sup\{|\chi_K T(\varphi)|: \|\varphi\| \leq 1\}$ is the mass of the current $T$ on the compact set $K$. For further details, see [10] and [20].

It turns out, that this property fully characterizes holomorphic chains in the space of currents. The following theorem is due to Harvey and Schiffman [20], [22].

**Theorem 1.** *Let $T$ be a closed current of bidimension $(p, p)$ and measure-type in a domain $D$ of $\mathbb{C}^n$ whose density exists and is integral at $\mathscr{H}_{2p}$-almost-every point of the support of $T$. Then $T$ is a holomorphic $p$-chain in $D$.*

The density condition can be replaced by the condition of local rectifiability of $T$ as in the original formulation of Harvey–Shiffman, however, rectifiability is perhaps a more complicated property. Each positive current is automatically of measure-type and moreover, for such currents, the Lelong number $n(T, a)$ (see §3.4) is defined at each point $a \in D$. The following theorem is due to Bombieri and Siu (see [43]).

**Theorem 2.** *Let $T$ be a positive closed current of bidimension $(p, p)$ in a domain $D$ in $\mathbb{C}^n$. Then, for any $c > 0$, the set $\{z \in D: n(T, z) \geq c\}$ is an analytic subset of $D$ of dimension $\leq p$.*

From this it follows, in particular, that each positive closed current of bidimension $(p, p)$, whose Lelong number is a natural number on an everywhere dense subset of the support, is a holomorphic $p$-chain.

The description of holomorphic chains in terms of currents is useful for applications, in particular, for problems on the behaviour of analytic sets under limiting processes and the removal of their singularities (see §5). In this way, one easily proves the following well known theorem of Bishop (see [6] and [20]).

**Theorem 3.** *Let $\{A_j\}$ be a sequence of purely $p$-dimensional analytic subsets of a complex manifold $\Omega$ having uniformly bounded volumes on each compact subset of $\Omega$, and suppose this sequence converges to a closed subset $A \subset \Omega$. Then, $A$ is also a purely $p$-dimensional analytic subset of $\Omega$.*

The convergence $A_j \to A$ here means that for each $z \in A$, there exists a sequence $z_j \in A_j$, such that $z_j \to z$ as $j \to \infty$, and $A_j \cap K$ is empty for any compact $K \subset \Omega \setminus A$ and all sufficiently large $j$. An analogous theorem is valid, of course, also for holomorphic chains (see [20]).

**4.2. Formulas of Poincaré–Lelong.** The zeros of a holomorphic function $f \not\equiv 0$ in a domain $\Omega$, counting multiplicities, define a positive holomorphic chain $D_f = \Sigma n_j [Z_j]$, where $Z_j$ are the irreducible components of the zero set $Z_f$, and $n_f$ is the multiplicity of the zeros of $f$ at points $z \in Z_j \cap \operatorname{reg} Z_f$ (it is constant for all such $z$). This chain is called the divisor of the function $f$ in $\Omega$. Analogously, one defines the *divisor of a meromorphic function*. In case $f = h/g$, where $h, g \in \mathcal{O}(\Omega)$, $h, g \not\equiv 0$ (this is always local), the divisor of $f$ is the difference $D_h - D_g$. If $\Omega$ is a domain in $\mathbb{C}$ and $f \in \mathcal{O}(\Omega), f \not\equiv 0$, then $\log|f| \in L^1_{\text{loc}}(\Omega)$ and the generalized Laplacian $\Delta \log|f|$ is equal to $c\Sigma n_j \delta_{a_j}$, where $a_j$ are the zeros of $f$, $n_j$ are their multiplicities, and $\delta_a$ is the delta-function at the point $a$ (i.e. the current $[a]$), and $c$ is an absolute constant. In the general case, $\log|f|$ is also locally integrable in $\Omega$ (for meromorphic $f \not\equiv 0$) and the following formula of Poincaré–Lelong holds.

**Theorem 1.** *Let f be a meromorphic function on a complex manifold Ω. Then*

$$D_f = \frac{i}{\pi}\partial\bar{\partial}\log|f|$$

*in the sense of currents on Ω.*

The assertion is local and its proof reduces to the one-dimensional case with the help of Fubini's Theorem and the representation as an analytic covering (see [17] and [20]).

If $A$ is a purely $p$-dimensional analytic subset of $\Omega$ and $\dim A \cap |D_f| < p$, then the function $\log|f|$ is also locally integrable on $A$ (this follows from the representation as a covering). Thus, on $\Omega$, the current

$$\langle A \log|f|, \varphi\rangle = \int_A (\log|f|)\cdot\varphi$$

is defined, and in this more general situation, an analogous formula holds.

**Theorem 2.** *Let f be a meromorphic function and T a holomorphic p-chain on the complex manifold Ω such that* $\dim|T| \cap |D_f| < p$. Then,

$$\frac{i}{\pi}\partial\bar{\partial}(T\cdot\log|f|) = T \wedge D_f$$

*in the sense of currents on Ω.*

The proof is technically rather difficult (see [17]); however, it fairly easily reduces to Theorem 1 and to the following non-trivial property on intersections: if $T_1$ and $T_2$ are holomorphic chains with proper intersection in the domain $D \subset \mathbb{C}^n$ and $T_{1\varepsilon}$ is the ε-regularization of $T_1$, then

$$\lim_{\varepsilon\to 0} T_{1\varepsilon} \wedge T_2 = T_1 \wedge T_2$$

(in the weak sense).

Analogous formulas of Poincaré–Lelong are valid also for holomorphic mappings $\Omega \to \mathbb{C}^k$ and for more general objects (see [17]).

**4.3. Characteristic Function. Jensen Formulas.** If $A$ is a purely $p$-dimensional analytic subset of the ball $B$: $|z| < R \le \infty$ in $\mathbb{C}^n$ and $A_r = A \cap \{|z| < r\}$, then the function $n(A, r) = \operatorname{vol} A_r/(c(p)r^{2p})$ is called the (unintegrated) counting function of the set $A$. For a holomorphic $p$-chain $T = \Sigma m_j A_j$ in $B$, the counting function is defined by additivity: $n(T, r) = \Sigma m_j n(A_j, r)$ (it can also be negative). The logarithmic average

$$N(T, r) = \int_0^r \frac{n(T, t)}{t}\,dt$$

is called the characteristic (or counting) function of the chain $T$. For analytic sets (and positive chains) it represents the growth of the volume of $A$ as $r \to \infty$ as compared to the growth of the $p$-dimensional ball $|z| < r$. Algebraic sets in $\mathbb{C}^n$ are completely characterized by the condition $N(A, r) \leq C \log(1 + r)$ (see §5.2). In the transcendental case, we define, as usual, the order of $A$ as

$$\overline{\lim_{r \to \infty}} \frac{\log N(A, r)}{\log r}.$$

Positive chains of finite order can be viewed as geometric analogues of entire functions of finite order.

For functions $f$, meromorphic in a ball $B$, the characteristic function of its divisor is, as in the one-dimensional case, closely connected to the average of $\log|f|$. The following "Jensen Formula" holds:

**Proposition 1.** *Let $f$ be meromorphic in the ball $|z| < R$ in $\mathbb{C}^n$ and $0 \notin |D_f|$. Then,*

$$N(D_f, r) = \int_{|z| = r} (\log|f|)\sigma - \log|f(0)|,$$

*for $r < R$, where*

$$\sigma = \frac{1}{2\pi^n} d^c \log|z| \wedge \omega_0^{n-1}$$

*is the normalized volume form on the sphere $|z| = r$.*

In particular, for holomorphic functions $f$ such that

$$\overline{\lim_{r \to R}} \int_{|z| = r} (\log|f|)\sigma < \infty,$$

we have the estimate $N(D_f, R) < \infty$, which can be rewritten in the form

$$\int_{D_f} (R - |z|)\omega^{n-1} < \infty.$$

For $n = 1$ and $R = 1$ this is the classical Blaschke condition $\Sigma(1 - |a_j|) < \infty$ for the zeros $a_j$ of the function $f$. In $\mathbb{C}^n$ this is also called the Blaschke condition. In more general domains of the form $D$: $\rho < 0$, where $d\rho \neq 0$ on $\partial D$, the growth of analytic sets near the boundary can be well estimated in terms of the defining function $\rho$. With the help of the Green Formula (§3.2), we obtain such an estimate, just as for the ball:

**Proposition 2.** *Let $D$: $\rho < 0$ be a bounded domain in $\mathbb{C}^n$, where $\rho \in C^2(\bar{D})$ and $d\rho \neq 0$ on $\partial D$. Let $f$ be a holomorphic function in the Nevanlinna class on $D$, i.e.*

$$\overline{\lim_{\delta \to 0}} \int_{\partial D_\delta} (\log^+|f|)\sigma_\delta < \infty,$$

*where $D_\delta = D \cap \{\rho < -\delta\}$ and $\sigma_\delta$ is the Euclidean volume form on $\partial D_\delta$. Then, the* *divisor of $f$ on $D$ satisfies the Blaschke condition:*

$$\int_{D_f} |\rho|\omega^{n-1} < \infty.$$

A more general situation arises in estimating the divisor of $f$ on an analytic subset $A \subset D$, i.e. the holomorphic chain $A \wedge D_f$. With the help of the averages $(\log|f|)_\varepsilon$, and of the formulas of Stokes and Poincaré–Lelong, one obtains

**Propositions 3.** *Let $A$ be a purely $p$-dimensional analytic subset of the ball $|z| < R$. Let $f$ be a function meromorphic in a neighbourhood of $A$ such that* $\dim A \cap |D_f| < p$ *and* $0 \notin A \cap |D_f|$. *Then*

$$N(A \wedge D_f, r) = \int_{bA_r} (\log|f|)\sigma_p - \frac{1}{\pi^p}\int_{A_r} (\log|f|)\omega_0^p - \mu_0(A)\cdot\log|f(0)|,$$

*where*

$$\sigma_p = \frac{1}{2\pi^p} d^c \log|z| \wedge \omega_0^{p-1}.$$

We remark that $\omega_0^p|A$ and $\sigma_p|bA_r$ are positive forms (i.e. the coefficients of proportionality of the corresponding volume forms are non-negative), and this allows one also to obtain various estimates depending on the growth of $f|A$.

**4.4. The Second Cousin Problem.** The equation of Poincaré–Lelong $i\pi^{-1}\partial\bar{\partial}\log|f| = D_f$ can be considered as an equation with an unknown function $f$ and a given right member, and then it becomes precisely the second Cousin problem of constructing a meromorphic function with a given divisor. The advantage of such a formulation is clear, particularly in connection with the great progress in solving $\bar{\partial}$-problems over the past 15–20 years. Whereas the classical theorems (in the Oka–Cartan theory) yield only conditions for the existence of solutions, with the help of the $\bar{\partial}$-problem, one can obtain solutions with estimates which depend on estimates for the right side. In this direction (in great part due to the efforts of the Lelong school), the study of divisors in $\mathbb{C}^n$ and of entire functions of several complex variables has progressed significantly (see [32], [40], and [44]).

The solution of so-called boundary $\bar{\partial}$-problems with estimates led to the solution of the problem on the Blaschke condition in $\mathbb{C}^n$. G.M. Khenkin [24] and Skoda [45] proved the following.

**Theorem.** *Let $D$: $\rho < 0$ be a strictly pseudoconvex bounded domain in $\mathbb{C}^n$ such that $H^1(D, \mathbb{R}) = 0$, and let $T$ be an arbitrary positive divisor (positive holomorphic $(n-1)$-chain) in $D$, satisfying the Blaschke condition:*

$$\int_T |\rho|\omega^{n-1} < \infty.$$

*Then there exists a holomorphic function $f$ in $D$ of Nevanlinna class such that $T = D_f$.*

Finally, we remark that all characterization theorems for holomorphic chains (see §4.1) were obtained by the method of solving the Poincaré–Lelong equation

$$\frac{i}{\pi}\partial\bar{\partial}\log|f| = T$$

with a given current on the right side (see [20], [22], and [43]).

**4.5. Growth Estimates for Traces on Linear Subspaces.** The Crofton Formulas (§3.3) show that the growth of the volume of an analytic set $A$ in $\mathbb{C}^n$ admits two-sided estimates in terms of the growth of the volume of the trace of $A$ on a "generic" linear subspace $L \ni 0$ of fixed dimension. This problem has been more thoroughly investigated for traces on hyperplanes. From the Jensen Formula, we have the following.

**Lemma 1.** *Let $A$ be a purely $p$-dimensional analytic subset of the ball $|z| < R$, $p \geq 1$, and $L^w$: $(z, w) = 0$. Then*

$$N(A \wedge L^w, r) - N(A, r) = \int_{bA_{r_0,r}} \log\frac{|(z,w)|}{|z||w|}\sigma_p - \frac{1}{\pi^p}\int_{A_{r_0,r}} \log\frac{|(z,w)|}{|z||w|}\omega_0^p,$$

*where $A_{r_0,r} = A_r \setminus \bar{A}_{r_0}$, $r_0 > 0$, and $\sigma_p$ is defined in §4.3.*

Since this relation depends only on $w/|w|$, we may integrate with respect to a measure on $\mathbb{P}_{n-1}$. In this context, the potential

$$u_\mu(z) = \int_{\mathbb{P}_{n-1}} \log\frac{|z||w|}{|(z,w)|}d\mu(w), \qquad z \in \mathbb{C}^n,$$

naturally arises. "Thin" sets for this problem are defined in terms of this potential. The following two-sided estimates are obtained in the work of Molzon, Shiffman, and Sibony [27].

**Theorem 1.** *Let $A$ be a purely $p$-dimensional analytic subset of the ball $|z| < R$ in $\mathbb{C}^n$, $p \geq 1$, and $\mu$ a probability measure on $\mathbb{P}_{n-1}$ whose potential is bounded: $u_\mu \leq c < \infty$. Then*

$$2ce^{-4pc}N(A, e^{-2c}r) \leq \int_{\mathbb{P}_{n-1}} N(A \wedge L^w, r)d\mu(w) \leq \left(\frac{c}{\log k} + 1\right)N(A, kr),$$

*for all $k > 1$ such that $kr \leq R$.*

The upper estimate for $N(A \wedge L^w, r)$ follows easily from Lemma 1. The finer lower estimate follows by averaging the following lemma of Gruman [18].

**Lemma 2.** *Let $A$ be a purely $p$-dimensional analytic subset of the ball $|z| < R$ and $v$ be a plurisubharmonic function in this ball of class $C^2$ and such that* $\log|z|$

$-c \le v(z) \le \log|z|$ on $A \setminus A_{r_0}$ *for some constants* $c, r_0 > 0$. *Then*

$$r^2 \int_{A_r} dd^c v \wedge \omega^{p-1} \ge 4c \int_{A_{r_0,r'}} \omega^p,$$

*where* $r' = re^{-2c}$ *and* $r_0 \le r \le R$.

A compact set $E \subset \mathbb{P}_{n-1}$ is, by definition, of positive logarithmic *capacity* if there exists a probability measure $\mu$, concentrated on $E$, whose potential is (uniformly) bounded. For example, all non-pluripolar compact sets and even some thinner sets are of positive capacity (for more details, see [27]). From Theorem 1, for example, one obtains the following consequences for purely $p$-dimensional analytic subsets $A \subset \mathbb{C}^n$, $p \ge 1$.

1) If the set $A \cap L^w$ is algebraic for each $w$ lying in a compact set $E \subset \mathbb{P}_{n-1}$ of positive logarithmic capacity, then $A$ is also an algebraic set.

2) If $\mu$ is a probability measure on $\mathbb{P}_{n-1}$ with bounded potential, then $\operatorname{ord} A \wedge L^w = \operatorname{ord} A$ for $\mu$ – a.e. $w \in \mathbb{P}_{n-1}$ (here $\operatorname{ord} A$ is the order of growth of $A$ in $\mathbb{C}^n$, see §4.3).

For the traces on linear subspaces of codimension $> 1$, the picture is not yet completely clear. Logarithmic capacity is clearly not suitable, however, an obvious substitute is not apparent. The following theorem was proved by Gruman [18].

**Theorem 2.** *Let* $A$ *be a purely* $p$*-dimensional analytic subset of the ball* $|z| < R$ *in* $\mathbb{C}^n$, $p \ge 1$, *and* $E$ *a Borel set in* $G(q, n)$ *of positive* $dL$*-volume*, $p + q \ge n$. *Then, there exist positive constants* $c_1, c_2$ *such that*

$$c_1 N(A, c_2 r) \le \int_E N(A \cap L, r)\, dL, \qquad r_0 \le r \le R.$$

Theorems on reciprocal estimates of the growth of analytic sets and of their traces on subspaces can be viewed as geometric analogues of theorems from the Nevanlinna theory of value distribution. In fact, such theorems arose from generalizing Nevanlinna theory to the multi-dimensional case.

**Notes.** For the results of the first four sections see the paper by Harvey in [20] and Griffiths–King [17]. A relatively simple proof of the theorem of Khenkin–Skoda for the ball is in Rudin's book [37]. The estimates of §4.5 are proved in the paper by Molzon–Shiffman–Sibony [27].

## §5. Analytic Continuation and Boundary Properties

**5.1. Removal of Metrically Thin Singularities.** The border of a $p$-dimensional analytic set has (metric) dimension $2p - 1$, and hence it is intuitively understandable that a set of lesser dimension cannot be an authentic obstruction for a purely $p$-dimensional analytic set. From the theorem on the analyticity of a generalized analytic covering (§1.5) one easily obtains the following theorem of Shiffman [41].

**Theorem 1.** *Let $E$ be a closed subset of a complex manifold $\Omega$, and let $A$ be a purely $p$-dimensional analytic subset of $\Omega \backslash E$. If the Hausdorff measure $\mathscr{H}_{2p-1}(E) = 0$,* then the closure *$\bar{A}$ of the set $A$ in $\Omega$ is a purely $p$-dimensional analytic subset of $\Omega$.*

The important particular case, when $E$ is an analytic subset of $\Omega$ of dimension less than $p$, constitutes the well-known theorem of Remmert–Stein (see, for example, [19]).

Sets of positive $(2p - 1)$-measure may serve as obstructions for $A$, however, there is here the following curious effect of "infectiousness" of the continuation (see [10]).

**Theorem 2.** *Let $M$ be a connected $(2p - 1)$-dimensional $C^1$-submanifold of a complex manifold $\Omega$, and let $A$ be a purely $p$-dimensional analytic subset of $\Omega \backslash M$. Suppose that in the neighbourhood of some point $z^0 \in M$, the set $\bar{A}$ is analytic. Then $\bar{A}$ is an analytic subset of $\Omega$.*

The proof is obtained in a rather standard fashion with the help of analytic coverings and canonical defining functions.

**5.2. Removal of Pluripolar Singularities.** Obstructions of higher dimension can also be removed for analytic sets under additional conditions on $E$ and on the "holes" through which $A$ continues. The following lemma was proved by Bishop [6].

**Lemma 1.** *Let $E$ be a closed totally pluripolar subset of a bounded domain $D = D' \times D''$ in $\mathbb{C}^n$, with $D' \subset \mathbb{C}^p$, and $A$ a purely $p$-dimensional analytic subset of $D \backslash E$ having no limit points on $D' \times \partial D''$. Suppose that there is a nonempty domain $U'$ in $D'$ such that $\bar{A} \cap (U' \times D'')$ is analytic. Then the set $\bar{A} \cap D$ is an analytic subset of $D$.*

A set $E \subset D$ is said to be *totally pluripolar* if there exists a plurisubharmonic function $\varphi \not\equiv -\infty$ in $D$ such that $E = \{z \in D \colon \varphi(z) = -\infty\}$.

From this one easily obtains the following theorem of Thullen on the "infectiousness" of continuation through analytic singularities.

**Proposition 1.** *Let $S$ be an analytic $p$-dimensional subset of a complex manifold $\Omega$ and $A$ a purely $p$-dimensional analytic subset of $\Omega \backslash S$. Suppose there is an open subset $U$ of $\Omega$ which meets every irreducible component of $S$ and such that the set $\bar{A} \cap U$ is analytic in $U$. Then $\bar{A}$ is an analytic subset of $\Omega$.*

The following theorem, essentially proved by Bishop [6], on the continuation through pluripolar singularities, also follows from the lemma without difficulty.

**Theorem 1.** *Let $E$ be a closed locally totally pluripolar subset of a complex manifold $\Omega$, and let $A$ be a purely p-dimensional analytic subset of $\Omega \backslash E$. Then, if the Hausdorff measure $\mathscr{H}_{2p}(A \cap K) < \infty$ for each compact $K \subset \Omega$ or if $\mathscr{H}_{2p}(\bar{A} \cap E) = 0$, then $\bar{A}$ is an analytic subset of $\Omega$.*

In applications, the first condition has been particularly useful. Not long ago, El Mir [12] proved an analogous theorem for positive closed currents having locally finite mass in $\Omega$. An important special case is when $E$ is a proper analytic subset. For example, if $A$ is an analytic subset of $\mathbb{C}^n$ such that $N(A, r) \leq c \log(1 + r)$, so that the Fubini–Study volume of the set $A$ in $\mathbb{P}_n \supset \mathbb{C}^n$ is bounded, and if the obstruction is the analytic set $E = \mathbb{P}_n \backslash \mathbb{C}^n$, then, by the theorem, $\bar{A}$ is an analytic and hence algebraic subset of $\mathbb{P}_n$. In this way, from Bishop's theorem, we obtain metric criteria, already mentioned in §4.3, in order for an analytic set to be algebraic.

**5.3. Symmetry Principle.** For the continuation of analytic sets through $\mathbb{R}$-analytic manifolds, a major role is played by a method analogous to the symmetry principle for holomorphic functions. In this direction, one has, for example, the following result (Alexander [2], Becker [5]).

**Proposition 1.** *Let $D$ be a domain in $\mathbb{C}^n$ and $A$ a purely p-dimensional analytic subset of $D \backslash \mathbb{R}^n$. If $p > 1$ or $p = 1$ and $A$ is symmetric with respect to $\mathbb{R}^n$, then $\bar{A} \cap D$ is an analytic subset of $D$.*

Any *totally real* $n$-dimensional $\mathbb{R}$-analytic *manifold* $M$ in $\mathbb{C}^n$ is locally biholomorphic to a domain in $\mathbb{R}^n \subset \mathbb{C}^n$. Hence an analogous symmetry principle is valid also for such obstructions $M$. If $M$ is given by a system of $\mathbb{R}$-analytic equations, then, an anti-holomorphic symmetry with respect to $M$ can be written explicitly in a neighbourhood of $M$ with the help of these equations. This symmetry preserves analyticity of sets and for $\mathbb{R}^n$ has the form $z \mapsto \bar{z}$. If, for example, $M = \Gamma$, the distinguished boundary $|z_1| = \ldots |z_n| = 1$ of the unit polydisc $U$ in $\mathbb{C}^n$, then the symmetry with respect to $\Gamma$ is the anti-holomorphic mapping $z \mapsto z^* = (1/\bar{z}_1, \ldots, 1/\bar{z}_n)$, and the following theorem of Shiffman [42] holds.

**Proposition 2.** *Let $A$ be a purely one-dimensional analytic subset of the unit polydisc $U$ in $\mathbb{C}^n$ such that $\bar{A} \subset U \cup \Gamma$, where $\Gamma$ is the distinguished boundary of $U$, and let $A^* = \{z^* : z \in A\}$. Then $\bar{A} \cup A^*$ is an algebraic subset of $\mathbb{C}^n$.*

If we don't require symmetry of the set $A$, then using the anti-holomorphic symmetry, we can obtain a continuation to some neighbourhood of $M$. In this direction, we have

**Theorem 1.** *Let $M$ be a $\mathbb{R}$-analytic submanifold of a complex manifold $\Omega$, and let $A$ be a purely p-dimensional analytic subset of $\Omega \backslash M$. If the CR-dimension of $M$ (i.e. the $\mathbb{C}$-dimension of the complex tangent space to $M$) is no greater than $p - 1$, then $A$ extends analytically to some neighbourhood of the set $\bar{A}$ in $\Omega$.*

The proof reduces to continuation through totally real manifolds with the help of appropriate traces and the following analogue of the Hartogs Lemma for analytic sets (Rothstein [34]).

**Lemma 1.** *Let $A$ be a closed subset of a domain $U = U' \times U''$ in $\mathbb{C}^n$, with $U' \subset \mathbb{C}^m$ and let $V''$ be a subdomain of $U''$. Suppose that:* 1) *$A \cap (U' \times V'')$ is a purely p-dimensional analytic subset of $U' \times V''$;* 2) *$A_c = A \cap \{z' = c\}$ is a purely q-dimensional or empty analytic subset of $c \times U''$ for each $c \in U'$; and* 3) *each irreducible component of the set $A_c$ meets $c \times V''$. Then $A$ is an analytic subset of $U$.*

The lemma is also perhaps valid without supposing that $A$ is closed.

Let us say a few words concerning the analytic continuation of sets. If $A$ and $\tilde{A}$ are purely $p$-dimensional locally analytic sets in $\Omega$ with $A \subset \tilde{A}$, and if each irreducible component of $\tilde{A}$ has a $p$-dimensional intersection with $A$, then $\tilde{A}$ is called an *analytic extension of the set A*. We say that $A$ extends analytically to $U$ if there exists an analytic subset $\tilde{A}$ of $U$ which is an analytic extension of the set $A \cap U$. The analytic continuation of sets has the following useful property of localness which is not difficult to prove.

**Lemma 2.** *Let $E$ be a closed subset of a complex manifold $\Omega$ and let $A$ be a purely p-dimensional analytic subset of $\Omega \backslash E$. Suppose that for each point $\zeta \in \bar{A} \cap E$, there is a neighbourhood $U_\zeta \ni \zeta$ in $\Omega$ to which $A$ extends analytically. Then $A$ extends analytically to some neighbourhood of $\bar{A}$ in $\Omega$.*

**5.4. Obstructions of Small *CR*-Dimension.** In the presence of a rich complex structure for a set $E$, an analytic subset of $\Omega \backslash E$ can approach $E$ with uncontrolled "wiggling" (as, for example, the graph of $w = e^{1/z}$ along the line $z = 0$ in $\mathbb{C}^2$), and in order to continue through $E$, it is necessary to impose additional conditions (as, for example, in Bishop's Theorem, §5.2). If, however, the complex structure of $E$ is sufficiently poor, then, from all indications, $E$ cannot be an obstruction for an analytic subset of $\Omega \backslash E$ of sufficiently high dimension. The following theorem was proved by the author [8].

**Theorem 1.** *Let $M$ be a $C^1$-submanifold of a complex manifold $\Omega$ and let $A$ be a purely p-dimensional analytic subset of $\Omega \backslash M$, $p > 1$. Suppose the complex tangent space $T_\zeta^{\mathbb{C}}(M)$ has (complex) dimension $< p - 1$ for each $\zeta \in \bar{A} \cap M$, with the possible exception of a set of zero $(2p - 1)$-measure. Then $\bar{A}$ is an analytic subset of $\Omega$.*

We remark that the border $bA$ of an analytic set $A$ of dimension $p$ (if it exists) has $CR$-dimension $p - 1$. Thus, the $CR$-structure of $M$ in the theorem is poorer than the possible $CR$-structure of a border. For positive closed currents of bidimension $(p, p)$ in $\Omega \backslash M$ (for twice smooth $M$) an analogous theorem was proved by El Mir [12].

**5.5. Continuation Through Pseudoconcave Hypersurfaces.** One of the central theorems on the continuation of holomorphic functions of several variables is the theorem on the continuation through a strongly pseudoconcave hypersurface (see [39], [50]). This remarkable property also carries over to analytic sets of dimension greater than 1 (see [33]).

**Theorem 1.** *Let $D \subset G$ be domains in $\mathbb{C}^n$ such that $\Gamma = (\partial D) \cap G$ is hypersurface of class $C^2$ whose Levi form has at least q negative eigenvalues, and let A be a purely p-dimensional analytic subset of D, where $p + q > n$. Then, there exists a neighbourhood $U \supset \Gamma$ in G such that A extends analytically to $D \cup U$. In particular, if $\Gamma$ is strictly pseudoconcave (i.e. $q = n - 1$), then each analytic subset of D of pure dimension $\geq 2$ extends analytically to a neighbourhood of $D \cup \Gamma$.*

Here, the main particular case is when $p = 2$ and $\Gamma$ is strictly pseudoconcave. The general case reduces to this one via the "Hartogs Lemma" of §5.3. We remark that Theorem 1 in §5.4, follows from this Theorem 1, in the case where $M$ is of class $C^2$, for then, $M$ can be represented as the intersection of "sufficiently pseudoconcave" hypersurfaces. Generally speaking, one-dimensional analytic sets do not continue through pseudoconcave hypersurfaces. It is easy to present appropriate examples.

From Theorem 1 and §5.3, one can deduce the following geometric analogue of the removability of compact singularities for functions of several complex variables.

**Theorem 2.** *Let D be a domain in $\mathbb{C}^n$ (or on a Stein manifold $\Omega$), K a compact subset of D, and A a purely p-dimensional analytic subset of $D \backslash K$, where $p \geq 2$ and the closure in D of each irreducible component of the set A is non-compact (in $\Omega$). Then A extends analytically to D.*

One can proceed with a proof by contradiction. By Lemma 2 of §5.3, there exists a minimal compact $E \subset K$ such that $A$ extends analytically to $D \backslash E$. Let $\xi \in E$ be such that $|\xi| = \max_E |z|$. Then, the hypersurface $|z| = |\xi|$ is strictly pseudoconcave, as the boundary of $D \cap \{|z| > |\xi|\}$, and hence $A$ extends through $\xi$ by Theorem 1. However this contradicts the minimality of $E$.

**5.6. The Plateau Problem for Analytic Sets.** The border of a $p$-dimensional analytic set (with border) is everywhere outside of a closed set of $(2p - 1)$-measure zero a $(2p - 1)$-dimensional manifold of $CR$-dimension $p - 1$ (such an odd-dimensional *manifold* is said to be *maximally complex*). A remarkable theorem of Harvey and Lawson [21] asserts that, for $p \geq 2$, this infinitesimal condition is essentially a characterization. This is a geometric analogue of the fact that each $CR$-function on the boundary of a bounded domain in $\mathbb{C}^n$, having connected complement, extends holomorphically to the interior of the domain. This is the theorem of Severi, see [39].

A closed subset $M$ of a complex manifold $\Omega$ is called a *maximal complex cycle* if there exists on $M$ a closed subset $\Sigma$ of zero $(2p - 1)$-measure such that $M \backslash \Sigma$ is an oriented maximally complex $(2p-1)$-dimensional manifold of locally finite $(2p-1)$-measure in $\Omega$, $p \geq 2$, and the current of integration on $M$ is closed in $\Omega$. For example, we may take any maximally complex submanifold of $\Omega$. The following theorem (Harvey, Lawson [21]) solves the problem of the existence of a complex-analytic "film" having a given compact boundary in $\mathbb{C}^n$.

**Theorem 1.** *Let $M$ be a compact maximally complex cycle of dimension $2p-1>1$ in $\mathbb{C}^n$ (or on a Stein manifold). Then there exists a unique holomorphic $p$-chain $T$ in $\mathbb{C}^n \setminus M$ having finite mass and compact support whose boundary is $M$, more precisely, $dT = -[M]$. Moreover, there is a set $E$ on $M$ of zero $(2p-1)$-measure such that for each point of $M \setminus E$, in the neighbourhood of which $M$ is of class $C^k$, $1 \leq k \leq \infty$, there exists a neighbourhood in which $|T| \cup M$ is a $C^k$-submanifold (with or without boundary).*

For $p = 1$ the border of $A$ satisfies an infinite set of orthogonality relations for holomorphic $(1,0)$-forms. For the correspondingly (already non-local) defined one-dimensional maximally complex cycles, the theorem on "soap films" is valid also for $p = 1$ (see [20]), but this case was already studied long ago (see [48]). We introduce one of the possible generalizations of Theorem 1 in the spirit of Stolzenberg [48].

**Theorem 2.** *Let $K$ be a polynomially compact set in $\mathbb{C}^n$ and $M$ a maximally complex cycle of dimension $2p \div 1 > 1$ in $\mathbb{C}^n \setminus K$ such that $M \cup K$ is compact. Then, there exists an analytic purely $p$-dimensional subset $A$ of $\mathbb{C}^n \setminus (M \cup K)$ such that $A \cup M \cup K$ is compact and $M \subset \bar{A}$. Moreover, outside of some closed subset of $M$ of $(2p-1)$-measure zero, $\bar{A}$ is a submanifold with border in $\mathbb{C}^n \setminus K$ having the same smoothness as $M$.*

We remark that a polynomially convex compact set can be approximated by polynomial polyhedra which when lifted to a space of higher dimension become convex polyhedra. The proof of the theorem of Harvey–Lawson in [20] carries over to half-spaces almost word for word, and the gluing of films in various half-spaces is also feasible by this scheme.

In the above described "complex Plateau problem", one obtains a film in general with singularities (as in the classical case). For smoothness criteria for the film, see the work of Yau [54]. Here one finds additional conditions in (not so easy) terms of the cohomology of the tangential Cauchy–Riemann complex. We introduce a result in this direction having a simple formulation and due to M.P. Gambaryan (see [10]).

**Proposition 1.** *If a maximally complex manifold $M$ lies on the boundary of a strictly pseudoconvex domain $D$ in $\mathbb{C}^n$ and $A$ is a purely $p$-dimensional analytic subset of $\mathbb{C}^n \setminus M$ such that $M = \bar{A} \setminus A$ and $\bar{A}$ is compact, then $\operatorname{sng} A$ consists of finitely many points and $\bar{A} \setminus \operatorname{sng} A$ is a submanifold with boundary in $\mathbb{C}^n \setminus \operatorname{sng} A$. Moreover, if $2p > n$ and there exists an $(n-p)$-subspace $L \subset \mathbb{C}^n$ such that $L \cap T_\xi M = 0$, for each $\xi \in M$, then $A$ is a complex submanifold of $\mathbb{C}^n \setminus M$.*

**5.7. Boundaries of One-Dimensional Analytic Sets.** In the theorem of Harvey and Lawson, there is the assertion concerning the regularity of the boundary behaviour of a $p$-dimensional analytic set A along a $(2p-1)$-dimensional manifold $M$ of class $C^k$, $k \geq 1$. Along manifolds of higher dimension, the behaviour of $A$, is general, uncontrolled. However, it turns out that this is fundamentally influenced by the $CR$-dimension of $M$, and if the latter is no more than $p-1$, then one can expect smoothness of $\bar{A}$ on the boundary, almost the same as in the theorem of Harvey–Lawson. The first difficult case is for $p = 1$ and $M$ a totally-real manifold. The following theorem was proved by the author [9].

**Theorem 1.** *Let $D$ be a domain in $\mathbb{C}^n$, $M$ a totally-real submanifold of $D$ of class $C^k$, $k > 1$, and $A$ a purely one-dimensional analytic subset of $D \setminus M$. Then, $bA \equiv \bar{A} \cap M$ is locally rectifiable. Moreover, on $bA$ there is a closed (possibly empty) subset $E$ of length zero such that each point $\xi \in (bA) \setminus E$ is either removable ($\bar{A}$ is an analytic subset in the neighbourhood of $\xi$), or is a border point ($\bar{A} \cap U_\xi$ is a manifold with border of smoothness $k-0$ in $U_\xi$), or a two-sided singularity ($\bar{A} \cap U_\xi$ is the union of two bordered manifolds whose borders are smooth of order $>1$ and have tangent cones at the point $\xi$ which constitute a complex plane). Moreover, the removable singularities and the border points form an open everywhere dense subset of $bA$.*

A manifold $S$ has smoothness of order $k \in \mathbb{R}_+$ if $S \in C^{[k]}$, where $[k]$ is the integral part of $k$, and the tangent space to $S$, considered as a point of the corresponding Grassmannian, satisfies a Hölder condition with exponent $k - [k]$. Smoothness of order $k - 0$ means that $S \in C^k$, if $k$ is non-integral and $S \in C^{k'}$ for each $k' < k$, if $k$ is integral. The loss of smoothness from $k$ to $k - 0$ actually takes place as is seen for example in the graphs of conformal mappings of plane domains. From the proof of the theorem, one obtains, for example, the following result on the regularity of suspended discs (used in many multi-dimensional problems).

**Proposition 1.** *Let $M \subset \mathbb{C}^n$ be a totally-real manifold of class $C^k$, $k > 1$, and $f$: $\Delta \to \mathbb{C}^n$ a holomorphic mapping of the unit disc $\Delta \subset \mathbb{C}$ such that the boundary values of $f$ along an open arc $\gamma \subset \partial\Delta$ lie in $M$. Then $f \in C^{k-0}(\Delta \cup \gamma)$.*

Results on the regularity of the boundary of one-dimensional analytic sets should apparently lead to analogous results for $p$-dimensional analytic sets in the neighbourhood of manifolds of $CR$-dimension $\leq p-1$ (with the help of the "Hartogs Lemma" from §5.3). However, this is not so simple, and the question of the boundary behaviour in this situation remains open.

**5.8. A Perspective and Prospects.** At the present time there are already many analogous theorems for holomorphic functions (mappings) on the one hand and for analytic sets on the other, which present the theory of analytic sets as a geometric variant of the theory of functions of several complex variables.

Adding a geometric viewpoint to analytic results and the reformulation and proof of the analogous assertions for analytic sets undoubtedly enriches both directions of multi-dimensional complex variables even more. One should remark however on a characteristic particularity of such a development. If one manages to find a correct (without obvious counterexamples) geometric formulation, then the subsequent proof is rather transparent (although perhaps technically very complicated as, for example, in the theorem of Harvey–Lawson). For this reason, at the present time, there are efforts in progress to formulate correct hypotheses by studying typical examples and the "inner geometry" of classical theorems of complex analysis. For example, concerning the removal of singularities, there is not yet an analogue to the theorem on the removal of singularities for holomorphic functions under the conditions of Hölder continuity, nor to the connection between the growth of a function along removable singularities and a metric characterization of singularities etc. Analogues of the strong results of recent years on the boundary properties of holomorphic mappings of strictly pseudoconvex domains also have not yet been obtained. One of the few results of this program for analytic sets is the following theorem of S.I. Pinchuk [31].

**Theorem 1.** *Let $D$ be a domain in $\mathbb{C}^{2n}$ and $M$ a submanifold of $D$ of the form $\Gamma_1 \times \Gamma_2$, where $\Gamma_j$ are strictly pseudoconvex $\mathbb{R}$-analytic hypersurfaces in $\mathbb{C}^n$. Then, each purely $n$-dimensional subset $A$ of $D \setminus M$ extends analytically to some neighbourhood of $M$.*

Here, the $CR$-dimension is completely sufficient to "block" $A$, however, this is not apparent from the powerful Levi form of the manifold $M$. This phenomenon is, as yet, completely uninvestigated. Part of the hypotheses can certainly be weakened but which and how is so far unknown.

**Notes.** The results of this chapter are essentially found in the articles mentioned in the text. An exception is the results of §§5.7 and 5.8 which are proved in the author's book [10]. The theme of this chapter is perhaps the most current one in the theory of analytic sets. Among the works of recent years we draw attention to [30], [31], [9], [12], [21], and [46].

# References*

1. Ajzenberg, L.A., Yuzhakov, A.P.: Integral representations and residues in multidimensional complex analysis. Nauka, Novosibirsk 1979; Engl. transl.: Am. Math. Soc., Providence, R.I. 1983. Zbl. 445.32002
2. Alexander, H.: Continuing 1-dimensional analytic sets. Math. Ann. *191*, No. 2, 143–144 (1971). Zbl. 211,102

* For the convenience of the reader, references to reviews in Zentralblatt für Mathematik (Zbl.), compiled using the MATH database, have, as far as possible, been included in this bibliography.

3. Alexander, H.: Volumes of images of varieties in projective space and in Grassmannians. Trans. Am. Math. Soc. *189*, 237–249 (1974). Zbl. 285.32008
4. Alexander, H., Taylor, B.A., Ullman, J.L.: Areas of projections of analytic sets. Invent. Math. *16*, No. 4, 335–341 (1972). Zbl. 238.32007
5. Becker, J.: Continuing analytic sets across $\mathbb{R}^n$, Math. Ann. *195*, No. 2, 103–106 (1972). Zbl. 223.32012
6. Bishop, E.: Condition for the analyticity of certain sets. Mich. Math. J. *11*, No. 4, 289–304 (1964). Zbl. 143,303
7. Chirka, E.M.: Currents and some of their applications. In the Russian translation of [20]. Mir, 122–154, Moscow 1979
8. Chirka, E.M.: On removable singularities of analytic sets. Dokl. Akad. Nauk SSSR *248*, 47–50 (1979); Engl. transl.: Sov. Math., Dokl. *20*, No. 5, 965–968 (1979). Zbl. 466.32007
9. Chirka, E.M.: Regularity of the boundaries of analytic sets. Mat. Sb., Nov. Ser. *117*, 291–336 (1982); Engl. transl.: Math. USSR, Sb. *45*, No. 3, 291–335 (1983). Zbl. 525.32005
10. Chirka, E.M.: Complex analytic sets. Nauka, Moscow 1985 [Russian]. Zbl. 586.32013
11. Draper, R.N.: Intersection theory in analytic geometry. Math. Ann. *180*, No. 3, 175–204 (1969). Zbl. 157,405 (Zbl. 167,69)
12. El Mir, H.: Sur le prolongement des courants positifs fermés. Acta Math. *153*, No. 1–2, 1–45 (1984). Zbl. 557.32003
13. Federer, H.: Geometric measure theory. Springer-Verlag, New York 1969. Zbl. 176,8
14. Fischer, G.: Complex analytic geometry. Lect. Notes Math. *538*, Springer-Verlag, New York 1976. Zbl. 343.32002
15. Griffiths, P.: Complex differential and integral geometry and curvature integrals, associated to singularities of complex analytic varieties. Duke Math. J. *45*, No. 3, 427–512 (1978). Zbl. 409.53048
16. Griffiths, P., Harris, J.: Principles of algebraic geometry. Wiley, New York 1978. Zbl. 408.14001
17. Griffiths, P., King, J.: Nevanlinna theory and holomorphic mappings between algebraic varieties. Acta Math. *130*, No. 3–4, 145–220 (1973). Zbl. 258.32009
18. Gruman, L.: La géométrie globale des ensembles analytiques dans $\mathbb{C}^n$. Lect. Notes Math. *822*, 90–99, Springer-Verlag, New York 1980. Zbl. 446.32007
19. Gunning, R.C., Rossi, H.: Analytic functions of several complex variables. Prentice Hall, Englewood Cliffs, N.J. 1965. Zbl. 141,86
20. Harvey, R.: Holomorphic chains and their boundaries. Proc. Symp. Pure Math. *30*, No. 1, 309–382 (1977). Zbl. 374.32002
21. Harvey, R., Lawson, H.P., Jr.: On boundaries of complex analytic varieties. I. Ann. Math., II. Ser. *102*, No. 2, 223–290 (1975). Zbl. 317.32017
22. Harvey, R., Shiffman, B.: A characterization of holomorphic chains. Ann Math., II. Ser. *99*, No. 3, 553–587 (1974). Zbl. 287.32008
23. Hervé, M.: Several complex variables. Local theory. Oxford University Press, London 1963. Zbl. 113,290
24. Khenkin, G.M.: H. Lewy's equation and analysis on a pseudoconvex manifold. II. Mat. Sb., Nov. Ser. *102*, 71–108 (1976); Engl. transl.: Math. USSR, Sb. *31*, No. 1, 63–94 (1977). Zbl. 358.35058
25. Lelong, P.: Intégration sur un ensemble analytique complexe. Bull. Soc. Math. Fr. *85*, No. 2, 239–262 (1957). Zbl. 79,309
26. Malgrange, B.: Ideals of differentiable functions. Oxford University Press, Oxford 1966, Zbl. 177,179
27. Molzon, R.E., Shiffman, B., Sibony, N.: Average growth estimates for hyperplane sections of entire analytic sets. Math. Ann. *257*, No. 1, 43–59 (1981). Zbl. 537.32009
28. Mumford, D.: Algebraic geometry. I. Complex projective varieties. Springer-Verlag, Berlin, Heidelberg, New York 1976. Zbl. 356.14002
29. Palamodov, V.P.: The multiplicity of a holomorphic transformation. Funkts. Anal. Prilozh. *1*, No. 3, 54–65 (1967); Engl. transl.: Funct. Anal. Appl. *1*, 218–226 (1968). Zbl. 164,92

30. Pinchuk, S.I.: On holomorphic mappings of real analytic hypersurfaces. Mat. Sb., Nov. Ser. *105*, 574–593 (1978); Engl. transl.: Math. USSR, Sb. *34*, No. 4, 503–519 (1978). Zbl. 389.32008
31. Pinchuk, S.I.: Boundary behaviour of analytic sets and algebroid mappings. Dokl. Akad. Nauk SSSR *268*, 296–298 (1983); Engl. transl.: Sov. Math., Dokl. *27*, No. 1, 82–85 (1983). Zbl. 577.32008
32. Ronkin, L.I.: Introduction to the theory of entire functions of several variables. Nauka, Moscow 1971; Engl. transl.: Am. Math. Soc., Providence, R.I. 1974. Zbl. 225.32001
33. Rothstein, W.: Zur Theorie der analytischen Mannigfaltigkeiten in Räumen von $n$ komplexen Veränderlichen. Math. Ann. *129*, No. 1, 96–138 (1955). Zbl. 64,80; *133*, No. 3, 271–280 (1957). Zbl. 77,289; *133*, No. 5, 400–409 (1957). Zbl. 84,72
34. Rothstein, W.: Zur Theorie der analytischen Mengen. Math. Ann. *174*, No. 1, 8–32 (1967). Zbl. 172,378
35. Rothstein, W.: Das Maximumprinzip und die Singularitäten analytischer Mengen. Invent. Math. *6*, No. 2, 163–184 (1968). Zbl. 164,382
36. Rudin, W.: A geometric criterion for algebraic varieties. J. Math. Mech. *17*, No. 7, 671–683 (1968). Zbl. 157,132
37. Rudin, W.: Function theory in the unit ball of $\mathbb{C}^n$, Springer-Verlag, Heidelberg 1980. Zbl. 495.32001
38. Santalo, L.A.: Integral geometry in Hermitian spaces. Am. J. Math. *74*, No. 2, 423–434 (1952). Zbl. 46,161
39. Shabat, B.V.: Introduction to complex analysis. Part II. Functions of several variables, 3rd ed. (Russian). Nauka, Moscow 1985. Zbl. 578.32001
40. Shabat, B.V.: Distribution of values of holomorphic mappings. Nauka, Moscow 1982; Engl. transl.: Amer. Math. Soc., Providence, R.I. 1985. Zbl. 537.32008
41. Shiffman, B.: On the removal of singularities of analytic sets. Mich. Math. J. *15*, No. 1, 111–120 (1968). Zbl. 165,405
42. Shiffman, B.: On the continuation of analytic curves. Math. Ann. *184*, No. 4, 268–274 (1970). Zbl. 176,380
43. Siu, Yum-Tong: Analyticity of sets associated to Lelong numbers and the extension of closed positive currents. Invent. Math. *27*, No. 1–2, 53–156 (1974). Zbl. 289.32003
44. Skoda, H.: Sous-ensembles analytiques d'ordre fini ou infini dans $\mathbb{C}^n$. Bull. Soc. Math. Fr. *100*, No. 4, 353–408 (1972). Zbl. 246.32009
45. Skoda, H.: Valeurs au bord pour les solutions de l'opérateur $d''$, et charactérisation des zéros des fonctions de la classe de Nevanlinna. Bull. Soc. Math. Fr. *104*, No. 3, 225–299 (1976). Zbl. 351.31007
46. Skoda, H.: Prolongement des courants, positifs, fermés de masse finie. Invent. Math. *66*, No. 3, 361–376 (1982). Zbl. 488.58002
47. Stoll, W.: The multiplicity of a holomorphic map. Invent. Math. *2*, No. 1, 15–58 (1966). Zbl. 158,84
48. Stolzenberg, G.: Uniform approximation on smooth curves. Acta Math. *115*, No. 3–4, 185–198 (1966). Zbl. 143,300
49. Stutz, J.: Analytic sets as branched coverings. Trans. Am. Math. Soc. *166*, 241–259 (1972). Zbl. 239.32006
50. Vladimirov, V.S.: Methods of the theory of functions of many complex variables Nauka, Moscow 1964; Engl. transl.: M.I.T. Press, Cambridge 1966. Zbl. 125,319
51. Whitney, H.: Local properties of analytic varieties. In: Differ. and Combinat. Topology. Princeton Univ. Press, 205–244, Princeton 1965. Zbl. 129,394
52. Whitney, H.: Tangents to an analytic variety. Ann. Math., II. Ser. *81*, No. 3, 496–549 (1965). Zbl. 152,277
53. Whitney, H.: Complex analytic varieties. Addison-Wesley Publ. Co., Reading, Massachusetts 1972, Zbl. 265.32008
54. Yau, St.: Kohn-Rossi cohomology and its application to the complex Plateau problem. I. Ann. Math., II. Ser. *113*, No. 1, 67–110 (1981). Zbl. 464.32012

# IV. Holomorphic Mappings and the Geometry of Hypersurfaces

A.G. Vitushkin

Translated from the Russian
by P.M. Gauthier

## Contents

## Introduction

The principal topic of this paper is non-degenerate (in the sense of Levi) hypersurfaces of complex manifolds and the automorphisms of such hypersurfaces. The material on strictly pseudoconvex hypersurfaces is presented most completely. We discuss in detail a form of writing the equations of the hypersurface which allows one to carry out a classification of hypersurfaces. Certain biholomorphic invariants of hypersurfaces are considered. Especially, we consider in detail a biholomorphically invariant family of curves called chains. A lot of attention is given to constructing a continuation of a holomorphic mapping.

The posing of the problem under consideration and the first concrete results go back to Poincaré [31]. His method of studying a hypersurface was to analyze its equations directly. In studying the classification of domains in $\mathbb{C}^2$, he formulated a series of concrete problems: the classification of real analytic hypersurfaces in terms of their defining equations, continuation of the germ of a mapping from one analytic hypersurface to another preassigned hypersurface, and others. Poincaré showed that a germ of a biholomorphic mapping of the sphere into itself extends to the whole sphere and moreover is a linear-fractional transformation. As a consequence the general form for an automorphism of the sphere was written out. Poincaré remarked that a certain family of series, of a special form having different coefficients of sufficiently high degree, yield pairwise non-equivalent hypersurfaces.

Segre [33], E. Cartan [10], and later Tanaka [36] worked out different approaches to the construction of a classification theory based on geometric methods. This topic acquired widespread popularity in the 70's following the work of Alexander [1], and Chern and Moser [11]. The paper of Alexander drew attention to itself by the clarity of its results. Therein, the above-mentioned result of Poincaré on the continuation of germs of mappings of the sphere, till then forgotten, is obtained anew. The paper of Chern and Moser has many levels. This paper develops both analytic and geometric methods and it gave rise to a lengthy cycle of works by other authors (Fefferman, Burns, Shnider, Diederich, Wells, Webster, Pinchuk, and others). The present chapter gives a survey of this theme over the past 10–15 years. In particular, we relate quite explicitly the results in this direction obtained in recent years in our common seminar with M.S. Mel'nikov at Moscow University (V.K. Bieloshapka, V.V. Ezhov, S.M. Ivashkovich, N.G. Kruzhilin, A.V. Loboda, and others).

Let us dwell on some of the results of the topic under consideration. In the geometric theory of Chern, the surface is characterized in terms of a special fibration. The base of the fibration is the surface itself, while the fiber is the stability group of a quadric (a quadric is the set of zeros of a real polynomial of order two; the stability group of a surface is the group of automorphisms, defined in the neighbourhood of a fixed point of the surface, which keep this point fixed). On this fiber bundle there is a finite set of differential forms which are invariant with respect to biholomorphic mappings of the base and which together uniquely determine the surface (see §10). The correspondence between these families of forms and surfaces in geometry is called a classification of surfaces.

In the frame of analytic methods, a surface is characterized in terms of special equations which Moser [11] calls the normal form of the surface (see §1). In the general situation, one and the same surface has, generally speaking, many different normal forms associated to it. The totality of all changes of coordinates, which bring the surface into normal form, forms a group which is isomorphic to the stability group of a quadric. This means that a stability group of an arbitrary surface is represented as a subgroup of the stability group of a quadric.

The class of surfaces with non-degenerate Levi form is divided into two types: in the first type we put quadrics and surfaces which are locally equivalent to them, and in the second we place all remaining surfaces, that is, surfaces not equivalent to quadrics. It turns out that the properties of surfaces of type I and the properties of surfaces of type II have little in common. In fact many properties of these subclasses are different. Thus, for example, the stability group for the sphere is non-compact, while for a strictly pseudoconvex surface of type II, it is compact [26]. On the other hand, from the point of view of local structure, almost everything is known about surfaces of type I. Consequently, interesting discoveries are, as a rule, in connection with surfaces of type II.

Among the results of a general nature which follow from the theory of Moser, one should mention foremost the theorem of V.K. Beloshapka [2] and A.V. Loboda [26]: if a surface is not equivalent to a quadric, its stability group can be embedded into the group of matrices preserving the Levi form of the surface (see §9). The dimension of the group of these matrices is not much smaller than the dimension of the stability group of a quadric. However, the group of these matrices is constructed essentially more simply than the group of a quadric. It is useful to compare the latter with what was said earlier concerning automorphisms in relation to the work of Chern and Moser.

The class of strictly pseudoconvex surfaces is of particular interest. A fundamental result for this class of surfaces is the theorem on the germ of a mapping. If a surface passing through the origin is real analytic, strictly pseudoconvex, and non-spherical (locally non-equivalent to a sphere), then the germ of a biholomorphic mapping, from one such surface to another and fixing the origin, has a holomorphic extension to a neighbourhood common for all such mappings. Moreover, a guaranteed size of the neighbourhood as well as for a constant estimating the norm of the continuation are determined in terms of the parameters of analyticity of the surface and the degree of non-sphericity (see §8).

The theorem on germs of mappings was the result of many years of work in our seminar. At first we expected to obtain its proof by purely analytic methods, more precisely by describing and estimating the dependent parameters of the stability groups. The resulting theorem of V.K. Beloshapka and A.V. Loboda turned out to be no less interesting in itself. An essential consequence was also the proof of compactness of the stability group ([4], [37]). With regards to the continuation of mappings, we succeeded at first, in this direction, to obtain only a few special cases of the theorem under consideration ([4], [37]; for statements, see §7.1 and §8.3). The obstruction was that the estimate for the dependent parameters automatically contained some additional quantities. Further progress required not only a perfecting of the analytic methods but also a fundamental investigation of a family of curves called chains ([38], [13], [12], [23], [14]). For such chains, the so-called circular form for the equations of the surface and the normal parametrization of a chain were introduced [38]. The result of this series of works was the theorem on the continuation of a mapping along a compact surface ([14], see §8.4). In [40], we succeeded finally in obtaining the necessary estimates on the size of the dependent parameters (see Lemma in §8.1) which allowed us to complete the solution of the problem on germs of mappings.

In closing we point out one more most promising result (N.G. Kruzhilin, A.V. Loboda). In the neighbourhood of each point of a strictly pseudoconvex, non-spherical surface, we may choose coordinates in which each local automorphism is a linear transformation (see §9.3). The question of the existence of analogous coordinates on surfaces whose Levi form is indefinite remains open.

This paper consists essentially of the paper [40] revised according to the requirements of the present series and containing complementary results on smooth surfaces (§10). Short proofs of the fundamental results are included.

## §1. The Normal Form for Representing a Hypersurface

In this section we consider a special form of representing a surface that was introduced by Moser. It is in a certain sense the simplest form, and so it has many applications.

**1.1. The Linear Normal Form.** Let $M$ be a real-analytic hypersurface in an $n$-dimensional complex manifold $X$. For every $\zeta \in M$ we can choose a local coordinate system $z=(z_1, \ldots, z_{n-1})$, $w=u+iv$, on $X$ for which $\zeta$ is the origin and $M$ is defined by $F_0(z, \bar{z}, w, \bar{w})=0$ where $dF_0(0) \neq 0$. We rotate $M$ about the origin until the $v$-axis is perpendicular to the surface. Then $M$ can be written in the form $v=F_1(z, \bar{z}, u)$, where $F_1(0)=0$ and $dF_1(0)=0$. Expanding $F_1$ as a series in $z, \bar{z}$ and isolating terms of the form $z_i \bar{z}_k$, we write the surface as

$$v=\langle z, z\rangle+F(z, \bar{z}, u), \tag{1.1.1}$$

where $\langle z, z\rangle=\sum_{i,k} a_{i,k} z_i \bar{z}_k$ is a Hermitian form and $F$ is a real-analytic function such that

$$F(0)=0, \quad dF(0)=0, \quad \left.\frac{\partial^2 F}{\partial z_1 \partial \bar{z}_k}\right|_0=0 \quad (i, k=1,2, \ldots, n-1).$$

The form $\langle z, z\rangle$ is called the Levi form of $M$ at the point $\zeta$. We shall consider hypersurfaces for which the Levi form is non-degenerate at all points.

If $n=1$, then the form $\langle z, z\rangle$ is not defined, and does not come into the equation of the surface (in this case, a one-dimensional curve). Therefore, by a change of coordinates any curve can be reduced to the form $v=0$. For $n \geq 2$ no biholomorphic change of coordinates can remove $\langle z, z\rangle$ from the equation of the surface. This means that the form of the surface cannot be significantly simplified.

By making an appropriate change of variables we may assume that

$$\langle z, z\rangle = \sum_{j=1}^{s} z_j\bar{z}_j - \sum_{j=s+1}^{n-1} z_j\bar{z}_j.$$

We define the differential operator $\Delta$ by the formula

$$\Delta = \sum_{j=1}^{s} \frac{\partial^2}{\partial z_j \partial \bar{z}_j} - \sum_{j=s+1}^{n-1} \frac{\partial^2}{\partial z_j \partial \bar{z}_j}.$$

Moser [11] showed that for each surface of the form (1.1.1) there is a biholomorphic map that sends this surface into one of the form

$$v = \langle z, z\rangle + \sum_{k,l \geq 2} F_{kl}(z, \bar{z}, u), \tag{1.1.2}$$

where $F_{kl}$ is a polynomial of degree $k$ in $z$ and $l$ in $\bar{z}$ with coefficients depending analytically on $u$, and $F_{22}$, $F_{32}$, and $F_{33}$ satisfy the conditions $\operatorname{tr} F_{22} = \operatorname{tr}^2 F_{32} = \operatorname{tr}^3 F_{33}$. The operator tr is a second order differential operator defined by the Levi form [11]. If

$$\langle z, z\rangle = \sum_{k=1}^{s} z_k\bar{z}_k - \sum_{k=s+1}^{n-1} z_k\bar{z}_k, \text{ then } \operatorname{tr}F_{kl} = \frac{1}{kl}\Delta F_{kl}.$$

Regarding surfaces of the form (1.1.2) we shall say that they and their equations have *linear normal* form (along with the linear form we introduce below the so-called circular form). Any biholomorphic map sending a surface of the form (1.1.1) into one of the form (1.1.2), that is, into a surface written in normal form, and leaving the origin fixed, will be called a *normalizing* map, or simply a *normalization* of the surface (or sometimes a *reduction* to normal form).

**1.2. The Initial Data of a Normalization.** We consider as an example the hyperquadric $v = \langle z, z\rangle$. Poincaré [31] and Tanaka [35] showed that any map defined and biholomorphic in a neighbourhood of the origin which sends this quadric into itself has the form

$$\begin{aligned} z^* &= \lambda U(z + aw)/\{1 - 2i\langle z, a\rangle - (r + i\langle a, a\rangle)w\}, \\ w^* &= \sigma\lambda^2 w/\{1 - 2i\langle z, a\rangle - (r + i\langle a, a\rangle)w\}, \end{aligned} \tag{1.2.1}$$

where $\sigma = \pm 1$, $\lambda > 0$ and $r$ are real number ($\sigma$ can take the value $-1$ only when the number of positive eigenvalues of $\langle z, z\rangle$ is equal to the number of negative ones), $a$ is an $(n-1)$-dimensional vector, and $U$ is an $(n-1) \times (n-1)$ matrix such that

$$\langle Uz, Uz\rangle = \sigma\langle z, z\rangle. \tag{1.2.2}$$

The converse is also true: if the set $\omega = (U, a, \lambda, \sigma, r)$ satisfies (1.2.2), then the corresponding map (1.2.1) sends the hyperquadric into itself. Thus, for hyperquadrics there are as many normalizations as there are sets $\omega = (U, a, \lambda, \sigma, r)$ satisfying (1.2.2).

Suppose that $H$: $z^* = f(z, w)$, $w^* = g(z, w)$ sends a surface of the form (1.1.1) into one of the same form, for example, into a surface defined in normal form. With every such map we associate a set $\omega = (U, a, \lambda, \sigma, r)$ defined by the system

$$\left.\frac{\partial f}{\partial z}\right|_0 = \lambda U, \qquad \left.\frac{\partial f}{\partial w}\right|_0 = \lambda U a$$
$$\left.\frac{\partial g}{\partial w}\right|_0 = \sigma\lambda^2, \qquad \left.\mathrm{Re}\frac{\partial^2 g}{\partial w^2}\right|_0 = 2\sigma\lambda^2 r. \tag{1.2.3}$$

We shall call this set the *set of initial data* of $H$. In particular if $H$ is a normalization, the set is called the set of initial data of the normalization.

**1.3. Moser's Theorem.** The main result in Moser's paper [11] can be stated as follows.

**Theorem.** *For each surface $M$ of the form* (1.1.1) *and any set of initial data $\omega$ there is a unique normalization $H_\omega(M)$ having initial data $\omega$. If $M$ is defined in linear normal form and $a = 0$, then $H_\omega$ has the form*

$$z^* = \lambda U z/(1 - rw), \quad w^* = \sigma\lambda^2 w/(1 - rw).$$

Here we must warn the reader who wishes to become acquainted with Moser's work: in [11] the initial data are introduced in terms of the inverse map of a normalization. It is more convenient for us to use a different definition. Let us explain this. A normalization is uniquely defined by the set of first derivatives and a single parameter, which is calculated from the second derivatives. When the initial data are defined in terms of the inverse map, this parameter is expressed in terms of the second derivatives of the map and the coefficients $\partial^2 F/\partial z_i \partial z_k$ (see (1.1.1)) of the normalized surface (it is not good that the definition of the parameter depends on the coefficients of the surface). Moser stated the theorem for surfaces that had been simplified beforehand (the holomorphic quadratic part of the series was removed by a substitution); in this case the formula turned out to be simple. In applications this is inconvenient because the preliminary processing has to be carried through the whole statement. These difficulties do not arise if the initial data are defined as we do here.

Below we discuss several applications of this theorem to the classification of hypersurfaces, to the estimation of the dimension of the automorphism group, and to the construction on hypersurfaces of a special family of curves called chains that is important in applications.

**1.4. The Classification of Hypersurfaces.** Let $M$ and $M^*$ be two hypersurfaces in $\mathbb{C}^n$ containing the origin. We shall say that $M$ and $M^*$ are *equivalent* if in a neighbourhood of the origin there is a biholomorphic change of variables sending $M$ into $M^*$. Clearly, if $M$ and $M^*$ are equivalent, then a normal form of

$M$ is at the same time a normal form of $M^*$. Conversely, if $M$ and $M^*$ can be represented by the same normal form, then they are equivalent.

The transition from one normal form of a surface to another is uniquely defined by the initial data (see §1.3). Therefore, the family of normal forms of a given surface has finite dimension. Since the normality of the form for a hypersurface involves no conditions on the coefficients of the polynomials $F_{kl}$ for $k>3$ and $l>3$, we see that the space of hypersurfaces not equivalent to one another has infinite dimension. As Wells [46] has observed, in the two-dimensional case, this fact was proved by Poincaré [31]. A detailed discussion of classification problems can be found in Fefferman's paper [18].

**1.5. Proof of Moser's Theorem.** For simplicity, we shall assume that the Levi form has the form:

$$\langle z, z\rangle = \sum_{j=1}^{n-1} \varepsilon_j z_j \bar{z}_j, \qquad \text{where } \varepsilon_j = \pm 1.$$

In this case we have for the polynomial $F_{kl}(z, \bar{z}, u)$, of degree $k$ in $z$ and $l$ in $\bar{z}$, that $\operatorname{tr} F_{kl} = (1/kl)\Delta F_{kl}$. We list several properties of the operator $\Delta$ which will be needed in the sequel:

a) If the matrix $U$ is such that $\langle Uz, Uz\rangle = \sigma\langle z, z\rangle$ (see §1.2), then from $\Delta F(z, \bar{z}, u) = 0$, it follows that

$$\Delta(Uz, \overline{Uz}, f(u)) = 0,$$

for each function $f$;

b) if $\Delta F(z, \bar{z}, u) = 0$, then $\Delta^2\langle z, z\rangle F(z, \bar{z}, u) = 0$;

c) each real polynomial $F_{22}(z, \bar{z}, u)$ can be written $F_{22} = \langle z, z\rangle({}^{\tau}z\,\Lambda\bar{z}) + N_{22}(z, \bar{z}, u)$, where $\Lambda(u)$ is a Hermitian matrix, that is ${}^{\tau}\Lambda = \bar{\Lambda}$, and $\Delta^2 N_{22} = 0$;

d) each real polynomial $F_{33}(z, \bar{z}, u)$ can be written

$$F_{33} = F(u)\langle z, z\rangle^3 + N_{33}(z, \bar{z}, u), \qquad \text{where } \Delta^3 N_{33} = 0;$$

e) each real polynomial $F_{32}(z, \bar{z}, u)$ can be written

$$F_{32} = \langle z, z\rangle^2 \langle z, F(u)\rangle + N_{32}(z, \bar{z}, u),$$

where $\Delta^2 N_{32}(z, \bar{z}, u) = 0$.

Let us fix a hypersurface $M$ given by (1.1.1) and a curve $\gamma(t)$: $z = p(t)$, $w = q(t)$, lying on $M$, where $t$ is a real parameter, and $p$ and $q$ are analytic functions such that $\gamma(0) = 0$ and $q'(0) \neq 0$. We show that there exists a biholomorphic mapping in the neighbourhood of the origin which sends $\gamma$ to the line $z = 0$, $v = 0$, and maps $M$ onto a hypersurface $M(\gamma)$ of the form

$$v = \langle z, z\rangle + \sum_{k, l \geq 2} F_{kl}(z, \bar{z}, u), \tag{1.5.1}$$

where $\Delta F_{22} = 0$ and $\Delta^3 F_{33} = 0$ (for the definition of $F_{kl}$, see §1.1). Moreover, imposing further conditions on $\gamma$, we obtain that $\Delta^2 F_{32} = 0$.

**Lemma.** *There exists a biholomorphic mapping (in some neighbourhood of the origin) which maps the curve $\gamma$ into the line $z = 0$, $v = 0$ and maps $M$ onto a hypersurface $M(\gamma)$ given by* (1.5.1). *Any biholomorphic mapping of a hypersurface given by* (1.5.1) *onto another such hypersurface, and leaving the line $z = 0$, $v = 0$ fixed, is a linear fractional transformation of the form*

$$z^* = \frac{\lambda U z}{1 - rw}, \quad w^* = \frac{\sigma \lambda^2 w}{1 - rw},$$

*where $\sigma = \pm 1$, $\lambda > 0$, $r$ is a real number, and*

$$\langle Uz, Uz \rangle = \sigma \langle z, z \rangle.$$

We construct the desired mapping $\varphi$ as the composition of six biholomorphic transformations $\varphi = \varphi_6 \varphi_5 \ldots \varphi_1$. In constructing these transformations, we shall, at each step, denote the old variables by $(z, w)$ and the new variables by $(z^*, w^*)$. We denote by $M_i$ the hypersurface obtained from $M$ by the transformation $\varphi_i \varphi_{i-1} \ldots \varphi_1$, and we write its equation as follows:

$$v = F^{(i)}(z, \bar{z}, u) \overset{\text{def}}{=} \sum_{k,l} F^{(i)}_{kl}(z, \bar{z}, u), \qquad i = 1, 2, \ldots, 6.$$

1. We define the transformation $\varphi_1$ via its inverse

$$\varphi_1^{-1}\colon z = z^* + p(w^*), \qquad w = q(w^*).$$

The transformation $\varphi_1$ maps the curve $\gamma$ to the line $z^* = 0$, $v^* = 0$, and the hypersurface $M$ *to* $M_1$: $v^* = F^{(1)}(z^*, \bar{z}^*, u^*)$, where

$$F^{(1)}_{11}(z^*, \bar{z}^*, 0) = \langle z^*, z^* \rangle, \; F^{(1)}(0, 0, u^*) = 0, \; u^*|_{\varphi_1(\gamma)} = t.$$

2. By the transformation $\varphi_2$: $z^* = z$, $w^* = w + g(z, w)$, $g(0, w) = 0$ we map $M_1$: $v = F^{(1)}(z, \bar{z}, u)$ to $M_2$: $v^* = F^{(2)}(z^*, w^*, u^*)$, and we show that $g$ can be uniquely chosen such that

$$F^{(2)}_{k0}(z^*, \bar{z}^*, u^*) \quad \text{and} \quad F^{(2)}_{0k}(z^*, \bar{z}^*, u^*) = 0; \qquad k = 1, 2, \ldots.$$

If in the equation for the hypersurface $M_2$, we write $z^*$ and $w^*$ as functions of $z$ and $w$, we obtain

$$F^{(2)}\left( z, \bar{z}, u + \frac{1}{2}(g(z, w) + \bar{g}(z, w) \right) = \frac{1}{2i}(g(z, w) - \bar{g}(\bar{z}, \bar{w})) + F^{(1)}(z, \bar{z}, u),$$

where $w = u + iF^{(1)}(z, \bar{z}, u)$. In this equation we set $\bar{z} = 0$. Then, from the condition $g(0, w) = 0$ it follows that the term $\bar{g}(\bar{z}, \bar{w})$ turns out to be zero. And since $F^{(2)}(z, 0, u) = \sum_k F^{(2)}_{k0}(z, \bar{z}, u)$, from the condition $\sum_k F^{(2)}_{k0}(z, \bar{z}, u) = 0$, we obtain that

$$0=\frac{1}{2i}g(z,u+iF^{(1)}(z,0,u))+F^{(1)}(z,0,u). \tag{1.5.2}$$

Set $\xi = u+iF^{(1)}(z,0,u)$. Since $F^{(1)}(z,0,u)|_{z=0}=0$, it follows from the implicit function theorem, that $u$ is a function of $z$ and $\xi$: $u=\xi+G(z,\xi)$, where $G(0,\xi)=0$. On the other hand, from (1.5.2) we have $0=(1/2i)g(z,\xi)+(1/i)(\xi-u)$, that is $u=\xi+(1/2)g(z,\xi)$. Thus, $g(z,w)=2G(z,w)$ is the desired function. Dropping the asterisks, we can rewrite the equation of the hypersurface $M_2$ in the form

$$v=F^{(2)}_{11}(z,\bar z,u)+\sum_{\substack{k,l\geq 1\\ k+l\geq 3}}F^{(2)}_{kl}(z,\bar z,u),$$

where $F^{(2)}_{11}(z,\bar z,0)=\langle z,z\rangle$.

3. We define $\varphi_3$ by its inverse transformation $\varphi_3^{-1}$: $z=c(w^*)z^*$, $w=w^*$. We choose $C(w^*)$ in such a way that the form $F^{(3)}_{11}$, appearing in the equation for the hypersurface $M_3=\varphi_3(M_2)$ is independent of $u$ and is equal to $\langle z,z\rangle$; namely, we choose $C$, satisfying the system

$${}^{\tau}C(u)H(u)\bar C(u)=J,\quad {}^{\tau}C(u)J=J\bar C(u),$$

where $H(u)$ is the matrix of the form $F^{(2)}_{11}(z,\bar z,u)$, and

$$J=\begin{pmatrix}\varepsilon_1 & & 0\\ & \ddots & \\ 0 & & \varepsilon_{n-1}\end{pmatrix}$$

is the form of $\langle z,z\rangle$ ($\varepsilon_i=\pm 1$). From this system, we get that $(C^{-1}(u))^2=HJ=E+L(u)$, where $L(0)=0$. Thus, for small $|u|$, there exists a unique $C(u)$, satisfying this equation and such that $C(0)=E$. Making the change of variables $z\mapsto C(u+iv)z$, $w\to w$, from the equation for $M_2$, we obtain the equation for $M_3$:

$$v={}^{\tau}z\,{}^{\tau}C(u)H(u)\bar C(u)\bar z+\sum_{\substack{k,l\geq 1\\ k+l\geq 3}}F^{(3)}_{kl}(z,\bar z,u)=\langle z,z\rangle+\sum_{\substack{k,l\geq 1\\ k+l\geq 3}}F^{(3)}_{kl}(z,\bar z,u).$$

4. We choose the mapping $\varphi_4$ such that $M_4=\varphi_4(M_3)$ is of the form:

$$v^*=\langle z^*,z^*\rangle+\sum_{k,l\geq 2}F^{(4)}_{kl}(z^*,\bar z^*,u^*).$$

To this end, we rewrite $M_3$ in the form

$$v=\langle z,z\rangle+\sum_{i=1}^{n-1}(z_i\bar A_i(z,u)+\bar z_iA_i(z,u))+\sum_{k,l\geq 2}F^{(3)}_{kl}(z,\bar z,u),$$

where

$$\sum_{i=1}^{n-1}(z_i\bar A_i(z,u)+\bar z_iA_j(z,u))=\sum_{j=1}^{\infty}(F^{(3)}_{1j}+F^{(3)}_{j1}).$$

We define $\varphi_4$ as follows: $z^* = z+f(z, w)$, $w^* = w$, where $f(z, w)$ is the vector with coordinates $(\varepsilon_1 A_1(z, w), \ldots, \varepsilon_{n-1} A_{n-1}(z, w))$, and $\varepsilon_i = \pm 1 (i = 1, 2, \ldots, n-1)$ are the coefficients of the form $\langle z, z\rangle$. Replacing $z^*$ by $z+f(z, w)$ in the form $\langle z, z\rangle$, we have

$$\langle z^*, z^*\rangle = \sum_{i=1}^{n-1} \varepsilon_i z_i^* \bar{z}_i^* = \sum_{i=1}^{n-1} \varepsilon_i (z_i + A_i(z, w))(\bar{z}_i + \bar{A}_i(z, w)) =$$

$$= \langle z, z\rangle + \sum_{i=1}^{n-1} (z_i \bar{A}_i(z, u) + \bar{z}_i A_i(z. u)) + \ldots .$$

That is,

$$\langle z, z\rangle + \sum_{i=1}^{n-1} (z_i \bar{A}_i(z, u) + \bar{z}_i A_i(z, u)) = \langle z^*, z^*\rangle + \ldots .$$

Since, for each $i$, the function $A_i(z, w)$ is a sum of polynomials having degree at least 2 in $z$, then, expressing the variables $(z, w)$ in terms of $(z^*, w^*)$ in the equation of $M_3$, we see that $M_4$ has the desired form.

5. We write the change of variables $\varphi_5^{-1}: z = V(w^*)z^*, w = w^*$, where $\langle V(u)z, V(u)z\rangle = \langle z, z\rangle$. We choose $V$ in such a way that $M_5$ has the form.

$$v^* = \langle z^*, z^*\rangle + \sum_{k, l \geq 2} F_{kl}^{(5)}(z^*, \bar{z}^*, u),$$

where $\Delta F_{22}^{(5)} = 0$. Because of property c) of the operator $\Delta$, the polynomial $F_{22}$ can be written in the form

$$F_{22}(z, \bar{z}, u) = \langle z, z\rangle ({}^{\tau}z \Lambda \bar{z}) + N_{22}(z, \bar{z}, u).$$

Since

$$z = V(u^* + iv^*)z^* = V(u^*)z^* + V'(u^*)(i\langle z^*, z^*\rangle + \ldots)z^*,$$

then making the substitution $\varphi_5^{-1}$ in the equation for $M_4$ and removing asterisks, we obtain for $M_5$:

$$v = \langle z, z\rangle + \langle z, z\rangle (i\langle V'z, Vz\rangle - i\langle Vz, V'z\rangle) + \langle z, z\rangle ({}^{\tau}z^{\tau}V\Lambda \overline{Vz})$$

$$+ N_{22}(Vz, \overline{Vz}, u) + \ldots .$$

Since $\Delta N_{22} = 0$, the equation $\Delta F_{22}^{(5)} = 0$ is equivalent to the equation

$$i\langle V'z, Vz\rangle - i\langle Vz, V'z\rangle + ({}^{\tau}z^{\tau}V\Lambda J^{-1} J\overline{Vz}) = 0,$$

which, in turn, is equivalent to $V(u)$ satisfying the equation $V'(u) = (1/2)J\Lambda(u)V(u)$. Let us recall that $J = J^{-1}$ is the matrix of the form $\langle z.z\rangle$. The function

$$V(u) = \exp\left(\frac{i}{2} J \int \Lambda(t)dt\right)$$

satisfies this equation and also

$$\langle V(u)z, V(u)z\rangle = \langle z, z\rangle.$$

Thus, we have constructed the desired transformation $\varphi_5$.

6. For $\varphi_6$ we choose a transformation of the form $z^* = \sqrt{Q'(w)}\,Uz$, $w^* = \sigma Q(w)$, where $\langle Uz, Uz\rangle = \sigma\langle z, z\rangle$, $Q(0)=0$, $\overline{Q(u)}=Q(u)$, and $Q'(0)>0$. The function $Q$ will be chosen in such a way that for the hypersurface

$$M_6\colon v^* = \langle z^*, z^*\rangle + \sum_{k,l\geq 2} F_{kl}^{(6)}(z^*, \bar{z}^*, u^*),$$

the conditions $\Delta F_{22}^{(6)} = 0, \Delta^3 F_{33}^{(6)} = 0$ are satisfied.

From the definition of $\varphi_6$ we have

$$v^* = \operatorname{Im}\sigma\left(Q(u)+ivQ'(u)-\frac{v^2}{2}Q''(u)-\frac{1}{6}iv^3Q'''+\ldots\right)=\sigma Q'v-\frac{\sigma}{6}Q'''v^3+\ldots,$$

$$\langle z^*, z^*\rangle = \left[\left(Q'+ivQ''-\frac{v^2}{2}Q'''+\ldots\right)\left(Q'-ivQ''-\frac{v^2}{2}Q'''+\ldots\right)\right]^{1/2}\langle z, z\rangle=$$

$$= Q'\sigma\langle z, z\rangle-\frac{1}{2}\left(Q'''-\frac{(Q'')^2}{Q'}\right)v^2\sigma\langle z, z\rangle+\ldots.$$

From this, we obtain

$$F_{22}^{(6)}(z^*, \bar{z}^*, u^*)=(Q')^2F_{22}^{(6)}(Uz, \bar{U}\bar{z}, \sigma Q(u))+\ldots$$

$$F_{32}^{(6)}(z^*, \bar{z}^*, u^*)=(Q')^{5/2}F_{32}^{(6)}(Uz, \bar{U}\bar{z}, \sigma Q(u))+\ldots$$

$$F_{33}^{(6)}(z^*, \bar{z}^*, u^*)=(Q')^3F_{33}^{(6)}(Uz, \bar{U}\bar{z}, \sigma Q(u))+\ldots$$

After a change of variables, $M_5$ can be written in the form

$$\sigma Q'v = \sigma Q'\langle z, z\rangle+(Q')^2F_{22}^*(Uz, \bar{U}\bar{z}, \sigma Q(u))+(Q')^{5/2}F_{32}^*(Uz, \bar{U}\bar{z}, \sigma Q(u))+$$
$$+(Q')^3F_{33}^*(Uz, \bar{U}\bar{z}, \sigma(Q(u))-\tfrac{1}{6}(2Q''-3(Q'')^2/Q')\sigma\langle z, z\rangle^3+\ldots,$$

that is,

$$F_{22}^*(z, \bar{z}, u)=\frac{\sigma}{Q'}F_{22}^{(5)}(U^{-1}z, \bar{U}^{-1}\bar{z}, Q^{-1}(u)),$$

$$F_{32}^*(z, \bar{z}, u)=\frac{\sigma}{(Q')^{3/2}}F_{32}^{(5)}(U^{-1}z, \bar{U}^{-1}\bar{z}, Q^{-1}(u)), \tag{1.5.2}$$

$$F_{33}^*(z, \bar{z}, u)=\frac{\sigma}{(Q')^3}\left[Q'F_{33}^{(5)}(U^{-1}z, \bar{U}^{-1}\bar{z}, Q^{-1}(u))+\tfrac{1}{6}(2Q'''-3(Q'')^2/Q')\right]\langle z, z\rangle^3.$$

From properties a) and d) of the operator $\Delta$, we have $\Delta F_{22}^* = 0$. The condition $\Delta^3 F_{33}^* = 0$ is satisfied if and only if $Q$ satisfies the equation

$$Q'F(u)+\frac{1}{6}\left(2Q'''-3\frac{(Q'')^2}{Q'}\right) = 0, \quad Q(0) = 0, \quad Q'(0)>0,$$

where $F$ is a real function such that

$$F_{33}(z, \bar{z}, u) = F(u)\langle z, z\rangle^3 + N(z, \bar{z}, u),$$

with $\Delta^3 N = 0$. The substitution $Q' = (S(u))^{-2}$ changes this equation to the linear equation $S'' = (3/2)F(u)S$, and hence, it is not difficult to see that the initial equation has a solution monotonic in $u$. Thus, $\varphi_6$ and consequently $\varphi$, are defined.

From the above calculations, it follows that any transformation which maps the hypersurface (1.5.1) into another such hypersurface, and which maps the line $z = 0$, $v = 0$ into itself, is of the form: $z = \sqrt{Q'(w)}\, Uz$, $w^* = \sigma Q(w)$. In this case, $Q$ satisfies the above differential equation in which $F(u) = 0$ and hence has the form stated in the theorem.

We now show that for a particular choice of the curve $\gamma$, the condition $\Delta^2 F_{32} = 0$ will be satisfied. From formula (1.5.2) it is clear that if $\Delta^2 F_{32} = 0$ for some choice of parameter on $\gamma$, then after a change of parameter, this condition will still be satisfied. We show that the condition $\Delta^2 F_{32} = 0$ is equivalent to $\gamma(t)$ satisfying a certain differential equation of order two. The inverse mapping of $\varphi_5\varphi_4 \ldots \varphi_1$ has the form:

$$z = p(w^*) + T(w^*)z^* + \ldots, \qquad w = q(w^*) + \ldots.$$

For definiteness, we shall assume that the original parameter $t$ is the value of the $u$-coordinate. In the equation for the original hypersurface $M$, if we substitute the above expressions for $z$ and $w$, we obtain that the term $F_{32}^{(5)}$, in the equation for the hypersurface $M_5$, has the form

$$F_{32}^{(5)} = \langle z, Bp''\rangle\langle z, z\rangle^2 + K_{32},$$

where the matrix $B$ as well as the coefficients of the polynomial $K_{32}$ depend analytically on $p, \bar{p}, p'$, and $\bar{p}'$, and for small $|u|$, the matrix $B$ is nonsingular. Thus, the condition $\Delta^2 F_{32}^{(5)} = 0$ is equivalent to $\gamma$ satisfying an ordinary differential equation of order two. A geometric solution of this equation is uniquely defined by the direction of a tangent vector at the origin.

Thus, the transformation $\varphi$ constructed with the indicated choice of the curve $\gamma$ is a normalization mapping. Here, the direction of the curve $\gamma$, the real numbers $Q'(0)$ and $Q''(0)$ and the matrix $U$ can be chosen arbitrarily provided $\langle Uz, Uz\rangle = \sigma\langle z, z\rangle$. It is not hard to convince oneself that these parameters can be so chosen that the normalizing mapping $\varphi$ has any preassigned set of initial data $\omega$.

We now show the uniqueness of a normalizing mapping having a given set of initial data. Let $\varphi$ and $\varphi^*$ be two normalizations of $M$ having $\omega$ as set of initial data.

Let us represent $\varphi^*$ in the form $\varphi^* = R(\varphi)$. By the lemma we proved in this section, $R$ is a fractional linear transformation. Since $\varphi$ and $\varphi^*$ have the same initial data, the set of initial data of the transformation $R$ is $(E, 0, 1, 1, 0)$, and hence, $R$ is the identity, i.e., $\varphi^* = \varphi$. The theorem is proved.

## §2. The Standard Normalization

Here we present a standard procedure that associates with any hypersurface a normal form, which is in a certain sense the most natural one, and we discuss certain properties and applications of this form.

**2.1. Definition of the Standard Normalization.** For a hypersurface $M$ the normalization $H_e(M)$ with the set of initial data $e = (E, 0, 1, 1, 0)$ (see §1.2) is called the *standard normalization* of $M$. We write $H_e^{-1}$ in the form $z = \varphi(z^*, w^*)$, $w = \psi(z^*, w^*)$. It is easy to see that $\varphi$ and $\psi$ satisfy the relations

$$\left.\frac{\partial \varphi}{\partial z^*}\right|_0 = E, \quad \left.\frac{\partial \varphi}{\partial w^*}\right|_0 = 0, \quad \left.\frac{\partial \psi}{\partial z^*}\right|_0 = 0, \quad \left.\frac{\partial \psi}{\partial w^*}\right|_0 = 1, \quad \left.\frac{\partial^2 \psi}{\partial w^{*2}}\right|_0 = 0.$$

In other words, the tangent mappings of $H_e$ and $H_e^{-1}$ are the same, and so are the real parts of the distinguished second derivatives.

**2.2. Approximation of a Normalization by a Linear-Fractional Mapping.** Let $M$ and $M^*$ be two hypersurfaces defined in normal form, and $H_\omega$ be a mapping sending $M$ into $M^*$ with the set of initial data $\omega$. We consider the composition $H_e(R_\omega)$, where $R_\omega$ is a linear-fractional mapping with the set of initial data $\omega$ (see §1.2), and $H_e$ is the standard normalization of the surface $R_\omega(M)$. We write $H_e$ as

$$z^* = z + f^*(z, w), \quad w^* = w + g^*(z, w).$$

**Lemma.** *The mapping $H_e(R_\omega) = H_\omega$ and the corresponding functions $f^*$ and $g^*$ are such that the order of smallness (with respect to $(z, w)$) of $f^*$ at the origin is not less than third order, and of $g^*$ not less than fourth order.*

It follows from the lemma, in particular, that any normalizing mapping $H$ of a hypersurface $M$ can be represented as $H = H_e(R_\omega(H_e))$, where $H_e$ (the first mapping) is the standard normalization of $M$, $R_\omega$ is a linear-fractional map with the set of initial data $\omega$, and the last mapping in the composition is the standard normalization of $R_\omega(H_e(M))$.

It is proved in [11] that $f^*$ and $g^*$ have second order of smallness at the origin. The result in the present paper was obtained by Kruzhilin [23]. The assertion can be restated thus: $H_\omega - R_\omega$ is defined by functions with third order of smallness, and the function corresponding to $w$ has fourth order of smallness.

**2.3. Parametrization of Mappings.** We first show that any mapping $H$: $z^* = f(z, w)$, $w^* = g(z, w)$ sending a hypersurface $M$ of the form (1.1.1) into another such hypersurface $\tilde{M}$ can be represented as the composition of standard transformations and a linear-fractional transformation. Clearly, $H$ can be written as $H = H_e^{-1}(H^*(H_e))$, where $H_e$ is a standard normalization and $H^*$

sends $M^* = H_e(M)$ into $\tilde{M}^* = H_e(\tilde{M})$. By the lemma in §2.2, $H^* = H_e(R_{\omega^*})$, where $R_{\omega^*}$ is a linear–fractional transformation with the set of initial data $\omega^*$. Thus $H = (H_e^{-1}(H_e(R_{\omega^*}H_e))$, that is, $H$ is the composition of a standard reduction, a linear–fractional map, another standard reduction, and the inverse of the standard normalization. The representation of $H$ in this form is unique, and so it uniquely defines the set $\omega^*$. The elements of $\omega^*$ are called the *parameters* of the map, and $\omega^*$ the *set of parameters* of the map $H$. From the lemma in §2.2 and properties of $H_e$ it can be deduced that all the elements of $\omega^*$ apart from $r^*$ are the same as the corresponding elements of the set of initial data of $H$, that is, that the following system of equalities holds:

$$\left.\frac{\partial f}{\partial z}\right|_0 = \lambda^* U^*, \quad \left.\frac{\partial f}{\partial w}\right|_0 = \lambda^* U^* a^*, \quad \left.\frac{\partial g}{\partial w}\right|_0 = \sigma^* \lambda^{*2}.$$

In general, the equality $\operatorname{Re}\left.\frac{\partial^2 g}{\partial w^2}\right|_0 = 2\sigma^*\lambda^{*2}r^*$ does not hold. $\operatorname{Re}\left.\frac{\partial^2 g}{\partial w^2}\right|_0$ is expressed in terms of elements of $\omega^*$ and $\left.\frac{\partial^2 \tilde{F}}{\partial z_i \partial z_k}\right|_0$ $(i, k = 1, 2, \ldots, n-1)$, where $\tilde{F}$ is the function in the equation of $\tilde{M}$: $v = \langle z, z\rangle + \tilde{F}$. It turns out that if $\tilde{M}$ is a hypersurface of the form

$$v = \langle z, z\rangle + \tilde{F}, \tag{2.3.1}$$

where $\left.\frac{\partial^2 \tilde{F}}{\partial z_i \partial z_k}\right|_0 = 0$ $(i, k = 1, 2, \ldots, n-1)$, then $\operatorname{Re}\left.\frac{\partial^2 g}{\partial w^2}\right| = 2\sigma^*\lambda^{*2}r^*$. This means, in particular, that if $H$ is a normalizing mapping, then the set of its parameters completely coincides with the set of its initial data (see §1.2). Hence the sets of parameters and of initial data are the same for any mapping sending a hypersurface of the form (2.3.1) into the same hypersurface. The set of initial data $\omega$ for such a mapping and the set of initial data $\tilde{\omega}$ of the inverse mapping are expressed in terms of one another in the same way as the analogous sets of a linear–fractional mapping

$$\tilde{U} = U^{-1}, \quad \tilde{a} = -\frac{\sigma U a}{\lambda}, \quad \tilde{\lambda} = \frac{1}{\lambda}, \quad \tilde{r} = -\frac{\sigma r}{\lambda^2}. \tag{2.3.2}$$

## §3. Chains

We introduce on a hypersurface a family of curves called chains. As we shall see later, this concept works well in the study of holomorphic mappings of hypersurfaces. A family of chains was constructed in the two-dimensional case by E. Cartan [10], and in the general case by Chern and Moser [11].

**3.1. Definition of a Chain.** A *chain* on a hypersurface $M$ is a maximally continued curve which, after the hypersurface is reduced to normal form in a neighbourhood of any point on the curve, by a certain normalization, is defined locally by $z = 0$, $v = 0$.

Clearly, a chain is an analytic curve, which at all points is transversal to the complex tangent space. Geometrically the construction of such curves is not simple. Fefferman [17] constructed an example of a hypersurface with a nondegenerate Levi form, on which chains are constructed as spirals; a chain, while keeping close to the complex tangent and slowly decreasing the diameter of the loops, approaches a point of the surface. It is still not clear whether a chain can approach a point while touching the complex tangent space but without increasing its curvature.

We show that for any point $\zeta \in M$ and any direction transversal to the complex tangent space we can find a unique chain through this point and tangent to the given direction. We choose a coordinate system with $\zeta$ as origin and in which $M$ has the form $v = \langle z, z \rangle + \ldots$ (see (1.1.1)). For a hypersurface in this form we can regard a change of variables straightening a chain as a normalization. Conversely, any normalization distinguishes a chain. A normalization $H_\omega$ with the set of initial data $\omega = (U, a, \lambda, \sigma, r)$ can be represented as the composition of three transformations: the linear transformation $z^* = z + aw$, $w^* = w$, the standard reduction $H_e$ (see §2.1), and a certain transformation $H_{\omega*}$. The tangent map for $H_{\omega*}$ has the form $z^* = \lambda U z$, $w^* = \sigma\lambda^2 w$. $H_{\omega*}$ sends one normal form into another, and the second element in its set of initial data (the vector $a^*$) is zero. Consequently, by Moser's theorem, $H_{\omega*}$ is a linear-fractional transformation leaving the line $z = 0$, $v = 0$ fixed. Thus, all normalizations with the same second element in the set of initial data (the vector $a$) send one and the same curve into the line $z = 0$, $v = 0$. This curve is tangent to the direction $(-a, 1)$ at the origin; this means that in any direction there is no more than one chain. Such a chain exists, since the vector $a$ in the construction of the normalization can be chosen arbitrarily.

Thus, the chain through a fixed point of a hypersurface is uniquely defined by its direction at this point. For a hypersurface in the form $v = \langle z, z^* \rangle + \ldots$ the direction of the chain through the origin can be characterized by the vector $(a, 1)$, where $a = (a_1, \ldots, a_{n-1})$. If a hypersurface $M$ is defined by an equation $F(z, \bar{z}, w, \bar{w}) = 0$, then at $\zeta \in M$ we define the direction of the chain $\gamma$ through $\zeta$ by a vector $a_\zeta$ lying in the complex tangent space to the surface at this point. We define $a_\zeta$ as the projection onto the complex tangent space of the tangent vector to the chain at $\zeta$, normalized so that the length of its projection onto the normal to the complex tangent space is 1 (here we mean the normal lying in the tangent space of the surface). We see that the vectors $a$ and $-a$ define one and the same chain.

**3.2. Chains on Quadrics.** On the hyperquadric $v = \langle z, z \rangle$ any chain through the origin can be obtained from the line $z = 0$, $v = 0$ by an automorphism of the quadric (see (1.2.1)). Hence, a chain on a quadric is a circle (or a straight line). It is the intersection of the quadric with a complex line. The chain in the direction $(a, 1)$ has curvature $\chi = 2\langle a, a\rangle/[1 + |a|^2]^{1/2}$. The angle $\alpha$ between the direction of the chain and the complex tangent space is the same at all points of the chain, and is found from $\cot\alpha = |a|$.

In the sequel we shall need another type of hyperquadric, namely, hypersurfaces of the form $1 - |w|^2 = \langle z, z\rangle$. This hyperquadric and the hyperquadric $v = \langle z, z\rangle$ considered earlier are obtained from one another by a linear-fractional transformation. The mapping sending $v = \langle z, z\rangle$ into $1 - |w^*|^2 = \langle z^*, z^*\rangle$ has the form $z^* = 2z/(i + w)$, $w^* = (i - w)/(i + w)$.

The general form of an automorphism of the hypersurface $1 - |w|^2 = \langle z, z\rangle$ that leaves the point $z = 0$, $w = 1$ fixed is

$$z^* = \frac{\lambda U(z - a(w - 1))}{\delta}, \quad w^* - 1 = \frac{\sigma\lambda^2(w - 1)}{\delta},$$

where $\delta = 1 + \langle z, a\rangle + (1/2)(1 - \sigma\lambda^2 - \langle a, a\rangle + ir)(w - 1)$. As before, $\sigma = \pm 1$. $\lambda > 0$ and $r$ are real numbers, $a$ is an $(n - 1)$-dimensional vector, and $U$ is a matrix satisfying the condition

$$\langle Uz, Uz\rangle = \sigma\langle z, z\rangle.$$

On the hypersurface $1 - |w|^2 = \langle z, z\rangle$ the chain passing through the point $z = 0$, $w = 1$ is again a circle (or a line). The curvature of the chain with the direction $a$ is given by $\chi = (1 + \langle a, a\rangle)/[1 + |a|^2]^{1/2}$, and the angle $\alpha$ between the chain and the complex tangent space is found from $\cot\alpha = |a|$, as before. This condition defines a system of differential equations on the chain.

**3.3. The Linear Normal Parameter.** Suppose that a hypersurface is reduced to linear normal form so that a segment of a chain $\gamma$ goes into the line $z = 0$, $v = 0$. In this case the values of the coordinate $u$ define a certain parametrization on this segment of $\gamma$. By considering different reductions of $M$ to linear normal form, we obtain on each chain a certain family of parameters. Any parameter of this family will be called a *linear normal* parameter. As we can easily see these parameters are expressed in terms of one another by linear-fractional transformations of the real line, of the form $u^* = \sigma\lambda^2 u/(1 - nu)$ (see the theorem in §1.2).

For $s \neq (n - 1)/2$ the transformations preserve the orientation of the line. Recall that $s$ is the number of positive eigenvalues of the Levi form. Thus, for $s \neq (n - 1)/2$ a certain orientation is distinguished on the chains. A first order differential form that is invariant under biholomorphic transformations is defined in terms of normal families on hypersurfaces. If a hypersurface is given in linear normal form, then at the origin, it is written in the form $k\,du$, where $k$ is the norm of $F_{22}$, that is, the square root of the sum of squares of the coefficients of the polynomial written in symmetrized form. This form characterizes the nonsphericity of a hypersurface: if it vanishes on a set of positive measure, then the hypersurface is spherical [42].

## §4. The Equation of a Chain

If a hypersurface $M$ is defined by $v = \langle z, z\rangle + \ldots$ then $(z, u)$ can be regarded as a local system of coordinates on $M$. In the coordinates $(z, u)$, chains through the point $z = 0$ are the integral curves of a system of differential equations

$$\frac{\partial^2 z}{\partial u^2} = \Lambda\left(z, \frac{\partial z}{\partial u}, u\right)$$

with a real-analytic right hand side. Variants of this equation have been given by Moser [11], and Burns and Shnider [7]. There are also other interpretations of a family of chains (see [11], [15] and [17]). Fefferman's construction [17] is interesting. He constructed a special bundle (with the surface as base and a circle as a fibre) and a metric on this bundle that is invariant under biholomorphic transformations of the surface; projections of the light rays of this metric onto the surface are chains ([6] and [41]).

**4.1. Straightening a Hypersurface Along an Analytic Curve.** We explain how the equation of a chain arises. We fix on $M$ an analytic curve $\gamma$, which at all its points is transversal to the complex tangent space. We say that a mapping *straightens* $\gamma$ if it sends points of this curve into the line $z = 0$, $v = 0$. By analyzing the construction of a reduction to normal form, we can construct a mapping that straightens a given curve and sends the hypersurface into the form

$$v = \langle z, z\rangle + \sum_{k,l \geq 2} F(z, \bar{z}, u),$$

where $\operatorname{tr} F_{22} = 0$ and $\operatorname{tr}^3 F_{33} = 0$. It turns out that, in general, the condition $\operatorname{tr}^2 F_{23} = 0$ is not fulfilled. We call the constructed mapping a *normalization straightening* $\gamma$, or a *normalization of the surface along* $\gamma$. We emphasize that for a hypersurface of this form any mapping sending one hypersurface into another of the same form and leaving the line $z = 0$, $v = 0$ fixed is, as in the case of chains, a linear–fractional transformation.

Clearly, a curve is a chain if and only if there is a normalization straightening it, after which the condition $\operatorname{tr}^2 F_{23} = 0$ holds. This condition yields a system of differential equations on the chain.

**4.2. The Equation of a Chain in the Natural Parameter.** Let $M \subset \mathbb{C}^n$ be a hypersurface with a non-degenerate Levi form and defined by an equation $A(\zeta, \bar{\zeta}) = 0$. We denote by $\langle z, z\rangle_\zeta$ the Levi form of $M$ at the point $\zeta$. Let us state more precisely what we have in mind. We transfer the origin to $\zeta$ and by a unitary transformation reduce the hypersurface to the form $v = \langle z, z\rangle_\zeta + F(z, \bar{z}, u)$ (see (1.1.1)). This transformation is defined up to a unitary change of variables in the plane $w = 0$; this limitation is not essential, since we shall speak only about the value of $\langle z, z\rangle_\zeta$ on one or another vector in the complex tangent space to the surface at $\zeta$, and this is independent of the choice of coordinates. We can regard $\langle z, z\rangle_\zeta$ as the restriction of the form $\frac{1}{2}\left|\frac{\partial A}{\partial \zeta}\right|_\zeta\bigg|^{-1} \sum_{i,k} \frac{\partial^2 A}{\partial \zeta_i \partial \bar{\zeta}_k} d\zeta_i \wedge d\bar{\zeta}_k$ to the complex tangent space of the hypersurface.

Let $\gamma(s)$ be a chain on a hypersurface $M$ ($s$ is the natural parameter), $\zeta = \gamma(s)$ a point of the chain, $a_s$ a vector in the complex tangent space defining the direction of the chain at this point, and $\langle \zeta, \zeta\rangle_s$ the Levi form of $M$ at $\gamma(s)$.

In certain situations, when speaking of a hypersurface, it is necessary to characterize the parameters of analyticity of the functions defining the hypersurface. To this end, we introduce the class $M(\delta, m)$ of hypersurfaces $M$ satisfying the following conditions. The hypersurface $M$ has the form (1.1.1). The corresponding function $F(z, \bar{z}, u)$ is holomorphic in the polydisc $\{|z_k| < \delta (k = 1, 2, \ldots, n-1),\ |\bar{z}_k| < \delta (k = 1, 2, \ldots, n-1),\ |u| < \delta\}$. Here the variables $z_1, \ldots, z_{n-1}, \bar{z}_1, \ldots, \bar{z}_{n-1}$, and $u$ are considered as independent complex variables. Also, $|F(z, \bar{z}, u)| < m$ in this polydisc.

**Lemma.** *Let $M \in M(\delta, m)$, (see §7.1) and let $\gamma(s)$ be a chain on $M$. Then in a small neighbourhood of the origin $\gamma(s)$ satisfies the equation*

$$\gamma''(s) = \frac{2\langle a_s, a_s\rangle_s}{(1 + |a_s|^2)^{1/2}} i\gamma'(s) + \eta(\gamma, \gamma'),$$

*where $\eta(\gamma, \gamma')$ is a vector such that $|\eta(\gamma, \gamma')| < \eta^*(\delta, m)$, and $\eta^*$ is a function satisfying the condition $\lim_{m \to 0} \eta^*(\delta, m) = 0$ for every $\delta > 0$. The size of the neighbourhood is also determined by $(m, \delta)$. If $M$ is a quadric, then $\eta(s) \equiv 0$, and the curvature of the chain is $2\langle a_s, a_s\rangle_s/[1 + |a|^2]^{1/2} = \text{const}$. If $M$ is given in linear normal form and $\gamma(0) = 0$, then $\eta(0) = 0$.*

Let the direction $a_0$ be such that $|a_0|$ is sufficiently large and $\langle a_0, a_0\rangle \sim |a_0|^2$. In this case, by using the equation it can be shown that in a large interval of variation of $s$ the chain is close to a circle. Here we mean a circle that is a chain of the quadric $v = \langle z, z\rangle_0$ and is tangent to $\gamma(s)$ at the origin. The chain $\gamma(s)$ makes approximately $|a_0|$-many loops near this circle. For a hypersurface with a positive definite Levi form the condition $\langle a_0, a_0\rangle \sim |a_0|^2$ holds; therefore, if the curvature of a chain on such a hypersurface is large at a point, then near the corresponding circle it makes several loops.

**4.3. Derivation of the Equation.** Let $H$ be a normalization straightening a chain $\gamma$. We fix the set of initial data for $H$ as $\omega = (E, a, 1, 1, 0)$. We recall that $H = H_e(R_\omega(H_e))$ (see §2.2), where $R_\omega$ is a linear–fractional transformation with the same set of initial data, and $H_e$ is the standard normalization. By the lemma in §2.2, $H$ and $R_\omega(H_e)$ coincide up to terms of the third order of smallness at the origin. Therefore, the curves $H^{-1}(0, u)$ and $(R_\omega(H_e))^{-1}(0, u) = H_e^{-1}(R_\omega^{-1}(0, u))$ have the same curvature. Here $H_e^{-1}$ is the mapping inverse to the standard normalization of $M$. We first write out the curvature $\chi_0$ of the chain $\gamma^* = H_e(\gamma)$. The mapping $R_\omega^{-1}$ sends points of the line $z = 0$, $v = 0$ into $\gamma^*$. $R_\omega^{-1}(u)$ has the form $R_\omega^{-1}(u) = \xi_a u/[1 - i\langle a, a\rangle u]$, where $\xi_a$ is the vector $(-a_1, \ldots, -a_{n-1}, 1)$ (see §1.2 and the formula for the transition to an inverse mapping in §2.3). From the form of $R_\omega^{-1}$ we find that $\chi_0 = 2\langle a, a\rangle/[1 + |a|^2]^{1/2}$. Since $\gamma = H_e^{-1}(\gamma^*)$ and $H_e^{-1} = E + H^*$ (see the lemma in §7.4), we have $\gamma'' = ik\gamma' + \eta$, where $\eta$ is a quantity whose modulus is bounded by the second derivatives of $H_e$ (see lemma in §7.1), and so $\gamma(s)$ satisfies the above equation.

## §5. The Circular Normal Form

The parametrization of chains introduced in §3.3 is applicable to a study of local properties of hypersurfaces. As is clear from the transition formula, a function realizing a change of parameter has a singularity on a chain. Thus, both parameter and substitution are defined on only a part of the chain. This leads to an essential difficulty in studying chains in the large. By changing the form of representing a hypersurface we introduce a new parametrization, which is more convenient for a study of chains in the large.

**5.1. The Form of a Hypersurface in Circular Coordinates.** We consider the space $\tilde{\mathbb{C}}^n$ with the coordinate functions $z_1, z_2, \ldots, z_{n-1}, \rho, \Theta$, *where* $z_1, \ldots, z_{n-1}$ are complex and $\rho$, $\Theta$ are real coordinates. We can regard $z_1, \ldots, z_{n-1}, w = \rho e^{i\Theta}$ as local complex coordinates, that is, $\tilde{\mathbb{C}}^n$ has a natural complex structure, and so we can talk about holomorphic transformations of $\tilde{\mathbb{C}}^n$. We emphasize that the coordinates $(z, w)$ in $\tilde{\mathbb{C}}^n$ will always be understood as local coordinates. When speaking of one or another many-valued function defined in the coordinates $(z, w)$ we shall mean some continuous branch of it. Whether it is specific or arbitrary will be clear from the context.

We consider a class of hypersurfaces of the form

$$1 - \rho^2 = \langle z, z\rangle + F(z, \bar{z}, \Theta), \tag{5.1.1}$$

where $F$ is a real-analytic function defined in a neighbourhood of $z = 0$,

$$\Theta = \Theta_0, \text{ such that } F(0, 0, \Theta_0) = 0 \; dF(0, 0, \Theta_0) = 0, \text{ and } \left.\frac{\partial^2 F}{\partial z_j \partial \bar{z}_k}\right|_{(0,0,\Theta_0)} = 0.$$

A mapping from $\tilde{\mathbb{C}}^n$ into $\tilde{\mathbb{C}}^n$ sends any such hypersurface into one of the form

$$1 - \rho^2 = \langle z, z\rangle + \sum_{k,l \geq 2} \Phi_{kl}(z, \bar{z}, \Theta). \tag{5.1.2}$$

Here the $\Phi_{kl}$ are polynomials of degree $k$ in $z$ and $l$ in $\bar{z}$, with coefficients depending analytically on $\Theta$. It is assumed that their coefficients are defined on a common interval of the $\Theta$-axis, and for any $\Theta_0$ from this interval $\sum_{k,l\geq 2} \Phi_{kl}(z, \bar{z}, \Theta)$ as a function of $z$, $\bar{z}$, and $\Theta - \Theta_0$ is expanded as a power series in $z$, $\bar{z}$ and $\Theta - \Theta_0$ which converges in some neighbourhood of the point $z = 0$, $\bar{z} = 0$, $\Theta - \Theta_0 = 0$. The terms $\Phi_{22}$, $\Phi_{23}$, and $\Phi_{33}$ satisfy the conditions

$$\operatorname{tr} \Phi_{22} = 0, \quad \operatorname{tr}^2 \Phi_{23} = 0, \quad \operatorname{tr}^3 \Phi_{33} = 0$$

(for the definition of the operator tr see §1.1).

As regards hypersurfaces of this form we say that they and their equations are defined in *circular normal form*, and call the coordinates *circular normal* coordinates. The corresponding mapping will be called a normalization, as before. The circular normal form of a hypersurface can be obtained from the linear normal form. The mapping $R_0$:

$$z^* = \frac{2z}{i + w}, \quad w^* = \frac{i - w}{i + w} \tag{5.1.3}$$

sends a hypersurface of the form $v = \langle z, z \rangle + \ldots$ (see (1.1.1)) into one of the form $1 - |w^*|^2 = \langle z^*, z^* \rangle + \ldots$ (see (5.1.1)). The inverse mapping sends a hypersurface of the second type into one of the first. It can be shown that both direct and inverse mappings preserve the normality of the form, that is, a hypersurface defined in linear form is taken by $R_0$ into one defined in circular normal form, and conversely.

**5.2. The Initial Data for a Circular Normalization.** By analogy with the linear case we associate with any normalization $H$: $z^* = f(z, w)$, $w^* = g(z, w)$, a set of initial data $\omega = (U, a, \lambda, \sigma, r)$ defined by the system:

$$\left.\frac{\partial f}{\partial z}\right|_{(0,1)} = \lambda U, \qquad \left.\frac{\partial f}{\partial w}\right|_{(0,1)} = -\lambda U a,$$

$$\left.\frac{\partial g}{\partial w}\right|_{(0,1)} = \sigma \lambda^2 \quad \text{and} \quad \operatorname{Im} \left.\frac{\partial^2 g}{\partial w^2}\right|_{(0,1)} = -\sigma \lambda^2 r.$$

In contrast with the linear case, $r$ is defined in terms of the imaginary part of the second derivative, rather than the real part.

For any hypersurface and any set of initial data we can find a normalization with the given set of initial data, and this normalization is unique. This assertion is easily obtained from Moser's theorem (see §§1.2 and 1.3). The essential difference between linear and circular normal forms arises in the transition formulae: if a hypersurface is defined in circular normal form, then the normalization with $a = 0$ in the set of initial data $\omega$ turns out to be better in a certain sense than in the case of linear normal forms.

**5.3. The Form of a Substitution Preserving a Chain.** Any map $\tilde{R}$ that leaves the line $z = 0$, $\rho = 1$ fixed, and sends a hypersurface of the form (5.1.2) into a similar hypersurface, is a composition $\tilde{R} = R_0(R(R_0^{-1}))$, where $R$ is a linear–fractional transformation sending a hypersurface given in linear normal form into a similar hypersurface and leaving the line $z = 0$, $v = 0$ fixed (see the theorem in §1.3). Therefore, in complex coordinates $R$ is written as

$$z^* = \sqrt{Q'(w)}\, Uz, \quad w^* = Q(w),$$

where $U$ is such that $\langle Uz, Uz \rangle = \sigma\langle z, z \rangle$, and $Q(w)$ is a linear–fractional transformation sending $|w| = 1$ into itself.

It is clear from these formulae that a mapping sending one normal form into another is holomorphic in a neighbourhood of the whole line $z = 0, \rho = 1$. This property of circular coordinates gives a natural construction for the continuation of a mapping along a chain.

**5.4. Continuation of a Normalization Along a Chain.** Let $\gamma_1$ and $\gamma_2$ be two arcs of a chain $\gamma$ on a hypersurface $M$ that have a common point. Let $H_1$ and $H_2$ be two normalizations of $M$ straightening $\gamma$ (that is, sending $\gamma$ into the line $z = 0$, $\rho = 1$) defined in neighbourhoods of $\gamma_1$ and $\gamma_2$, respectively, and sending $M$ into a circular normal form. Then in some neighbourhood of the common point of these arcs we have $H_1 = \tilde{R}(H_2)$, where $\tilde{R}$ is a linear–fractional transformation (see §5.3). But since $\tilde{R}$ does not have singularities on the line $z = 0$, $\rho = 1$, $\tilde{R}(H_2)$ is a continuation of $H_1$ from $\gamma_1$ to $\gamma_2$.

**Theorem.** *Let $M$ be a hypersurface of a complex manifold that has a nondegenerate Levi form. Then every normalizing mapping that straightens $\gamma \subset M$ and sends $M$ into circular normal form can be continued indefinitely along $\gamma$. Any two normalizations $H_1$ and $H_2$ that straighten one and the same chain are connected by a relation $H_2 = \tilde{R}(H_1)$, where $\tilde{R}$ is a linear–fractional transformation* (*see* §5.3) [38].

Let us clarify the assertion of the theorem. We fix a chain $\gamma$ and a normalization $H$ at some point $x \in \gamma$ that straightens this chain. We denote by $H_\gamma$ the family of all normalizations at points of $M$ that can be obtained by continuing $H$ holomorphically along $\gamma$. The family $H_\gamma$ is called a *normalization along the chain* $\gamma$.

## §6. Normal Parametrization of a Chain

**6.1. The Circular Normal Parameter.** Let $H: X \to \tilde{\mathbb{C}}^n$ be a normalization of a hypersurface $M$ along a chain $\gamma \subset M$. The map defined by the corresponding family $H_\gamma$ sends $\gamma$ into the interval $(\Theta_H^-, \Theta_H^+)$ $(-\infty \leq \Theta_H^- \leq \Theta_H^+ \leq \infty)$ of the line $z = 0$, $\rho = 1$. The mapping $H^{-1}$ can be continued holomorphically to a neighbourhood of this interval, and so it associates with every point of this interval a point $H^{-1}(0, 1, 0)$ of $\gamma$. We observe that if one of the points $\Theta_H^-$, $\Theta_H^+$ is a finite point of the coordinate axis, then $H^{-1}$ cannot be automatically continued locally biholomorphically to any neighbourhood of this point.

The continued mapping, which we denote by the same symbol $H^{-1}$, sends the interval $(\Theta_H^-, \Theta_H^+)$ of the line $z = 0, \rho = 1$ onto the whole chain. The restriction of $H^{-1}$ to this interval can be regarded as a parametrization of $\gamma$: this mapping associates with each value of the parameter $\Theta_H = \Theta$ from $(\Theta_H^-, \Theta_H^+)$ the point $H^{-1}(0, 1, \Theta)$ of the chain.

Such a parametrization is called *normal*, and the corresponding parameter a *normal* parameter.

If a chain is not closed and has no multiple points, then the interval of values of the parameter covers the chain $\gamma$ univalently; if the chain is closed, then the covering is infinitely-valued. We recall for comparison that under a parametrization defined by linear normal form (see §3.3), the interval of values of the parameter covers, in general, only part of the chain.

As an example we consider normal parametrizations of chains of the quadric $1 - |w|^2 = \langle z, z \rangle$ passing through the point $z = 0, w = 1$. Setting $|w| = \rho$ we see that the hypersurface is defined in circular normal form. The branch of the mapping $\rho = |w|, \Theta = \operatorname{Arg} w, z^* = z$ sends the circle $\gamma_0: z = 0, \rho = 1$ into the line $z^* = 0, \rho = 1$. Therefore, $\gamma_0$ is a chain, and every branch of the argument of $w$ is a normal parameter on it, and so varies from $-\infty$ to $+\infty$. An arbitrary normal parameter on this circle can be obtained from those indicated above by a linear–fractional substitution sending the circle onto itself. Under a single circuit of $\gamma_0$ the normal parameter changes by $\pm 2\pi$.

An arbitrary chain $\gamma$ on the quadric in question can be obtained from $\gamma_0$ by a suitable automorphism (see §3.2), and so under a single circuit of $\gamma$ each normal parameter of it changes by $2\pi$. Similarly, under a single circuit of a chain on the quadric $v = \langle z, z \rangle$ the normal parameter changes by $2\pi$.

**6.2. The Formula for Changing a Parameter.** Let $P$ denote the projection of $\tilde{\mathbb{C}}^n$ into $\mathbb{C}^n$ that associates with the set $(z, \rho, \Theta)$ the point $(z, w = \rho e^{i\Theta})$. If $M \subset \tilde{\mathbb{C}}^n$ is a hypersurface of the form (5.1.1) then $P(M)$ can be written as $1 - |w|^2 = \langle z, z \rangle + \sum_{k,l \geq 2} \Phi_{k,l}(z, \bar{z}, \Theta)$, where $\Theta$ is a branch of $\operatorname{Arg} w$ on $|w| = 1$. The interval $(\Theta^-, \Theta^+)$ on the line $z = 0, \rho = 1$ is a chain on $M$. The projection $P$ winds this interval onto the circle $z = 0, |w| = 1$. The multiplicity of the covering, that is, the integer part $\left[\dfrac{|\Theta^+| - |\Theta^-|}{2\pi}\right]$, is independent of the choice of the normal parameter. In other words, the number of complete circuits made by the point $P(0, 1, \Theta)$ as $\Theta$ varies in $(\Theta^-, \Theta^+)$ is an invariant of the chain.

**Lemma.** *If $\Theta$ and $\Theta^*$ are two normal parameters on one and the same chain on a hypersurface $M$, then they are related by $e^{i\Theta^*} = \varphi(e^{i\Theta})$, where $\varphi(w)$ is a linear–fractional transformation sending $|w| = 1$ into itself. If $\Theta(x'') - \Theta(x')$*

$= 2\pi k$ *on some segment* $[x', x'']$ *of the chain, where* $k$ *is an integer, then* $\Theta^*(x'') - \Theta^*(x') = \pm 2\pi k$ *(see* [38] *and* [14]*).*

The above lemma is easily obtained from the formula in §5.3.

We mention another property of a normal parameter. If a normal parameter $\Theta$ changes by $2\pi$ on a segment of a chain, then for any other normal parameter $\Theta^*$ of this chain there is a point $\Theta_0$ of the same segment such that $\left|\frac{d\Theta^*}{d\Theta}\right|_{\Theta_0} \leq 1$ and $\left|\frac{d^2\Theta^*}{d\Theta^2}\right|_{\Theta_0} \leq \left|\frac{d\Theta^*}{d\Theta}\right|_{\Theta_0}$.

**6.3. The Initial Data of a Parametrization.** Now, having introduced the circular normal form, we turn again to hypersurfaces of the form $v = \langle z, z\rangle + \ldots$. When considering a hypersurface of this type and a normal parameter on the chains on this hypersurface, we have to speak of a circular normalization of a hypersurface of this form. In this case the normalizing map $H$: $z^* = f(z, w)$, $w^* = g(z, w)$ of $M$ can be represented by two compositions of the form $H = H_\omega(R_0)$ and $H = R_0(H_{\omega^*})$, where $R_0$ is a transformation sending a linear normal form into a circular one (see §6.1), $H_\omega$ is a circular normalization of $R_0(M)$ with some set of initial data $\omega = (U, a, \lambda, \sigma, r)$, and $H_{\omega^*}$ is a linear normalization of $M$ with a certain set $\omega^*$. In defining the initial data we saw to it that we can now say that $\omega = \omega^*$. We shall call the set $\omega$ the set of initial data of the normalization $H$. The elements of this set and the derivatives of the functions $f$ and $g$ are connected by the relations

$$\left.\frac{\partial f}{\partial z}\right|_0 = -2i\lambda U, \quad \left.\frac{\partial f}{\partial w}\right|_0 = -2i\lambda U a,$$

$$\left.\frac{\partial g}{\partial w}\right|_0 = 2i\sigma\lambda^2, \quad \operatorname{Im}\left.\frac{\partial^2 g}{\partial w^2}\right|_0 = -4\sigma\lambda^2 r.$$

Let $H$ be a circular normalization of the hypersurface $M$ that straightens a chain $\gamma$. This mapping defines a normal parameter $\Theta$ on $\gamma$. If $M$ is given as $v = \langle z, z\rangle + \ldots$ or $1 - |w|^2 = \langle z, z\rangle + \ldots$, and $\gamma$ passes through the origin or the point $z = 0$, $w = 1$, respectively, then a set of initial data for $H$ is determined. The parameter $\Theta$ is uniquely determined by the triple $\sigma, \lambda, r$ of this set. We call this triple the *set of initial data* of the parametrization.

On any analytic curve $\gamma$ lying on the hypersurface $M$ and transversal to the complex tangent space we can define in the same way as on chains a family of normal parametrizations (linear and circular). A normalization of $M$ that straightens the curve $\gamma$ (see §6.1) sends $M$ into a hypersurface of the form $v = \langle z, z\rangle + \sum_{k,l \geq 2} F_{kl}(z, \bar{z}, u)$. The transformation $R_0$ sends the resulting hypersurface into one of the form

$$1 - \rho^2 = \langle z, z\rangle + \sum_{k,l \geq 2} \Theta_{kl}(z, \bar{z}, \Theta).$$

The composition $H$ of these maps can be continued indefinitely along $\gamma$ and thus defines a normal parameter $\Theta$ on $\gamma$. We note that in contrast to the normalization of chains, when straightening arbitrary curves we cannot assert that a normalization with a given set of initial data is unique. But if we specify what curve is straightened by a given map, then the set of initial data $\omega = (U, a, \lambda, \sigma, r)$ uniquely determines the map. Thus the triple $(\sigma, \lambda, r)$ uniquely defines a parametrization of the curve just as in the case of chains.

**6.4. Normal Parametrization and the Continuation of Mappings.** Let there be defined a map $H$ of a hypersurface $M$ into a hypersurface $M^*$ that sends $x \in M$ into a point $x^* \in M^*$. Suppose a chain $\gamma \subset M$ passes through $x$ and let $\gamma^* = H(\gamma)$. Let $\varphi$ *and* $\varphi^*$ be normalizations sending $M$ and $M^*$ into linear normal form and straightening $\gamma$ and $\gamma^*$, respectively. We write $H$ as $H = \varphi^{*-1}(R(\varphi))$. Here, $R$ is a linear–fractional transformation leaving the line $z = 0$, $v = 0$ fixed (see the theorem in §1.3). If $\varphi$ and $\varphi^*$ are defined on large portions of $\gamma$ and $\gamma^*$, and $R$ on a large part of the line $z = 0$, $v = 0$, then the map $H$, which is defined, in general, in a small neighbourhood of $x$, can be continued to a large segment of $\gamma$. However as is clear from the formulae for the transformation of linear normal forms (see §3.9), $u = 1/r$ is a singular point. For large $r$ the mapping $R$ is holomorphic only in a small neighbourhood of the origin, and this turns out to be an obstacle to its continuation. If $\varphi$ and $\varphi^*$ are circular normalizations, then the corresponding $R$ is holomorphic on the entire line $z = 0$, $\rho = 1$. Circular normalizations can be continued indefinitely along chains. Thus, if, for example, the interval $(\Theta^{*-}, \Theta^{*+})$ of values of a normal parameter of a chain $\gamma$ is the whole line, then the map $H$ can be continued to the whole chain $\gamma$. A chain on the sphere in $\mathbb{C}^n$ is a circle. In this case $(\Theta^{*-}, \Theta^{*+})$ is the whole line, and thus a locally defined map of an arbitrary hypersurface into a sphere can be continued to the whole hypersurface.

It is known that a locally defined map of one hypersurface into another cannot, in general, be continued (see [9] and [3]). In these examples a chain $\gamma$, along which a mapping $H$ cannot be continued, and $\gamma^* = H(\gamma)$ are constructed such that the variation of a normal parameter on $\gamma$ is infinite, and on $\gamma^*$ finite.

**6.5. The Equation for Passing to a Normal Parameter.** We consider a family of hypersurfaces $M_\alpha$: $1 - |w|^2 = \langle z, z\rangle + \alpha\langle z, z\rangle^3$, where $\alpha \in R$. The circle $z = 0$, $|w| = 1$ is a chain on each hypersurface of the family. However, as Ezhov [12] has shown, its normal parametrizations are different for different $\alpha$, and $\alpha = 1/24$ is found to be critical in the sense that for $\alpha = 1/24$ the variation of a normal parameter on the chain is $2\pi$, while for $\alpha < 1/24$ it is infinite, and for $\alpha > 1/24$ it is strictly less than $2\pi$.

In general it turns out ([12], [23], and [14]) that an estimation of the variation of a normal parameter on a chain requires an analysis of $\Phi_{33}$. For hypersurfaces of the form

$$1-|w|^2=\langle z,z\rangle+\sum_{k,l\geq 2}\Phi_{kl}(z,\bar{z},t),$$

where $t=\arg w$, $\operatorname{tr}\Phi_{22}=0$, and $\operatorname{tr}^2\Phi_{23}=0$, the transition function $\Theta=g(t)$ from a parameter $t$ to a normal parameter $\Theta$ satisfies the equation

$$\alpha_n \operatorname{tr}^3\Phi_{33}(g')^2+2g'''g'-3(g'')^2+(g')^4-(g')^2=0,$$

where $\alpha=\binom{n}{3}$ ([12] and [14]).

An analysis of this equation and of the whole construction of the reduction enables us, in a number of cases, to give an estimate of the length of the interval of variation of a normal parameter. The hypersurfaces most studied are those with a positive definite Levi form. If a chain on such a hypersurface makes at some point a small angle with the complex tangent space, then it turns out that the interval of variation of a normal parameter on this chain is large. This will be discussed in more detail in §8.2. In particular, if the angle of inclination of the chain to the complex tangent space decreases to zero as a point moves along a chain, then the variation of any normal parameter on this chain is infinite (see [23]). The need for estimates of this sort arose in connection with problems of continuation of holomorphic maps ([38] and [14]). The estimation of the term $\Phi_{33}$ involves complicated calculations. A property of chains necessary for the continuation of mappings will be stated differently (see §§7.4 and 8.2). This will enable us to avoid laborious calculations.

## §7. The Non-Sphericity Characteristic of a Hypersurface

A connected hypersurface is called *spherical* if in a neighbourhood of each of its points it is equivalent to a quadric, that is, for a suitable choice of coordinates it can be written as $v=\langle z,z\rangle$. Otherwise a hypersurface is said to be non-spherical (when speaking of sphericity or non-sphericity it will always be assumed that the hypersurface is connected).

From the point of view of the local structure practically everything is known about spherical hypersurfaces: any normal form of such a hypersurface has the form $v=\langle z,z\rangle$, and the automorphism group of a quadric as been known from Poincaré's time (see §1.2). Therefore, as a rule, our future discussion will concern non-spherical hypersurfaces. It turns out that in a number of problems it is essential to know not only whether a hypersurface is spherical or not, but to have the possibility of characterizing the "magnitude" of its non-sphericity. For this purpose we introduce a special numerical characteristic.

**7.1. Estimate for the Radius of Convergence, and the Norms of the Defining Series.** In order to write estimates for the radius of convergence and for the norms, we introduce appropriate parameters for the hypersurface and the normalization mapping. We shall describe the analyticity of the hypersurface $M$ as before by a pair of positive numbers $\delta$ and $m$, writing $M \in M(\delta, m)$ (see section 4.2).

We characterize the set of initial data $\omega = (U, a, \lambda, \sigma, r)$ of a normalizing mapping by a number $v$, writing $|\omega| \leq v$ by which we mean the system of inequalities $\{1/v \leq \|U\| \leq v, |a| \leq v, 1/v \leq \lambda \leq v, |r| \leq v\}$.

**Lemma.** *Let $M \in M(\delta, m)$ and $|\omega| \leq v$. Then a normalizing mapping $H_\omega$ of a hypersurface $M$ is representable as $H_\omega = R_\omega + H(z, w)$, where $R_\omega$ is a linear-fractional transformation with initial data $\omega$ (see §1.2), and $H(z, w)$ is a mapping holomorphic in the polydisc.*

$$\{|z_k| < \delta^* (k = 1, 2, \ldots, n-1), |w| < \delta^*\}$$

*and satisfying $|H(z, w)| \leq m^*(|z|^2 + |w|^2)$ in this polydisc, where $\delta^* > 0$ and $m^* = m^*(\delta, m, v)$ is such that $\lim_{m \to 0} m^*(\delta, m, v) = 0$ for every $\delta > 0$, and $\delta^*, m^*$ depend only on $\delta$, $m$, $v$, and the matrix of $\langle z, z\rangle$. The hypersurface obtained as the result of the normalization belongs to $M(\delta', m')$ ($\delta'$ and $m'$ are also defined by $m$ and $v$).*

The assertion of the lemma is valid for linear and circular normalizations of hypersurfaces given in the form $v = \langle z, z\rangle + \ldots$ or in the form $1 - |w|^2 = \langle z, z\rangle + \ldots$. The proof is obtained from the fact that every normalizing mapping is a composition of standard normalizations, of their inverses, and a suitable linear-fractional transformation. The standard reduction is constructed quite concretely, and so the proof of the lemma reduces to purely technical estimates. This thankless task is carried out in [4]. From these estimates it is not difficult to obtain, in particular, the following result.

**Theorem.** *Let $M, M^* \in M(\delta, m)$ and let $H(\zeta)$ be a biholomorphic mapping defined in some neighbourhood of the origin that sends points of $M$ into points of $M^*$ and is such that $|\partial H/\partial\zeta|_0 < m_0$ and $|\partial^2 H/\partial\zeta^2|_0 < m_0$. Then $H$ extends holomorphically to the ball $|\zeta| < \delta^*$, and in this ball $|H(\zeta)| < m^*$, where $\delta^*$ and $m^*$ depend only on $\delta$, $m$, $m_0$, and the norm of the matrix of $\langle z, z\rangle$* ([4], [39]).

**7.2. The Non-Sphericity Characteristic.** Let $M$ be given by an equation $A(\zeta, \bar{\zeta}) = 0$. Fix $\zeta \in M$. By a unitary transformation we map $\zeta$ into the origin and $M$ into a hypersurface of the form $v = \langle z, z\rangle + F$ (see (1.1.1)). This transformation is determined to within a unitary change of variables in the plane $w = 0$. Let $U$ be such a unitary change of variables. Under $U$ the hypersurface $v = \langle z, z\rangle + F$ goes into a hypersurface of the same form, and we write it as $v = \langle z, z\rangle_U + F_U$. Then we carry out the standard normalization of this hypersurface and write the result as $v = \langle z, z\rangle_U + F_{U,e}$.

We associate with a point $\zeta$ of $M$ a number $N_{\delta,m}(\zeta, M)$, which we shall call the *non-sphericity characteristic* of $M$ at $\zeta$. $N_{\delta,m}$ is defined at $\zeta$ provided the corresponding hypersurface $v = \langle z, z\rangle + F_{U,e}$ belongs to $M(\delta, m)$ for all $U$. In this case we set

$$N_{\delta,m}(\zeta, M) = \max_{U} \max_{(z,\bar{z},u)} |F_{U,e}(z, \bar{z}, u)|,$$

where the maximum is taken over all $(z, \bar{z}, u)$ satisfying $|z_k| \le \delta/2$ $(k = 1, 2, \dots n-1)$, $|\bar{z}_k| \le \delta/2$ $(k = 1, 2, \dots, n-1)$, $|u| \le \delta/2$ (here $z, \bar{z}, u$ are regarded as independent complex variables).

For all points of the quadric $v = \langle z, z\rangle$ the non-sphericity characteristic is zero. If the characteristic is zero at a point $\zeta \in M$, then the standard normalization sends a neighbourhood of this point into a quadric, and therefore every non-sphericity characteristic is zero at all points of a neighbourhood of $\zeta$. The lemma in §7.1 ensures that this neighbourhood is rather large. By progressively changing the point and repeating the argument, we find that the characteristic is zero at all points of the hypersurface $M$.

Let $M$ be defined by $A(\zeta, \bar{\zeta}) = 0$, and let $M_0$ be the part of $M$ lying in the ball $|z|^2 + |w|^2 \le 1$. We shall assume that $M_0$ is compact and connected, and the matrix of its Levi form belongs to a compact set of non-singular matrices, which we assume is fixed. Fix two pairs of positive numbers $\delta, m$ and $\delta^*, m^*$. It is assumed that $N_{\delta,m}$ is defined at some point $\zeta_0$ of the hypersurface $M_0$ and that $N_{\delta^*,m^*}$ is defined at all points of $M_0$.

**Lemma.** *If $N_{\delta,m}(\zeta_0, M_0) = N_0 > 0$, then $N_{\delta^*,m^*}(\zeta, M_0) > N_0^*$ at each point $\zeta \in M_0$, where $N_0^* > 0$ is a function of $\delta$, $m$, $\delta^*$, $m^*$, and $N_0$.*

The family of hypersurfaces of the type $M_0$ satisfying the conditions of the lemma is compact. Therefore, assuming that the assertion of the lemma is not true, we can find a hypersurface of the type $M_0$ such that $N_{\delta,m}(\zeta, M_0) \ge N_0$ at some point $\zeta$ of this hypersurface and $N_{\delta^*,m^*}(\xi, M_0)$ vanishes at some other point $\xi$ which, as was mentioned above, is impossible.

We state a problem which arises in connection with the definition of the non-sphericity characteristic. It would be nice to have a definition of a measure of non-sphericity that is not linked to $(\delta, m)$. It is natural to try to define non-sphericity as the lower bound of the deviations of a given hypersurface from spherical hypersurfaces. The first question that arises in this connection can be stated thus, for example: we fix a non-spherical hypersurface and some compact part of it. Can this compact part be approximated with any accuracy (in the metric of deviations) by a spherical hypersurface?

**7.3. The Variation of the Characteristic Under a Mapping.** Suppose that $H$ sends a non-spherical hypersurface $M$: $v = \langle z, z\rangle + F(z, \bar{z}, u)$ of the form (1.1.1) into $\tilde{M}$: $v = \langle z, z\rangle + \tilde{F}(z, \bar{z}, u)$ with $H$ defined by $z \to \mu z$, $w \to \mu^2 w$, where $\mu \in R$. Then $\tilde{F}(z, \bar{z}, u) = \dfrac{1}{\mu^2} F(\mu z, \mu\bar{z}, \mu^2 u)$. If $M$ is given in normal form, then the expansion of $F$ begins with terms of degree at least 4. Hence $|\tilde{F}| \le \mu m$ for small $\mu$

and so $N_{\delta,m}(0, \tilde{M}) \to 0$ as $\mu \to 0$. Thus, a large "dilation" decreases the non-sphericity of a hypersurface.

We consider two examples of maps for which it turns out that the variation of the characteristic of a hypersurface can be estimated. Let $H$ be a mapping of $M$ into $\tilde{M}$ such that the first and second order partial derivatives of $H$ and its inverse at the origin are bounded by a constant $m_0$. We say that such mappings belong to the class $H(\delta, m, m_0)$. For fixed $\delta$ and $m$ the class of hypersurfaces $M(\delta, m)$ is compact. It turns out that $H(\delta, m, m_0)$ for fixed $m_0$ is also compact (see the theorem in §7.1). It follows from the compactness of these classes that if $N_{\delta,m}(0, M) > N > 0$, then for $\tilde{M} = H(M)$ $(H \in H(\delta, m, m_0))$ $N_{\delta,m}(0, \tilde{M})$ is at least $N^* > 0$, where $N^*$ is a function of $(\delta, m, m_0)$. Next suppose that $M$ and $\tilde{M}$ are hypersurfaces with a positive definite Levi form, and that $R$ sends $M$ into $\tilde{M}$ and has the form $z^* = \lambda Uz/(1 - rw)$, $w^* = \lambda^2 w/(1 - rw)$ (see §1.2 for the definition of $U$ and $\lambda$), where $\lambda < v$ and $|r| < v$. The group of matrices $U$ preserving a positive definite Levi form is compact, and so the first and second derivatives of $R$ for fixed $v$ and a fixed Levi form can be bounded from above. If we bound $\lambda$ from below by assuming, for example, that $1/\lambda < v$, then a similar bound holds also for the derivatives of $R^{-1}$. In this case, as is clear from the above example, the variation of the non-sphericity characteristic is bounded from both above and below. It was shown at the beginning of this section that under a large dilation, the non-sphericity characteristic decreases. Thus, without imposing any lower bound on $\lambda$, it can be said that $N_{\delta,m}(0, \tilde{M})$ cannot be much less than $N_{\delta,m}(0, M)$. Indeed, if $\lambda$ is small and the mapping $R$ significantly decreases the non-sphericity characteristic, then the composition of $R$ and the dilation $z^* = \frac{1}{\lambda} z$, $w^* = \frac{1}{\lambda^2} w$ considerably decreases the non-sphericity. The latter is impossible, because this composition is a linear-fractional transformation of the same form, for which $\lambda$ is equal to 1. The following assertion is easily derived from what has been said.

**Lemma.** *Let $M$ and $\tilde{M}$ be strictly pseudoconvex hypersurfaces of the class $M(\delta, m)$ for which the characteristic $N_{\delta,m}$ is defined and $N_{\delta,m}(0, M) > N > 0$. Let $H$ be a mapping sending $M$ into $\tilde{M}$ that is defined by the composition $H = H_1(R(H_2))$, where $H_1, H_2 \in H(\delta, m, v)$ and $R$ is a linear-fractional transformation of the form $z^* = \lambda Uz/(1 - rw)$, $w^* = \lambda^2 w/(1 - rw)$, where $\lambda < v$ and $|r| < v$. Then $N_{\delta,m}(0, M) > N^*$, where $N^* > 0$ is a function of $\delta, m, v, N$ and the norm of the matrix of the Levi form.*

Similarly we can observe the variation of the non-sphericity characteristic under mappings of the form $H_2^{-1}(R(H_1))$, where $H_1$ and $H_2$ are mappings from $\tilde{\mathbb{C}}^n$ into $\tilde{\mathbb{C}}^n$, and $R$ is a linear-fractional transformation sending a circular normal form into a circular norm form, and leaving the circle $z = 0$, $|w| = 1$ fixed.

**7.4. Chains on a Hypersurface Close to a Quadric.** Let $\gamma(\Theta)$ be a chain on a hypersurface $M$: $v = \langle z, z \rangle + \dots$ that passes through the origin, and let $\gamma_0(\Theta)$

be a chain on the quadric $v = \langle z, z\rangle$ that passes through the origin and is tangent to $\gamma(\Theta)$ there. Suppose for definiteness that $\gamma(0) = \gamma_0(0) = 0$. We assume that the parametrizations of $\gamma(\Theta)$ and $\gamma_0(\Theta)$ are compatible, that is, they have the same initial data (see §6.3). The function $\gamma_0(\Theta)$ is defined for all values of $\Theta$, and $\gamma(\Theta)$ is defined on some interval $(\Theta^-, \Theta^+)$ (see §6.1). We assume that $M \in M(\delta, m)$ is given in linear normal form, that $N_{\delta,m}(0, M)$ for this hypersurface is defined and is small and that the projection of the circle $\gamma_0$ onto the plane $w = 0$ belongs to the polydisc $|z_k| \leq \delta/2$ $(k = 1, \ldots, n-1)$. The latter implies that for small $\delta$ the modulus of the vector $a$ defining the direction of $\gamma_0$ and $\gamma$ (see §3.1) is sufficiently large.

Let $H$ and $R$ be normalizing mappings of $M$ and the quadric $v = \langle z, z\rangle$ that define a parameter $\Theta$ on $\gamma$ and $\gamma_0$. We shall assume that the set of initial data of $H$ and $R$ is

$$(E, -a, 1, 1, 0).$$

**Lemma.** *For each large $p$ and small $\varepsilon > 0$ we can find an $N > 0$ such that if $N_{\delta,m}(0, M) < N$, then $H^{-1}$ is defined and is holomorphic in a $\delta^*$-neighbourhood of the segment $z = 0, \rho = 1, -p \leq \Theta \leq p$, and $|H^{-1} - R^{-1}| < \varepsilon$ everywhere in this neighbourhood, where $\delta^* > 0$ is a function of $\delta, m, p$, and $|a|$.*

The substance of the lemma is that if the hypersurface $M$ is nearly spherical, then on a large part of the domain of $\Theta$ the point $\gamma(\Theta)$ is close to the corresponding point of a circle.

*Proof.* The mapping $R^{-1}$ can be regarded as a composition $R^{-1} = R_0^{-1}(R_\omega)$, where $R_0$ is a linear–fractional transformation sending the quadric $v = \langle z, z\rangle$ into the quadric $1 - |w|^2 = \langle z, z\rangle$ (see §§3.2 and 5.1), and $R_\omega$ is a linear–fractional transformation with initial data $\omega = (E, -a, 1, 1, 0)$ that sends $1 - |w|^2 = \langle z, z\rangle$ into itself. Since the parametrizations of $\gamma$ and $\gamma_0$ are compatible, $H^{-1} = H_e(R_0^{-1}(R_\omega))$, where $H_e$ is the standard normalization (see §2.1). If the characteristic $N_{\delta,m}(0, M)$ is small, then $N_{\delta',m'}(0, \tilde{M})$ is also small for $\tilde{M} = H(M)$ (see §§7.1 and 7.3) ($\delta'$ and $m'$ are defined by the triple $|a|, \delta, m$). Hence $N_{\delta'',m''}(0, M^*)$ is small for $M^* = R_0^{-1}(R_\omega(\tilde{M}))$ ($\delta''$ and $m''$ are again defined by the triple $|a|, \delta', m'$) (see §7.3). Thus, by the lemma in §7.1 the standard normalization $H_e$, which sends $M^*$ into $M$, is close to the identity in a large neighbourhood of the origin. From what has been said we obtain that $H^{-1}$ is close to $R^{-1}$ in a $\delta_0$-neighbourhood of $(0, 1, 0)$. Here $\delta_0$ is defined by the triple $(|a|, \delta, m)$, and the deviation $|H^{-1} - R^{-1}|$ is defined by the same triple and $N_{\delta,m}(0, M)$. When the latter decreases in the indicated neighbourhood, $|H^{-1} - R^{-1}|$ tends uniformly to zero. In particular, on the segment $(0, 1, -\frac{1}{2}\delta_0 \leq \Theta \leq \frac{1}{2}\delta_0)$ $\frac{d}{d\zeta}H^{-1}$ and $\frac{d^2}{d\zeta^2}H^{-1}$ are close to $\frac{d}{d\zeta}R^{-1}$ and $\frac{d^2}{d\zeta^2}R^{-1}$, respectively. Therefore, by the theorem in §7.1, $H^{-1}$ is holomorphic in a $\delta^*$-neighbourhood of this segment and $|H^{-1}| < m^*$. We may assume that $\delta^* < \delta_0$. At $(0, 1, \frac{1}{2}\delta^*)$ $H^{-1}$ and $R^{-1}$ are close in the metric of $C^2$, and so, by the theorem in §7.1, $H^{-1}$ extends holomorphically to a $\delta^*$-neighbourhood of this point. Since $\left|\frac{d}{d\zeta}R\right|$ and $\left|\frac{d^2}{d\zeta^2}R\right|$ are bounded on the whole line $z = 0$, $\rho = 1$, by passing successively along the chain of points $\left(0, 1, \pm\frac{k}{2}\delta^*\right)$ and using the proximity of $H^{-1}$ and $R^{-1}$ obtained at earlier stages, we continue $H^{-1}$ to a neighbourhood of the segment $(0, 1, -p \leq \Theta \leq p)$ of the line $z = 0$, $\rho = 1$. The size of this neighbourhood and the constant that bounds $|H^{-1}|$ are determined by $(|a|, \delta, m)$ and $N_{\delta,m}(0, M)$. The required accuracy $\varepsilon$ for fixed $|a|$, $\delta$, $m$, and $p/\delta^*$ is ensured by choosing the characteristic $N_{\delta,m}(0, M)$ sufficiently small.

**7.5. The Behaviour of a Chain Near Points of High Curvature.** We show that if a chain on a strictly pseudoconvex hypersurface has large curvature at some point, then on a large interval of variation of a normal parameter it is close to a circle, and on each loop (a single circuit along this circle) the variation of the normal parameter is nearly $2\pi$.

We recall that the curvature of a chain $\gamma$ at a point $x \in M$: $v = \langle z, z\rangle + \dots$ is approximately equal to $2\langle a, a\rangle/[1 + |a|^2]^{1/2}$ (see §4.2), where $a$ is the vector in the direction of $\gamma$ at $x$ (see §3.1). Therefore, for strictly pseudoconvex hypersurfaces the curvature $\chi$ of $\gamma$ at $x \in M$ is equivalent to $|a|$ for large $|a|$. The angle $\alpha$ between the chain and the complex tangent space is defined by $\cot\alpha = |a|$. Hence, for chains on a strictly pseudoconvex hypersurface, to say that the curvature is large or the angle is small is one and the same.

Let $\gamma(\Theta)$ be a chain on $M$: $v = \langle z, z\rangle + \dots$ that passes through the origin, and $\gamma_0(\Theta)$ a chain on the quadric $v = \langle z, z\rangle$ that passes through the origin and is tangent to $\gamma(\Theta)$ there. Let $a$ be the vector in the direction of $\gamma$ and $\gamma_0$ at their point of contact, $\gamma(0) = \gamma_0(0) = 0$, and let $H$ and $R$ be normalizations of $\gamma$ and $\gamma_0$ with initial data $\omega = (E, -a, 1, 1, 0)$ that define a parameter $\Theta$ on them. We shall denote by $a(\Theta)$ the vector in the direction of $\gamma$ at $\gamma(\Theta)$.

**Lemma.** *Let $M$ be a strictly pseudoconvex hypersurface from the class $M(\delta, m)$ given in linear normal form. Then for each $p > 0$ and $\varepsilon > 0$ we can find an $a_0 > 0$ depending only on $p$, $\varepsilon$, $\delta$, $m$, and the norm of the matrix of $\langle z, z\rangle$, such that when $|a| = |a(0)|$ is greater than $a_0$, the segment $[-p, p] \subset (\Theta^-, \Theta^+)$, and at all points $[-p, p]$ we have*

$$|\gamma(\Theta) - \gamma_0(\Theta)| < \frac{\varepsilon}{|a|}, \qquad |a(\Theta) - a| < \varepsilon|a|,$$

*and for H we have the estimates*

$$\left|\frac{d}{d\zeta}\bigg|_{\gamma(\Theta)} H\right| < c|a|^2, \qquad \left|\frac{d^2}{d\zeta^2}\bigg|_{\gamma(\Theta)} H\right| < c|a|^2,$$

$$\left|\frac{d}{d\zeta}\bigg|_{(0,1,\Theta)} H^{-1}\right| < \frac{c}{|a|}, \qquad \left|\frac{d^2}{d\zeta^2}\bigg|_{(0,1,\Theta)} H^{-1}\right| < \frac{c}{|a|},$$

*where c is an absolute constant.*

This assertion can be reformulated as follows. When a normal parameter varies by $2\pi$, the corresponding segment of a chain is geometrically close to a circle; the angle of inclination of the chain to the complex tangent space is almost constant on this segment; the derivatives of a normalizing mapping on this segment and the derivatives of its inverse cannot be much larger than $|a|^2 + |a|^{-1}$.

*Proof.* We introduce the substitution $z \to \mu z$, $w \to \mu^2 w$, and denote the resulting images of $M$, $\gamma$, and $\gamma_0$ by $M^*$, $\gamma^*$, and $\gamma_0^*$, respectively. We choose $\mu$ such that the radius of $\gamma_0^*$ is 1. Let $a^*$ be a vector in the direction of $\gamma^*$. The curvature of the circle $\gamma_0^*$ is given by $2\langle a^*, a^*\rangle/[1 + |a^*|^2]^{1/2}$ (see §3.2). Thus $[1 + |a^*|^2]^{1/2} = 2\langle a^*, a^*\rangle$; but $a^* = \mu a$ and so $1 + \mu^2|a|^2 = 4\mu^4\langle a, a\rangle^2$. Since the form $\langle z, z\rangle$ is positive definite, $\mu \sim 1/|a|$ and $|a^*| \sim 1$.

The above substitution sends $M: v = \langle z, z\rangle + F$ into $M^*: v = \langle z, z\rangle + F^*$, where $F^* = \frac{1}{\mu^2} F(\mu z, \mu\bar{z}, \mu^2 u)$. Since $M \in M(\delta, m)$ and the expansion of $F$ contains no terms of order less than 4, by taking $\mu^3$ out of the parentheses in $F(\mu z, \mu\bar{z}, \mu^2 u)$ and writing $\frac{1}{\mu^3} F(\mu z, \mu\bar{z}, \mu^2 u)$ as $\tilde{F}$ we obtain $|F^*| < \mu|\tilde{F}|$. If $\mu$ is small, then $M^* \in M(3, \mu)$. The value 3 is chosen so that $\gamma_0^*$ lies strictly within a large polydisc. For $\mu$ small, $M^*$ is close to a quadric, and so by the lemma in §7.4, $H^{-1}$ and $R^{-1}$ are close in a $\delta^*$-neighbourhood of the segment $[z = 0, \rho = 1, -\rho \le \Theta \le p]$. Thus, for any point $\gamma(\Theta)$ ($\Theta \in [-p, p]$), $H$ and $R$ are close in a large neighbourhood of $\gamma(\Theta)$. Consequently, the first and second derivatives of $H$ and $H^{-1}$ are close to the corresponding derivatives of $R$ and $R^{-1}$, and the latter are bounded by absolute constants.

By making the inverse substitution it is not difficult to obtain the inequalities in the lemma.

## §8. Strictly Pseudoconvex Hypersurfaces

In this section we shall prove a theorem that brings out an interesting property of biholomorphic maps that send a real-analytic strictly pseudoconvex non-spherical hypersurface in $\mathbb{C}^n$ into another such hypersurface. We obtain as corollaries several known results about this class of hypersurfaces.

**8.1. A Theorem on the Germ of a Mapping.** We formulate our main result in terms of normal forms; it is simpler to talk about the non-sphericity of a hypersurface and properties of maps in these terms.

Let $M$: $v = \langle z, z\rangle + F(z, \bar{z}, u)$ be a hypersurface of class $M(\delta, m)$ (see §7.1) defined in linear normal form. We characterize its non-sphericity by $N(M) = \max_{z,\bar{z},u} |F(z, \bar{z}, u)|$, where $z, \bar{z}, u$ are regarded as independent complex variables, and the maximum is take over a polydisc of radius $\delta/2$: $|z_k| \leq \delta/2$, $k = 1, 2, \ldots, n-1$, $|\bar{z}_k| \leq \delta/2$, $k = 1, 2, \ldots, n-1$, $|u| \leq \delta/2$.

We fix positive $\delta$, $m$, and $N$, and denote by $M_+$ the class of all strictly pseudoconvex non-spherical hypersurfaces $M \in M(\delta, m)$ of the form $v = \langle z, z\rangle + \ldots$ defined in linear normal form, for which $N(M) \geq N$. If $M$ and $M^*$ belong to the class $M_+$, then every biholomorphic map sending $M$ into $M^*$ can be regarded as a normalization of $M$ with a certain set of initial data $\omega = (U, a, \lambda, \sigma, r)$.

We characterize the properties of this map by a number $v$, writing as before $|\omega| \leq v$ (see §7.1).

**Lemma.** *For each positive $\delta$, $m$, and $N$ we can find a $v$ such that if a normalization $H_\omega$ of $M \in M_+$ sends it into $M^* \in M_+$, then $|\omega| \leq v$.*

The proof will be given at the end of this section. We remark that the essential part of the lemma is the estimate on the parameters $a$, $\lambda$, and $r$. An estimate on the norm of the matrix $U$ follows directly from the compactness of the group of unitary matrices. For indefinite forms the analogous group of matrices is clearly non-compact, and hence the assertion of the lemma cannot be generalized to the case of hypersurfaces with indefinite Levi form. In this situation, we might expect an estimate on the parameter $a$, $\lambda$, and $r$, which depends on the matrix $U$. In particular, for automorphisms, it is known that the parameters $a$, $\lambda$ and $r$ can be estimated in terms of the norm of the matrix $U$ and some parameters which determine the hypersurface [4].

We now state a fundamental consequence of the lemma.

**Theorem.** *If a biholomorphic map $H$ fixing the origin sends $M \in M_+$ into $M^* \in M_+$, then $H$ can be continued holomorphically to a $\delta^*$-neighbourhood of the origin, and $|H| < m^*$ in this neighbourhood, where $\delta^*$ and $m^*$ are functions of $\delta$, $m$, and $N$.*

The assertion follows easily from the above lemma and the lemma in §7.1 We now consider several corollaries of this theorem.

## 8.2. Properties of the Stability Group

**Corollary 1.** *The group of automorphisms of a hypersurface $M \in M_+$ that leave the origin fixed is compact* [37].

This follows immediately from the theorem. We remark that the group of a quadric, and therefore of any spherical surface, is non-compact. For hypersurfaces with indefinite Levi form, this group is, in general, non-compact even when the hypersurface is non-spherical. For example, the stability group of the hypersurface $v = \langle z, z\rangle + \langle z, z\rangle^4$ is isomorphic to the group of matrices preserving a given form $\langle z, z\rangle$, and so is non-compact if $\langle z, z\rangle$ is indefinite.

**Corollary 2.** *If $M \in M_+$, then by using a biholomorphic change of variables leaving the origin fixed, we can go over to new coordinates in which all the automorphisms of the hypersurface that leave the origin fixed are linear transformations.*

This follows from the compactness of the group in question (see Corollary 1) and Bochner's theorem [5]: close to any point of a real-analytic hypersurface we can find a coordinate system in which all the transformations of any compact subgroup of the stability group of this hypersurface with centre at the fixed point are linear. In the next section a stronger assertion will be proved regarding the stability group of hypersurfaces from the class $M_+$.

### 8.3. Compactness of the Group of Global Automorphisms

**Corollary 3.** *Let $M$ and $M^*$ be non-spherical compact real-analytic strictly pseudoconvex hypersurfaces of some complex manifolds. Then each biholomorphic map of $M$ into $M^*$ can be continued holomorphically to a neighbourhood $\Omega$ of $M$, which is common for all these maps. Here the family of all such maps is equicontinuous on $\Omega$* [37].

The corollary is obtained from the theorem on germs and the lemma in §7.2. It means, in particular, that the group of global automorphisms of a hypersurface with the listed properties is compact. From this it follows in turn that if a domain bounded by such a hypersurface is defined in a complex manifold, then its automorphism group is compact ([47] and [32]) for in this situation the automorphisms are extended holomorphically to the boundary of the domain (see [36] and [16] or [30] and [27]). We note that the reverse implication from the compactness of groups and the symmetry principle to Corollary 3 is impossible, since here we obtain more, namely, the uniformity of the estimates of the size of the domain of holomorphy.

### 8.4. Continuation of a Germ of a Mapping Along a Compact Hypersurface

**Corollary 4.** *Suppose $M$ and $M^*$ satisfy the conditions of Corollary 3, and let $H$ be a germ of a biholomorphic mapping from $M$ into $M^*$, that is, a biholomorphic mapping defined in a neighbourhood of some $x \in M$ and sending points of $M$ into points of $M^*$. Then $H$ can be continued holomorphically along any path lying on $M$ and starting at $x$.*

This assertion for the case of hypersurfaces from $\mathbb{C}^n$, that is hypersurfaces given by a single chart, was proved by Pinchuk [33], and in the general case by Ezhov, Kruzhilin, and the author ([13] and [14]). The assertion, just as Corollary 3, follows from the theorem on germs and the lemma in §7.2. In view of the uniformity of the estimates, the map can be continued to any point of the first hypersurface by a finite chain of expansions of the defining series.

### 8.5. Classification of Coverings

**Corollary 5.** *Let $M$ and $M^*$ satisfy the conditions of Corollary* 3. *Then if $M$ and $M^*$ are locally equivalent* (*see* §1.4), *their universal coverings are globally equivalent* (*the complex structure of the coverings is obtained by lifting the structure of the original hypersurfaces*).

This is a reformulation of Corollary 4. As simple examples show the global equivalence of $M$ and $M^*$ does not, in general, follow from their local equivalence, and so we have to speak of the equivalence of their coverings. Of course, if $M$ and $M^*$ are simply-connected, then their global equivalence follows from their local equivalence. The latter is also true for spherical simply-connected hypersurfaces (this follows easily from Poincaré's theorem stated in the preamble).

### 8.6. Discussion of Examples of Non-Continuable Mappings

**Corollary 6.** *If $M$ satisfies the conditions of Corollary* 3, *then its universal covering $\tilde{M}$ is a surface which can be said to be maximal in the following sense: if a hypersurface $\mathfrak{M}$ is connected, and there is a locally biholomorphic map on $\tilde{M}$ sending $\tilde{M}$ into $\mathfrak{M}$, then $H(\tilde{M}) = \mathfrak{M}$.*

If the hypersurface is spherical, then its universal covering does not, in general, have to be a maximal hypersurface. Similarly, if the Levi form of a hypersurface is indefinite, then its covering does not have to be a maximal hypersurface. These hypersurfaces are used in the known examples of non- continuable local maps.

We recall these examples. Burns and Shnider [7] conceived a hypersurface (in $\mathbb{C}^2$) that is real-analytic, compact, strictly pseudoconvex, spherical, but not simply-connected, and a non-continuable local map from a sphere into this hypersurface. Beloshapka [3] constructed a non-continuable local map of a non-spherical hypersurface with an indefinite Levi form into another such hypersurface.

We recall that the question of why local maps are not always continuable was discussed also in §6.4.

**8.7. Mapping of Hypersurfaces with Indefinite Levi Form.** The continuability of holomorphic maps of strictly pseudoconvex hypersurfaces is accounted for by the compactness of the group of local automorphisms. Maps sending a hypersurface with indefinite Levi form into another such hypersurface also have the continuation property. But the reason here is different. When the Levi form is indefinite, then in different directions the hypersurface is convex on different sides. For any side we can construct a one-parameter family of analytic discs whose boundaries lie on the hypersurface; under a variation of the parameter the boundaries of the discs contract to a point, and the discs fill out part of a neighbourhood of this point that lies on one side of the hypersurface. Therefore, every function holomorphic on a hypersurface (and consequently a map) can be continued to all the discs of this family. There is an extensive literature devoted to hulls of holomorphy of hypersurfaces, and the removal of singularities of holomorphic maps (see, for example, [19], [34], [44], [21], [20], [45]).

We state one of these results. Suppose that in $\mathbb{C}P^n$ there are defined two compact hypersurfaces with $C^2$-smoothness, and that their Levi forms are indefinite at all points. Then any locally biholomorphic map of one hypersurface onto the other can be continued holomorphically to the whole of $\mathbb{C}P^n$, and so it turns out to be a linear map (see [21]).

As regards applications of the technique of normal forms in this context, so far we may speak only about possibilities. Although the group of local automorphisms of hypersurfaces with indefinite Levi form is non-compact, perhaps we might expect certain elements of the set of initial data of automorphisms of a non-spherical hypersurface to form in the aggregate over the whole group a compact set. We shall discuss some of these results in the next section.

**8.8. Proof of the Lemma in §8.1.** We fix $M$, $M^*$, and $H$, which maps $M$ into $M^*$. Let $\gamma(\Theta)$ and $\gamma_0(\Theta)$ be chains on $M$ and on the quadric $v = \langle z, z \rangle$ with the same vector $b$ defining their direction at the origin (see §3.1). We assume that the parameterizations of these chains are compatible (see §7.4) and that $\gamma(0) = \gamma_0(0) = 0$. Let $b^*$ be a vector defining the direction of $\gamma^*$. We show that $|b^*|$ cannot be much greater than $|b|$. Naturally, we can assume that $|b^*| > |b|$ (it is unnecessary to force an open door). Furthermore we assume that $|b|$ is a large number. Then by the lemma in §7.5 the interval $[0, 2\pi]$ belongs to the domain of definition of $\gamma(\Theta)$ and $\gamma^*(\Theta)$; the corresponding segments $\tilde{\gamma} = \gamma|_{[0,2\pi]}$ and $\tilde{\gamma}^* = \gamma^*|_{[0,2\pi]}$ are close to the circles $\gamma_0$ and $\gamma_0^*$; both circles and these segments of chains lie inside a polydisc of radius $\delta/2$ with centre at the origin, and $||b(\Theta)| - |b|| < \varepsilon|b|$ for $\Theta \in [0, 2\pi]$.

We define $\mu$ by $|b^*| = (1/\mu)|b|$, and make the substitution $z \to \mu z$, $w \to \mu^2 w$. From $M^*$, $\gamma^*$, $\tilde{\gamma}^*$, and $b^*(\Theta)$ we obtain similar sets $M^{**}$, $\gamma^{**}$, $\tilde{\gamma}^{**}$, and $b^{**}(\Theta)$. By the lemma in §7.5 when $|b|$ is large, $\tilde{\gamma}^{**}$ lies in a polydisc of radius $\delta/2$ (since $|b^{**}| = |b|$), $\tilde{\gamma}^{**}$ is close to $\tilde{\gamma}_0^{**}$, and $|b^{**}(\Theta)|$ is close to $|b^{**}(0)|$ at all points of $\tilde{\gamma}^{**}$. By taking $|b|$ sufficiently large and so obtaining all we need for what follows, we also assume that $|b| \leq b_0$, where $b_0$ is a function of $\delta$ and $m$.

We denote by $H_1$ and $H_2$ normalizations of $M$ and $M^{**}$ with initial data $\omega_1 = (E, -b, 1, 1, 0)$ and $\omega_2 = (E, -b^{**}, 1, 1, 0)$, respectively, that send $M$ and $M^{**}$ into circular normal form. These normalizations straighten the chains $\gamma$ and $\gamma^{**}$. We shall assume that the parameter $\Theta$ on the chains $\gamma(\Theta)$ and $\gamma^*(\Theta)$ is defined by these normalizations. We consider a map $H^{**}$ sending $M$ into $M^{**}$ that is the composition of a dilation followed by $H$.

The idea of the proof that $|b^*|$ cannot be much bigger than $|b|$ consists in showing that $\mu$ is small for large $|b^*|$, and so the hypersurface $M^{**}$ is close to spherical. On the other hand, by the lemma in §6.2, $\tilde{\gamma}^{**} = H^{**}(\tilde{\gamma})$, and thus it is possible to find an $x_1 \in \tilde{\gamma}$ at which $H^{**}$ has derivatives not large in modulus, and at the same time sends $x_1$ into $x_2 \in \tilde{\gamma}^{**}$. But by the lemma in §7.3, this is impossible since $M$ has large non-sphericity at $x_1$ and $M^{**}$ is almost spherical at $x_2$. We note that if $N(M) > N$, then $N_{\delta,m}(0, M) > N$ ($N(M)$ is introduced here only to avoid binding the assertion to an awkward definition). We have $|b^{**}| = \mu|b^*| = |b|$. If $\mu$ is small, then $M^{**}$ is more spherical than $M^*$. More precisely, the non-sphericity characteristic $N_{\delta,m}$ of the hypersurface $M^{**}$ is defined and $N_{\delta,m}(0, M^{**}) < N^*$, where $N^* > 0$ is such that $N^* \to 0$ as $\mu \to 0$.

We represent the mapping $H^{**}$ as $H^{**} = H_2^{-1}(R(H_1))$, where $R \overset{\text{def}}{=} H_2(H^{**}(H_1^{-1}))$. The mapping $R$ sends a normal form into a normal form and preserves the orientation of the line $z = 0$, $\rho = 1$. Thus, when $R$ is written in complex coordinates, it turns out to be a linear-fractional transformation (see §5.2). It sends the disc $z = 0$, $|w| \le 1$ into itself. For any linear-fractional transformation $R$ sending the disc $z = 0$, $|w| \le 1$ into itself there is a $\Theta_0 \in [0, 2\pi]$ such that

$$\left|\frac{\partial R}{\partial w}\right|(0, 1, \Theta_0)\right| \le 1 \quad \text{and} \quad \left|\left|\frac{\partial^2 R}{\partial w^2}\right|(0, 1, \Theta_0)\right| \le \left|\left|\frac{\partial R}{\partial w}\right|(0, 1, \Theta_0)\right|.$$

The minimum point of the derivative of the map satisfies these conditions. Thus, at $(0, 1, \Theta_0)$ the mapping $R$ has the initial data $(U(\Theta_0), 0, 1, \lambda(\Theta_0), r(\Theta_0))$, and so knowing the general form of $R$ (see §5.2) it is not difficult to see from the inequalities

$$\left|\left|\frac{\partial R}{\partial w}\right|(0, 1, \Theta_0)\right| \le 1, \qquad \left|\left|\frac{\partial^2 R}{\partial w^2}\right|(0, 1, \Theta_0)\right| \le \left|\left|\frac{\partial R}{\partial w}\right|(0, 1, \Theta_0)\right|$$

that each of $\lambda$, $|r|$ does not exceed 1 in this case. We explain this: earlier, the initial data of a normalization were defined only at the origin (see §5.2). Here we are speaking of initial data at $(0, 1, \Theta_0)$, having in mind the quantities obtained in the image and the inverse image of corresponding points after transfer to the origin.

The lemma in §7.5 gives an upper bound for the first and second derivatives of the mapping $H_1$ at points of $\tilde{\gamma}$, and of $H_1^{-1}$ at points of the segments $z = 0$, $\rho = 1$, $0 \le \Theta_0 \le 2\pi$. Similarly, the derivatives of $H_2$ on $\tilde{\gamma}^*$ and of $H_2^{-1}$ on the segment $z = 0$, $\rho = 1$, $0 \le \Theta_0 \le 2\pi$ can be estimated from above. Thus, having chosen $N_{\delta',m'}$ we can use the lemma in §7.3.

We choose $\delta'$ and $m'$ so that $N_{\delta',m'}$ is defined for all points of $M$ and $M^{**}$ lying inside a polydisc of radius $\delta/2$, and in particular for $x_1 \in \gamma(\Theta_0) \in M$ and $x_2 \in \gamma^{**}(\Theta_0) \in M^{**}$. The pair $\delta', m'$ is defined by the pair $\delta, m$. By the lemma in §7.2, $N_{\delta',m'}(x_1, M)$ can be estimated from below if $\delta, m$, and $N$ are fixed beforehand. Hence, by the lemma in §7.3, $N_{\delta',m'}(x_2, M^*)$ has an estimate from below. Therefore, by the lemma in §7.2, $\mu = |b|/|b^*|$ cannot be arbitrarily small; more precisely, it is bounded from below for any positive $\delta, m$, and $N$ regardless of the choice of $M$ and $M^*$. From this it follows that for any vector $b$ with a sufficiently large modulus, for example $|b| = b_0$, we have $|b^*| \leq cb_0$, where $c$ is a function of $\delta, m$, and $N$. Similarly, for $H^{-1}$ we obtain that if $|b^*| = b_0$, then $|b| \leq cb_0$.

Let $\omega = (U, a, \lambda, \sigma, r)$ be the set of initial data of the mapping $H$ in question. We estimate $|a|$ and $\lambda$. From the definition of $\omega$ (see §2.1) and the vectors $b$ and $b^*$ we have $b^* = \sigma U \lambda^{-1}(b + a)$. By substituting in this an arbitrary vector $b$ such that $|b| = b_0$ we obtain that $|\sigma U \lambda^{-1}(b + a)| \leq cb_0$. Similarly, if $|b^*| = b_0$, then $|\sigma \lambda U^{-1} b^* - a| \leq cb_0$. Substituting in the second inequality a vector $b^*$ such that $\lambda U^{-1} b^*$ has the same direction as $-a$, we obtain $|a| \leq cb_0$. Similarly, using the first inequality we obtain $1/\lambda \leq c$. Hence, for any $M$ and $M^*$ and $H$ sending $M$ into $M^*$ we have obtained upper estimates for $|a|, \lambda, 1/\lambda$. An estimate for the norm of the matrix $U$ follows from the compactness of the group of unitary matrices.

We now estimate the parameter $r$. The map $H$ defines a change of parameter on a segment of the chain $\tilde{\gamma}^*$ in exactly the same way as in the above construction $H^{**}$ defined a reparametrization of a segment of $\tilde{\gamma}^{**}$. Again we use the formula for a change of parameter. We represent $H$ as $H^{*-1}(R^*(H_1))$, where $H^*$ is a normalization of $M^*$ straightening $\gamma^*$, and $R^* = H^*(H(H_1^{-1}))$ is a linear-fractional transformation sending a normal form for $H_1(M)$ into the normal form for $H^*(M^*)$. We define the initial data $\omega^*$ and $\omega_1$ by $\omega^* = (E, -b^*, 1, 1, 0)$, and as before, $\omega_1 = (E, -b, 1, 1, 0)$. As before we shall assume that $|b|$ is large, for example, $|b| = b_0$. By the lemma in §7.5 the first and second derivatives of $H_1$ and $H^{*-1}$ can be estimated from above. Thus $r$ is bounded above by $r^* = \left|\frac{\partial^2 R}{\partial w^2}\right|_{(0,\,1,\,0)}\right|$, $b_0$ and $\lambda$. Therefore, if $r$ is large, then so is $r^*$. But $R^*$ (in complex coordinates) is a linear-fractional transformation sending the circle $z = 0$, $w = 1$ onto itself. Thus if $r^*$ is large, then we can find a $\Theta_0 \in [0, 2\pi]$ such that $\left|\frac{\partial R^*}{\partial w}\right|_{(0,\,1,\,\Theta_0)}\right|$, is large. By the lemma in §7.5 the derivatives of $H^*$ at all points of $\tilde{\gamma}^*$ and the derivatives of $H_1^{-1}$ at all points such that $z = 0$, $\rho = 1$, $0 \leq \theta \leq 2\pi$ can be estimated from above. Hence $\left|\frac{\partial}{\partial \zeta}\right|_{\gamma(\Theta_0)} H\right|$ is large. This is impossible. We explain why. Close to $x_1$ we choose coordinates with $x_1$ as origin and such that the $u$-axis is perpendicular to the complex tangent space; we construct the standard normalization of $M$ at $x_1$. Similarly, close to $x_2 = H(x_1)$ we choose linear normal coordinates for $M^*$. Then we can find $\delta', m'$, and $N'$ depending only on $\delta$, $m$, and $N$, and such that $N_{\delta', m'}(0, M) > N'$ and $N_{\delta', m'}(0, M^*) > N'$. By rewriting $H$ in the new coordinates, we can assume that $\left|\frac{d}{d\zeta}\right|_0 H\right|$ remains large. But we already know that the parameters $a$ and $\lambda$ for such maps are bounded above, and so $\left|\frac{d}{d\zeta}\right|_0 H\Big|$ cannot be large. Hence we also have an estimate on $r$.

## §9. Automorphisms of a Hypersurface

Can we associate with a hypersurface a coordinate system in which all the automorphisms of the hypersurface are linear–fractional transformations?

**9.1. Estimation of the Dimension of the Stability Group.** Let $M$ be a hypersurface with a non-degenerate Levi form. We fix a point $\zeta \in M$ and estimate the dimension of the stability group (the group of automorphisms of $M$ defined in a neighbourhood of $\zeta$ and leaving this point fixed). Let us introduce a coordinate system with origin at $\zeta$ in which $M$ has normal form. Then every automorphism can be regarded as a normalization of the hypersurface. A normalization is uniquely determined by a set of initial data $\omega = (U, a, \lambda, \sigma, r)$ and every such set gives an automorphism of a quadric (see §1.2). Hence, the set of all normalizations can be regarded as a group, which is isomorphic to the stability group of the quadric. Therefore, the stability group of $M$ is isomorphic to a subgroup of the stability group of a quadric.

Let us now calculate the dimension of the stability group of a quadric. The dimension of the group in the parameters $\sigma$, $\lambda$, $r$ is 2, in $a$ it is $2(n-1)$, and in the matrix $U$ it is $(n-1)^2$. Thus the dimension of the stability group of a quadric is $n^2 + 1$ and consequently, for an arbitrary hypersurface with a non-degenerate Levi form the dimension does not exceed $n^2 + 1$. This bound, as we shall see later, is exact only for spherical hypersurfaces, that is, those equivalent to a quadric.

**9.2. Parametrization of the Group of Automorphisms.** A spherical hypersurface is written in normal coordinates in the same way as a quadric. Using such a representation, for any set $\omega$ satisfying (1.2.2) we can find an automorphism of the hypersurface with the given set of initial data.

A different situation arises when $M$ is non-spherical. In this case the initial data $(U, a, \lambda, \sigma, r)$ corresponding to automorphisms are no longer independent. Beloshapka and Loboda ([2] and [26]) have proved the following result.

**Theorem.** *Let $M$ be a real-analytic non-spherical hypersurface having a non-degenerate Levi form and given in normal form. Then for any automorphism the initial data $a$, $\lambda$, $\sigma$, $r$ are uniquely defined by the matrix $U$. In other words, the stability group of $M$ is isomorphic to a subgroup of the group of matrices preserving the Levi form of this hypersurface.*

This assertion can be reformulated in geometrical terms. A local automorphism of a non-spherical hypersurface is completely defined by the restriction of the tangent map (at a fixed point) to the complex tangent.

It follows from this theorem that the dimension of the group of local automorphisms of a non-spherical hypersurface does not exceed $(n-1)^2$.

The very simple example of a hypersurface $v = \langle z, z\rangle + \langle z, z\rangle^4$ shows that the highest possible dimension is actually realized.

We emphasize that although for large $n$ the number of initial data of automorphisms that turn out to be dependent is relatively small, these parameters have significantly complicated the study of automorphism groups.

**9.3. Linearization of the Group.** Consider the stability group of a strictly pseudoconvex hypersurface. Kruzhilin and Loboda [25] have shown that the stability group of such a hypersurface is linearizable in the following sense.

**Theorem.** *If a real-analytic hypersurface is strictly pseudoconvex and non-spherical, then in a neighbourhood of any of its points we can choose normal coordinates in which every automorphism leaving this point fixed is written as*

$$z^* = Uz, \quad w^* = w,$$

*where $U$ is a matrix preserving the Levi form.*

The initial data $(U, a, \lambda, r)$ of automorphisms of a quadric are not connected with each other by any relations, and so the stability group of a spherical hypersurface is non-linearizable. The possibility of linearization in the non-spherical case is suggested by a theorem of Bochner [5]: for every compact subgroup of the stability group of a real-analytic hypersurface we can find a linearizing coordinate system. However, a direct application of this theorem is not acceptable, since in the case in question we need a system of normal coordinates.

Let us recall the question stated at the beginning of this section. For hypersurfaces with a positive definite Levi form the answer turns out to be positive. If a hypersurface is spherical, then all normal coordinates have the required property. For a non-spherical hypersurface a suitable coordinate system is assured by the theorem of Kruzhilin and Loboda. The question of the linearization of the stability group of a hypersurface with indefinite Levi form remains open. The theorem of §9.2 gives one possible approach to this problem (see also the parameter estimates in 8.1).

From this point of view the works of Burns, Shnider and Wells [9], and Webster [43] are also interesting. In these works, a linearization of the system of coordinates is constructed under certain restrictions on the type of hypersurface.

**9.4. Proof of the Kruzhilin–Loboda Theorem.** Suppose $M$ contains the origin and is given in normal form. We denote by $\Phi$ the stability group of the hypersurface at the origin. First of all, we show that there exists a chain $\gamma$ in $M$,

$\gamma(0) = 0$, whose points are fixed for each transformation $\varphi \in \Phi$. Denote by $p$ the vector tangent to $M$ at the point $(0, 0)$ and having coordinates $z = 0$, $w = 1$. We set $p_\varphi = (d\varphi|_0)p$, where $d\varphi|_0$ is the differential of the automorphism $\varphi$ at $(0, 0)$.

By Corollary 1 of Theorem 8.1, $\Phi$ is compact and therefore there is a finite positive measure on $\Phi$, invariant with respect to the group action (the Haar measure). Let us denote by $\mu$ this measure and by $h$ the average of $p_\varphi$ with respect to this measure

$$h = \int_{\varphi \in \Phi} p_\varphi d\mu.$$

Each automorphism $\varphi$ can be considered as a normalization with a certain set of initial data $\omega = (U, a, \lambda, \sigma, r)$. Since $M$ is strictly pseudoconvex, $\sigma = 1$. Consequently, for each $\varphi$, the projection of $p_\varphi$ on the $u$ axis is numerically equal to $2\sigma\lambda^2$, hence positive. Hence, the vector $h$ also has a positive projection on the $u$ axis and therefore, it clearly is transversal to the complex tangent space to $M$.

Since $h$ is obtained by averaging $p_\varphi$, for each $\psi \in \Phi$ we have

$$(d\psi|_0)h = (d\psi|_0) \int_{\varphi \in \Phi} p_\varphi d\mu = \int_\varphi (d\psi|_0)p_\varphi d\mu = \int_\varphi (d\psi \circ \varphi|_0)p d\mu =$$

$$= \int_\varphi p_{\psi \circ \varphi} d\mu = \int_\varphi p_\varphi d\mu = h.$$

By Bochner's Theorem [21], there is a system of coordinates $(z^*, w^*)$, $w^* = u^* + iv^*$, in which each automorphism $\varphi \in \Phi$ is linear. We shall suppose that $h$ has coordinates $(0, 1)$ in the coordinates $z^*$, $w^*$ and $M$ is given by the equation $v^* = G(z^*, \bar{z}^*, u^*)$, where $G(0) = 0$, $dG|_0 = 0$.

In this system of coordinates, each $\varphi \in \Phi$ is a linear transformation which leaves the hyperplane $w^* = 0$ invariant as well as the point $z^* = 0$, $w^* = 1$. It follows that the transformation $\varphi$ leaves the coordinate $w^*$ invariant.

Consider a chain $\gamma(t)$ on $M$ such that $\gamma(0) = 0$ and $(d\gamma/dt)(0) = h$. Since automorphisms of $M$ send chains to chains and $\gamma$ is the only chain passing through the origin in the direction of the vector $h$, $\varphi(\gamma) = \gamma$ for each $\varphi \in \Phi$. And since the projection of $\gamma$ on the $u$-axis is one-to-one for $t$ small, $\varphi(\gamma(t)) = \gamma(t)$ and the points of $\gamma$ remains fixed under the automorphism $\varphi$.

Let us now bring the equation of $M$ into a linear normal form such that the chain $\gamma$ is carried to the line $z = 0$, $v = 0$. Each local automorphism $\varphi$ fixes $\gamma$ and hence is a linear–fractional transformation of a completely concrete type (see Theorem 1.3). Since all points of $\gamma$ are fixed, $\lambda = 1$ and $r = 0$, and so $\varphi$ is a unitary transformation of the variable $z$. This completes the proof of the theorem.

**9.5. Proof of the Beloshapka–Loboda Theorem.** We shall write the equation for the hypersurface in the form

$$v = \langle z, z \rangle + \sum_{\nu = k}^{\infty} F_\nu(z, \bar{z}, u)$$

where $F_\nu$ are homogeneous polynomials of degree $\nu$, i.e. such that

$$F_\nu(tz, t\bar{z}, t^2u) = t^\nu F_\nu(z, \bar{z}, u).$$

Let $M$ and $M^*$ be hypersurfaces given in normal form

$$v = \langle z, z \rangle + F(z, \bar{z}, u), \quad v = \langle z, z \rangle + F^*(z, \bar{z}, u)$$

respectively, and let $h = (f, g)$ be a biholomorphic mapping of $M$ to $M^*$ in a neighbourhood of the origin. In this case we have the identity

$$(-\operatorname{Im} g + \langle f, f \rangle + F^*(f, \bar{f}, \operatorname{Re} g))|_{w = u + i\langle z, z \rangle + iF(z, \bar{z}, u)} = 0. \qquad (9.5.1)$$

This is an analytic equation with the property that if the variables are bound by the equation for $M$, then the values are bound by the equation for $M^*$.

We mention several consequences of this identity which, in case $M = M^*$ is not spherical, allow us to express the parameters $(a, \lambda, \sigma, r)$ via the matrix $U$ and thus obtain the proof of the theorem in question. In the formulations, we shall set $\lambda U = c$ and $\sigma\lambda^2 = \rho$ for brevity.

(1) We have the identity (see (4.1.2))

$$\langle Uz, Uz \rangle = \sigma \langle z, z \rangle,$$

which allows us to express $\sigma$ via $U$.

(2) If $F_k$ is the first non-zero term of $F$, then $\lambda^{k-2} F_k^*(Uz, \bar{U}\bar{z}, \sigma u) = \sigma F_k(z, \bar{z}, u)$. This identity allows us to express $\lambda$ in terms of $U$ and $\sigma$.

In order to write down the succeeding relations, we introduce a notation. If $P$ is a polynomial or a matrix, we shall denote by $[P]$ an ordering of the coefficients of $P$.

(3) We have the relation (see [2], [26]):

$$\alpha([F_k])\begin{pmatrix} a \\ \xi_1 \end{pmatrix} = A(\lambda, \sigma, [U], [F_{k+1}], [F_{k+1}^*]),$$

where $\alpha$ is a square matrix depending linearly on $[F_k]$, with $\det \alpha = 0$ if and only if $F_k = 0$; $\xi_1$ is a group of variables, depending on the coefficients of $h$; and $A$ is a polynomial of degree $k-1$ in $\lambda$ whose coefficients depend linearly on $[U]$, $[F_{k+1}]$, and $[F_{k+1}^*]$.

(4) We have the relation

$$B([F_k])\begin{pmatrix} r \\ \xi_2 \end{pmatrix} = B(\lambda, a, \sigma, [U], [F_{k+1}], [F_{k+1}^*]), \beta[F_{k+2}], [F_{k+2}^*]),$$

where $\beta([F_k])$ is a square matrix depending linearly on $[F_k]$, whose determinant vanishes only for $F_k = 0$; $\xi_2$ is a group of variables depending on the coefficients of $h$; and $B$ is a polynomial in $\lambda$ and $a$ of degree $k$ in $\lambda$ and degree 2 in $a$ whose coefficients depend linearly on $[U]$, $[F_{k+1}]$, $[F_{k+2}]$, $[F_{k+1}^*]$, $[F_{k+2}^*]$.

Let us return to equation (9.5.1). Along with the mapping $h$, we consider the mapping $\tilde{h}$ which is the automorphism of the hyperquadric $v = \langle z, z \rangle$ given by the formulas from (1.2) where as initial data we choose the initial data of $h$.

Analogous to (9.5.1), we obtain the identity

$$(-\operatorname{Im}\tilde{g}+\langle\tilde{f},\tilde{f}\rangle)|_{w=u+i\langle z,z\rangle}\equiv 0. \tag{9.5.2}$$

Substracting (9.5.2) from (9.5.1), denoting by $\Phi$ the left member of the resulting identity, and identifying terms of both sides, we obtain:

$$\Phi_s(z,\bar{z},u)\equiv 0, \qquad s=0,1,\ldots.$$

Let

$$h(z,w)=\sum_{s=0}^{\infty}h_s(z,w)$$

be the power series expansion of $h$ in $z$ and $w$, where

$$h_s(tz,t^2w)=t^s h_s(z,w), \quad h_s=(f_s,g_s).$$

Further, suppose $F\not\equiv 0$, i.e.

$$F=\sum_{s=p}^{\infty}F_s, \quad F^*=\sum_{s=p^*}^{\infty}F_s^*, \quad \text{where } F_p, F_{p^*}^*\neq 0.$$

Here, we formally allow $p^*=\infty$, but it will be shown below (see (9.5.8) that $p^*=p$. We set $k=\min(p,p^*)$. The $s$-th term of (9.5.1) allows us to define $f_{s-1}$ and $g_s$ (see [2] and [26]), but for $s=0,1,\ldots,k-1$, the $s$-th term of (9.5.1) agrees with the $s$-th term of (9.5.2). Thus, we conclude that

$$f_s=\tilde{f}_s, \quad \text{for } s=0,1,\ldots,k-2 \tag{9.5.4}$$

$$g_s=\tilde{g}_s, \quad \text{for } s=0,1,\ldots,k-1.$$

Our immediate goal is to calculate $\Phi_k$, $\Phi_{k+1}$ and $\Phi_{k+2}$. For the calculation it is convenient to introduce the following notation. In calculating terms of weight $s$, we shall write $\alpha\to\beta$, meaning that all terms of weight $s$ which appear in $\alpha$ also appear in $\beta$. Furthermore, we denote $\Delta h=h-\tilde{h}=(\Delta f,\Delta g)$, $\tilde{w}=u+i\langle z,z\rangle$, $\varphi=C^{-1}f$, and $\psi=\rho^{-1}g$. We represent $\Phi(z,\bar{z},u)$ in the following form:

$$\begin{aligned}\Phi(z,\bar{z},u)=&\operatorname{Re}i(g|_{w=\tilde{w}+iF}-g|_{w=\tilde{w}})+\operatorname{Re}i\Delta g|_{w=\tilde{w}}+\\&+(\langle f,f\rangle|_{w=\tilde{w}+iF}-\langle f,f\rangle|_{w=\tilde{w}})+2\operatorname{Re}(\Delta f,f)+(\langle\Delta f,\Delta f\rangle|_{w=\tilde{w}})+\\&+F_k^*(f,\tilde{f},\operatorname{Re}g)+F_{k+1}^*(f,\tilde{f},\operatorname{Re}g)+F_{k+2}^*(f,\tilde{f},\operatorname{Re}g)+\ldots.\end{aligned}$$

In this representation we shall write the terms in the form: (I), (II), (III), etc.

We need explicit formulae for the first few components of the mapping $\tilde{h}$ and its derivative $(\partial\tilde{h})/(\partial w)$. These can be obtained immediately from the formulas for representing quadrics, namely:

$$\begin{aligned}&\tilde{\varphi}_0=0,\ \tilde{\varphi}_1=z,\ \tilde{\varphi}_2=2i\langle z,a\rangle z+aw,\\&\tilde{\varphi}_3=-4\langle z,a\rangle^2 z+(r+i\langle a,a\rangle)wz+2i\langle z,a\rangle aw,\\&\tilde{\psi}_0=0,\ \tilde{\psi}_1=0,\ \tilde{\psi}_2=w,\ \tilde{\psi}_3=2i\langle z,a\rangle w,\\&\tilde{\psi}_4=-4\langle z,a\rangle^2 w+(r+i\langle a,a\rangle)w^2,\end{aligned}$$

$$\left(\frac{\partial\tilde{\varphi}}{\partial w}\right)_0 = a, \quad \left(\frac{\partial\tilde{\varphi}}{\partial w}\right)_1 = 2i\langle z, a\rangle a + (r + i\langle a, a\rangle)z, \tag{9.5.5}$$

$$\left(\frac{\partial\tilde{\psi}}{\partial w}\right)_0 = 1, \quad \left(\frac{\partial\tilde{\psi}}{\partial w}\right)_1 = 2i\langle z, a\rangle,$$

$$\left(\frac{\partial\tilde{\psi}}{\partial w}\right)_2 = 2(r + i\langle a, a\rangle)w - 4\langle z, a\rangle^2.$$

Calculation of $\Phi_k$:

$$(\mathrm{I}) = \mathrm{Re}\, i\left(\frac{\partial g}{\partial w}\bigg|_{w=\tilde{w}} i(F_k + \ldots) + \ldots\right) \to -\rho F_k.$$

Here, we have used the fact that

$$\left(\frac{\partial g}{\partial w}\right)_0 = \left(\frac{\partial \tilde{g}}{\partial w}\right)_0 = \rho \qquad \text{(see (9.5.5))},$$

$$(\mathrm{II}) \to \mathrm{Re}\, i\Delta g|_{w=\tilde{w}} = \rho\, \mathrm{Re}\, i\Delta\psi_k|_{w=\tilde{w}}.$$

$$(\mathrm{III}) = 2\mathrm{Re}\left(\left\langle f, \frac{\partial f}{\partial w}\right\rangle i(F_k + \ldots) + \ldots\right)\bigg|_{w=\tilde{w}} \to F_k 2\mathrm{Re}\, i\left\langle f, \frac{\partial f}{\partial w}\right\rangle\bigg|_{w=\tilde{w}}.$$

However, $\langle f, (\partial f/\partial w)\rangle$ does not contain terms of degree 0, and so (III) $\to 0$.

$$(\mathrm{IV}) \to 2\mathrm{Re}\langle \Delta f_{k-1}, \tilde{f}_1\rangle|_{w=\tilde{w}} = 2\rho\, \mathrm{Re}(\Delta\varphi_{k-1}, z\rangle|_{w=\tilde{w}}.$$

$$(\mathrm{V}) \to F_k^*(f_1, \tilde{f}_1, \mathrm{Re}\, g_2) = F_k^*(Cz, \bar{C}\bar{z}, \rho u).$$

Summing, we obtain:

$$\Phi_k = \rho\, \mathrm{Re}(i\Delta\psi_k + 2\langle\Delta\varphi_{k-1}, z\rangle)|_{w=\tilde{w}} + F^*(Cz, \bar{C}\bar{z}, \rho u) - \rho F_k(z, \bar{z}, u).$$

If we denote $\mathrm{Re}(i\Delta\varphi_{s-1} + 2\langle\Delta\varphi_{s-1}, z\rangle)|_{w=\tilde{w}}$ by $L_s(h)$, we have

$$L_k(h) = F_k(z, \bar{z}, u) - \rho^{-1} F_{k*}(Cz, \bar{C}\bar{z}, \rho u). \tag{9.5.6}$$

In [2], [26] it is shown that the equation $L_s(h) = G \pmod R$ has a unique solution for $\Delta\varphi_{s-1}$, $\Delta\psi_s$ in the class of mappings $(\Delta\varphi, \Delta\psi)$ for which the initial data is $(E, 0; 1, 1, 0)$. $R$ is the collection of equations written in the normal form. The right member of (9.5.6) belongs to $R$. Thus, $L_k(h) = 0 \pmod R$ and by the above assertion, we conclude that

$$\Delta\varphi_{k-1} = 0, \quad \Delta\psi_k = 0, \tag{9.5.7}$$

and also

$$F_k^*(Cz, \bar{C}\bar{z}, \rho u) = \rho F_k(z, \bar{z}, u). \tag{9.5.8}$$

Calculation of $\Phi_{k+1}$:

$$(\mathrm{I}) = \mathrm{Re}\, i\left(\frac{\partial g}{\partial w} i(F_k + F_{k+1} + \cdots) + \ldots\right)\Bigg|_{w=\tilde{w}} \to$$

$$\to -\rho F_{k+1}\mathrm{Re}\left(\frac{\partial \psi}{\partial w}\right)_0 - \rho F_k \mathrm{Re}\left(\frac{\partial \tilde{\psi}}{\partial w}\right)_1 = -\rho F_{k+1} - \rho F_k 2\mathrm{Re}\, i\langle z, a\rangle.$$

$(\mathrm{II}) \to \rho\,\mathrm{Re}\, i\Delta\psi_{k+1}|_{w=\tilde{w}}$.

$$(\mathrm{III}) = 2\mathrm{Re}\, i\left\langle f_0, \frac{\partial f}{\partial w}\right\rangle\Bigg|_{w=\tilde{w}} (F_k + F_{k+1} + \ldots) + \ldots \to$$

$$-\rho F_k 2\mathrm{Re}\, i\left(\left\langle \tilde{\varphi}, \frac{\partial \varphi}{\partial w}\right\rangle\right)_1 - \rho F_{k+1} 2\mathrm{Re}\, i\left(\left\langle \tilde{\varphi}, \frac{\partial \tilde{\varphi}}{\partial w}\right\rangle\right)_0,$$

$$\left(\left\langle \tilde{\varphi}_0, \frac{\partial \varphi}{\partial w}\right\rangle\right)_1 = \left\langle \tilde{\varphi}_1, \left(\frac{\partial \tilde{\varphi}}{\partial w}\right)_0\right\rangle = \langle z, a\rangle; \qquad \left(\left\langle \tilde{\varphi}, \frac{\partial \varphi}{\partial w}\right\rangle\right)_0 = 0.$$

Thus, $(\mathrm{III}) \to -\rho F_k 2\mathrm{Re}\, i\langle z, a\rangle$, and

$(\mathrm{IV}) \to \rho 2\mathrm{Re}\langle \Delta\varphi_k, z\rangle|_{w=\tilde{w}}$.

$(\mathrm{V}) = F_k^*(f, \bar{f}, \mathrm{Re}\, g) = \rho F_k(\varphi, \tilde{\varphi}, \mathrm{Re}\,\psi) = \rho F_k(z + \varphi_2 + \ldots,$

$\bar{z} + \bar{\varphi}_2 + \ldots, u + \mathrm{Re}\,\psi_3 + \ldots) = \rho(F_k + dF_k(\varphi_2 + \ldots, \bar{\varphi}_2 + \ldots,$

$$\mathrm{Re}\,\psi_3 + \ldots) + \ldots)|_{(z,\bar{z},u)} \to \rho 2Re\left(\frac{F_k}{\partial z}(2i\langle z, a\rangle z + \tilde{w}a) + i\langle z, a\rangle \tilde{w}\frac{\partial F_k}{\partial u}\right).$$

$(\mathrm{VI}) = F_{k+1}^*(f, \bar{f}, \mathrm{Re}\, g) \to F_{k+1}^*(Cz, \bar{C}\bar{z}, \rho u)$.

Summing, we obtain:

$$\Phi_{k+1} = \rho L_{k+1}(h) + F_{k+1}^*(Cz, \bar{C}\bar{z}, \rho u) - \rho F_{k+1}(z, \bar{z}, u) +$$

$$+ 2\rho\mathrm{Re}\left(-2i\langle z, a\rangle F_k + 2i\langle z, a\rangle \frac{\partial F_k}{\partial z}(z) + \tilde{w}\frac{\partial F_k}{\partial z}(a) + i\langle z, a\rangle \tilde{w}\frac{\partial F_k}{\partial u}\right).$$

Thus,

$$L_{k+1}(h) + T(F_k, a) = F_{k+1}(z, \bar{z}, u) - \rho^{-1} F_{k+1}^*(Cz, \bar{C}\bar{z}, \rho u), \tag{9.5.9}$$

where

$$T(F, a) = \mathrm{Re}\left(-2i\langle z, a\rangle F + 2i\langle z, a\rangle \frac{\partial F}{\partial z}(z) + \tilde{w}\frac{\partial F}{\partial z}(a) + i\langle z, a\rangle \tilde{w}\frac{\partial F}{\partial u}\right).$$

We remark the functional $T$ is real linear in each of its variables.

Calculation of $\Phi_{k+2}$.

$$(\mathrm{I}) = \mathrm{Re}\, i\left(\frac{\partial g}{\partial w}\Bigg|_{w=\tilde{w}} i(F_k + F_{k+1} + F_{k+2} + \ldots) + \ldots\right) \to$$

$$\to -\rho F_{k+2} - \rho F_{k+1} 2\mathrm{Re}\, i\langle z, a\rangle - \rho F_k 2\mathrm{Re}((r + i\langle a, a\rangle)\tilde{w} - 2\langle z, a\rangle^2).$$

$$(\mathrm{II}) \to \rho \operatorname{Re} i \Delta \psi_{k+2}|_{w=\bar{w}},$$

$$(\mathrm{III}) \to 2\rho \operatorname{Re} i\left(\left\langle \varphi, \frac{\partial \varphi}{\partial w}\right\rangle\right)_2 - \rho F_{k+1} 2\operatorname{Re} i\left(\left\langle \varphi, \frac{\partial \varphi}{\partial w}\right\rangle\right)_1$$

$$- \rho F_{k+2} 2\operatorname{Re} i\left(\left\langle \varphi, \frac{\partial \varphi}{\partial w}\right\rangle\right)_0,$$

$$\left\langle \varphi, \frac{\partial \varphi}{\partial w}\right\rangle_2 = \left\langle \varphi_2, \left(\frac{\partial \varphi}{\partial w}\right)_0\right\rangle + \left\langle \varphi_1, \left(\frac{\partial \varphi}{\partial w}\right)_1\right\rangle =$$

$$= 2i\langle z, a\rangle^2 + w\langle a, a\rangle - 2i\langle z, a\rangle\langle a, z\rangle + (r - i\langle a, a\rangle)\langle z, z\rangle,$$

$$\left(\left\langle \varphi, \frac{\partial \varphi}{\partial w}\right\rangle\right)_1 = \left\langle \varphi_1, \left(\frac{\partial \varphi}{\partial w}\right)_0\right\rangle = \langle z, a\rangle,$$

$$\left(\left\langle \varphi, \frac{\partial \varphi}{\partial w}\right\rangle\right)_0 = 0.$$

Thus,

$$(\mathrm{III}) \to \rho F_k 2\operatorname{Re}(2\langle z, a\rangle^2 - 2\langle z, a\rangle\langle a, z\rangle) - \rho F_{k+1} 2\operatorname{Re} i\langle z, a\rangle.$$

$$(\mathrm{IV}) \to 2\rho \operatorname{Re}\langle \Delta\varphi_{k+1}, z\rangle|_{w=\bar{w}} + 2\rho \operatorname{Re}\langle \Delta\varphi_k, \bar{\varphi}_2\rangle|_{w=\bar{w}}.$$

$$(\mathrm{V}) = F_k^*(f, \bar{f}, \operatorname{Re} g) = \rho F_k(\varphi, \bar{\varphi}, \operatorname{Re}\psi) =$$

$$= \rho F_k(z + \varphi_2 + \ldots, \bar{z} + \bar{\varphi}_2 + \ldots, u + \operatorname{Re}\psi_3 + \ldots) \to$$

$$\to -\rho dF_k(\varphi_3, \bar{\varphi}_3, \operatorname{Re}\psi_4) + \tfrac{1}{2} d^2 F_k(\varphi_2, \bar{\varphi}_2, \operatorname{Re}\psi_3).$$

In order to isolate the coefficient of $r$, we write the first term in detail:

$$\rho \operatorname{Re}\left(2\frac{\partial F_k}{\partial z}(\varphi_3) + \frac{\partial F_k}{\partial u}\psi_4\right) = \rho \operatorname{Re}\left(2\frac{\partial F_k}{\partial z}(-4\langle z, a\rangle^2 z + \right.$$

$$\left. + (r + i\langle a, a\rangle)\bar{w}z + 2i\langle z, a\rangle\bar{w}a\right) +$$

$$+ \frac{\partial F_k}{\partial u}(-4\langle z, a\rangle^2\bar{w} + (r + i\langle a, a\rangle)w^2).$$

Hence, the coefficient of $r$ is

$$\rho \operatorname{Re}\left(2\bar{w}\frac{\partial F_k}{\partial z}(z) + \bar{w}^2\frac{\partial F_k}{\partial u}\right).$$

$$(\mathrm{VI}) = F_{k+1}^*(f, \bar{f}, \operatorname{Re} g) = F_{k+1}^*(Cz + f_2 + \ldots, \bar{C}\bar{z} + \bar{f}_2 + \ldots, \rho u + \operatorname{Re} g_3 + \ldots) =$$

$$= F_{k+1}^* + dF_{k+1}^*(f_2 + \ldots, \bar{f}_2 + \ldots, \operatorname{Re} g_3 + \ldots)|_{(Cz, \bar{C}\bar{z}, \rho u)} \to$$

$$\to dF_{k+1}^*(f, \bar{f}, \operatorname{Re} g_3)|_{(Cz, \bar{C}\bar{z}, \rho u)}.$$

$$(\mathrm{VII}) \to F_{k+2}^*(Cz, \bar{C}\bar{z}, \rho u).$$

Summing, we obtain:

$$\rho^{-1}\Phi_{k+2} = L_{k+2}(h) + \rho^{-1}F^*_{k+2}(CZ, \bar{C}\bar{Z}, \rho u) - F_{k+2}(z, \bar{z}, u) + \qquad (9.5.10)$$
$$+ P_1(\rho^{-1}F^*_{k+1}(CZ, \bar{C}\bar{Z}, \rho u), a) + F_{k+1}P_2(a) + rP_3(F_k) +$$
$$+ P_4(F_k, a) + P_5(F_k, a, a) + P_6(\Delta\varphi_k, a),$$

where each of the expressions $P_1, \ldots, P_6$ is real linear in each of its variables.
We notice that

$$P_3(F) = \operatorname{Re}\left( -2wF + 2w\frac{\partial F}{\partial z}(z) + \tilde{w}^2\frac{\partial F}{\partial u}\right).$$

Thus, the proof is complete.

## §10. Smooth Hypersurfaces

Some of the results stated above, regarding real-analytic hypersurfaces, have been developed subsequently for other classes of surfaces. There is also a series of works on real-analytic manifolds of high codimension (see Tanaka [36], Moser and Webster [28], A.E. Tumanov and G.M. Khenkin [22]). Limitations of space force us to leave the exposition of these works to authors of other papers.

**10.1. Invariant Structures on a Smooth Hypersurface.** As in the case of a real-analytic hypersurface, on a smooth hypersurface, there exists an invariant family of curves called a chain and on each chain a certain class of distinguished parametrizations. On real-analytic hypersurfaces, geometrically, these curves coincide with the chains introduced in §3. However on such surfaces, the parametrization differs from those described above (see §§3, 6). In addition to families of chains, other invariant structures can be defined on hypersurfaces with the help of the Cartan-Chern theory and these can be useful in studying mappings of hypersurfaces.

In [11] Chern constructed a certain principal bundle $Y$ over a hypersurface $M$ having a non-degenerate Levi form. On $Y$ he defined a connection form $\pi$ with the property that two hypersurfaces $M_1$ and $M_2$ are $CR$-diffeomorphic if and only if there exists a diffeomorphism of the bundle $Y_1$ over $M_1$ onto the bundle $Y_2$ over $M_2$ which preserves the connection form. The connection form $\pi$ corresponds to the curvature form, $\Omega = d\pi - \pi \wedge \pi$.

Let $\omega$ be a real 1-form on $M$ which is non-vanishing and which at each point of $M$ annihilates the complex tangent space to $M$. It is known that locally on $M$ we can find smooth complex 1-forms $\omega^1, \ldots, \omega^{n-1}$ which at each point are a basis for the (1, 0)-forms on the complex tangent space and for some real 1- form

$\varphi$, we have

$$(10.1.1) \qquad d\omega = i\sum_{i,j} g_{ij}\omega^i \wedge \bar{\omega}^j + \omega \wedge \varphi,$$

where $\{g_{ij}\}$ is the Levi-form.

From the definition of the bundle $Y$ in [11], the choice of the collection of forms $(\omega, \omega^1, \ldots, \omega^{n-1}, \varphi)$ satisfying (10.1.1) gives a section $\sigma: M \to Y$. The forms $\sigma^*(\pi)$, $\sigma^*(\Omega)$ are smooth forms on $M$ and hence can be represented via $\omega$, $\omega^j$, and $\bar{\omega}^j$. The coefficients in these representations consist of variables defined on $M$ and depending on the choice of forms satisfying (10.1.1). One such variable is the so-called pseudoconformal curvature tensor $S = (S_{\alpha\beta\gamma\delta})$ $(1 = \alpha, \beta, \gamma, \delta \leq n-1)$. For $n \geq 2$, if $S \equiv 0$ on $M$, it follows that the curvature form $\Omega$ vanishes on $M$ and this means that $M$ is locally $CR$-diffeomorphic to the $(2n-1)$-dimensional quadric (see [11] with its appendix and [7]). A point $M$ at which $S = 0$, is called an *umbilical point*.

If a real-analytic hypersurface $M$ is given by equations in normal form

$$v = \langle z, z \rangle + \sum_{k, l \geq 2} F_{kl}(z, \bar{z}, u),$$

and if for the form $\omega$, we take $i\partial(v - \langle z, z \rangle - \Sigma F_{kl})$, and if the forms $\omega^1, \ldots, \omega^{n-1}$ coincide at the origin with $dz^1, \ldots, dz^{n-1}$, then the components of the tensor $S$, corresponding to this choice of forms, are the same as the coefficients of the polynomial $-4F_{22}$ (assuming that the polynomial is written in symmetric form, i.e. its coefficients are symmetric with respect to a permutation of variables).

Set

$$\|S\|^2 = \sum_{\alpha_1\beta_1\gamma_1\delta_1\alpha_2\beta_2\gamma_2\delta_2} g^{\alpha_1\bar{\alpha}_2} g^{\beta_1\bar{\beta}_2} g^{\gamma_1\bar{\gamma}_2} g^{\delta_1\bar{\delta}_2} S_{\alpha_1\beta_1\gamma_1\delta_1} S_{\alpha_2\beta_2\gamma_2\delta_2},$$

where $(g^{\alpha\bar{\beta}})$ is the inverse matrix of $(g_{\alpha\bar{\beta}})$. The following fact was noticed and exploited by Webster.

**Lemma.** The form $\|S\|\omega$ has continuous coefficients and is independent of the choice of forms $(\omega, \omega^j, \varphi)$ satisfying (10.1.1).

Thus, on each hypersurface there exists an invariant real 1-form $\theta = \|S\|\omega$, which vanishes only at those points where $\|S\| = 0$. In particular, if $M$ is strictly pseudoconvex, then $\theta$ vanishes only at umbilical points.

We restrict our attention now to the open set of those points of $M$ where $\|S\| \neq 0$. On this set, $\theta$ is a smooth 1-form. At each point the form $d\theta$ gives a non-degenerate real 2-form, on the complex tangent space, which corresponds to a certain Hermitian metric $ds^2$. Moreover, at each point of $M$, there is a unique tangent vector $x$, transversal to the complex tangent space, and such that $\theta(x) = 1$ and $d\theta(x, 0) = 0$. Thus, on $M$, an invariant vector field is defined. Using this field and the metric on the complex tangent space, we can construct an invariant indefinite metric tensor on $M$. In case the hypersurface is strictly pseudoconvex, this tensor can be chosen positive and such that it yields an invariant Riemannian metric on $M$.

The above invariants, defined on the set of non-umbilical points, were used by Webster [41] to construct a special normal form for a real-analytic hypersurface in the neighbourhood of a non-umbilical point. A system of coordiantes corresponding to this normal form is defined uniquely up to unitary transformations.

In [8] Burns and Shnider, using these invariants, showed that in the space of plurisubharmonic functions, there exists a family of second category such that for functions of this family, the different level sets are not $CR$-diffeomorphic.

In case $n=2$, the tensor $S$ is identically zero on any hypersurface and in its place one considers the variable denoted by $Q$ in [11]. For a real-analytic hypersurface given in normal form, $Q$ is precisely (up to a multiplicative constant) the coefficient in $F_{42}$. If $Q \equiv 0$, then $M$ is spherical. Points at which $Q = 0$ are also called umbilical points. The form $|Q|^{(1/2)}\omega$ turns out to be invariant and permits us to define an invariant vector field and an invariant metric on the set $\{Q \neq 0\}$. These invariants were constructed by $E$. Cartan [10]. Moreover, for $n = 2$, we can invariantly choose a basis vector, uniquely up to sign, for the complex tangent space. If the hypersurface is given in normal form, then this corresponds to choosing the coordinate $z^1$ so that the coefficient in $F_{42}$ is equal to one. Thus, for $n = 2$, on hypersurfaces without umbilical points, we have a smooth field of repères, and the question of $CR$-equivalence of two such hypersurfaces reduces to the possibility of mapping one hypersurface to another while preserving this field of repères.

The existence of such a repère field on $M$ means that we may uniquely fix a section $\sigma$: $M \to Y$. Thus, the coefficients in the expression of the forms $\sigma^*(\pi)$ and $\sigma^*(\Omega)$ turn out to be invariant functions on $M$. Making use of this, $E$. Cartan [10] defined nine functions on a three-dimensional hypersurface such that two hypersurfaces are in general $CR$-diffeomorphic if and only if there exists a diffeomorphism which identifies these nine-tuples.

The invariants which we have considered are defined and non-trivial only for non-umbilical points. However, in studying mappings of hypersurfaces, it is not always possible to avoid considering umbilical points. Thus, those invariants, which can be continuously extended to the whole hypersurface, present particular interest. As we already mentioned, the form $\theta$ is such an invariant and Webster [41] used the continuity of its coefficients to show that the component of the identity in the group of $CR$-diffeomorphisms of a compact hypersurface $M$ is compact if $M$ is not everywhere locally $CR$-diffeomorphic to the sphere.

The Hermitian metric $ds^2$ defined on the complex tangent space also extends continuously to umbilical points. However, the invariant vector field and consequently the invariant metric tensor do not have this property. On the other hand, the $(2n-1)$-dimensional measure on $M$, corresponding to the Riemannian metric in the strictly pseudoconvex case, is so constructed that, in any coordinate system its density with respect to real measure extends continuously to all points. This was used by Burns and Shnider [8] in studying proper mappings of strictly pseudoconvex domains. In addition, in the same paper, Burns and Shnider constructed a certain semi-metric on $M$ with the help of $ds^2$. We shall discuss this in the next section.

**10.2. Compactness of the Group of Global Automorphisms.** Let $M$ be a smooth strictly pseudoconvex hypersurface. The Hermitian metric $ds^2$ is defined in the complex tangent spaces at non-umbilical points and extends continuously to zero at umbilical points. Thus, we define the length of any piecewise smooth curve on $M$ whose tangent vectors lie in the complex tangent bundle to $M$. For a pair of points $x, y \in M$, we define the number $q(x, y)$ as the minimal length of all such curves joining $x$ and $y$. It is not difficult to see that:

1. $q: M \times M \to \mathbb{R}^1$ is a semi-metric on $M$, i.e. satisfies the triangle inequality,
2. If the point $x$ is non-umbilical, and $y \neq x$, then $q(x, y) > 0$,
3. If $d: M \times M \to \mathbb{R}^1$ is an arbitrary smooth metric on $M$, then for each compact set $K \subset M$, there is a $c$ such that for $x, y \in K$, we have $q(x, y) \leq c(d(x, y))^{1/2}$.

N.G. Kruzhilin [23] showed that from the existence of such a semi-metric on $M$ it immediately follows that the group of global $CR$-diffeomorphisms is compact provided $M$ is a compact smooth hypersurface, not everywhere locally $CR$-diffeomorphic to the sphere. Indeed, let $V$ be the open subset consisting of non-umbilical points of $M$. For some $\varepsilon > 0$, the set $V_\varepsilon = \{x \in V: q(x, M \setminus V) \geq \varepsilon\}$ is a non-empty compact subset of $V$. Each automorphism of $M$ is an isometry of $V$, with respect to the invariant smooth Riemannian metric on $V$ introduced above. This isometry maps $V_\varepsilon$ to $V_\varepsilon$. Since the group of such isometries is compact, it turns out that the group of $CR$-automorphisms of $M$ is also compact.

**10.3. Chains on Smooth Hypersurfaces.** The definition of a chain on a hypersurface has no connection with umbilical points. For this reason, chains also prove to be useful in studying mappings of hypersurfaces having umbilical points. However, to obtain similar applications, we should clarify some properties of chains. Just as in the real-analytic case, it turns out that, on strictly pseudo-convex hypersurfaces, chains which form small angles with the complex tangent spaces are similar to chains on spheres. This is related to the geometric behaviour of a chain as well as to its parametrization. We recall that on each chain, there is a distinguished family of parameters and we pass from one set of such parameters to another via a fractional-linear transformation. A segment of a chain, forming a small angle with the complex tangent space, and for which one of the distinguished parameters varies from o to $\pi$ is close to a circle of small radius. As before, we shall say that such a segment performs one loop. From the definition of chains and there parametrizations in terms of the bundle $Y$ over $M$

and its connection form $\pi$, one can give estimates on the above-mentioned closeness. We shall not introduce these estimates here. Let us merely remark that, using these estimates, one can prove the following assertion:

**Lemma.** *Let $x \in M$. Then each point of $M$ sufficiently close to $x$ can be joined to $x$ by a chain segment making less than one loop* (*N.G. Kruzhilin* [24]).

**10.4. Properties of the Stability Group.** Let $M$ be a smooth strictly pseudoconvex hypersurface and $x \in M$. As in the real-analytic case, to each local automorphism $f$, defined in a neighbourhood of the point $x$, we may associate a set of initial data $\omega = (U, a, \lambda, r)$. For this we will suppose that the hypersurface is given by an equation in which the terms of small degree satisfy the same condition as the corresponding terms of an equation given in normal form. As initial data of an automorphism of a hypersurface, we take the parameters of the corresponding automorphism of the hyperquadric.

**Theorem.** *If $M$ is not CR-diffeomorphic to the sphere near $x$, and $f$ and $g$ are local automorphisms of $M$ with initial data $\omega_f = (U_f, a_f, \lambda_f, r_f)$ and $\omega_g = (U_g, a_g, \lambda_g, r_g)$ and if $U_f = U_g$, then $f = g$ (N.G. Kruzhilin [24]).*

To prove the theorem, it is sufficient to show that if $U_f$ is the identity matrix $E$, for a local automorphism $f$, then $\lambda_f = 1$, $a_f = 0$, $r_f = 0$ and, consequently, $f$ is the identity mapping.

The proof follows if we show that the following assertions hold for an automorphism of a hypersurface which is not equivalent to the sphere:

1. $\lambda_f = 1$,
2. from $a_f = 0$, it follows that $r_f = 0$,
3. from $U_f = E$, it follows that $a_f = 0$.

The proof of these assertions is based on Lemmas 10.1 and 10.3. If some automorphism does not satisfy one of the properties 1, 2, 3, then we consider a chain segment, which starts at $x$ and completes no less than one loop. It turns out that if this segment lies in a sufficiently small neighbourhood of the point $x$, then on the segment, we may define iterations of the mapping $f$ or of the inverse mapping $f^{-1}$. The length of the image of the segment, after these iterations, tends to zero (in the proof of 1 and 3- at the expense of making the angle small between the chain and the complex tangent space, and in the proof of 2- at the expense of changing the parameters of the chosen quantities). However, by Lemma 10.1, the integrals of the invariant form $\theta$ along the segment or along its image coincide. Since $\theta$ has continuous coefficients, this means that $\theta = 0$ at each point of the original segment. Lemma 10.3 shows that all points sufficiently close to $x$ are umbilical and hence, $M$ is locally $CR$-diffeomorphic to the sphere.

In contrast to the real-analytic case (see 9.2) the condition of strict pseudoconvexity turns out, in the case of a smooth hypersurface, to be essential, as the following example shows.

Let $M$ be the hyperquadric in the space $\mathbb{C}^3(z_1, z_2, w = u + iv)$, given by the equation $v = z_1\bar{z}_2 + z_2\bar{z}_1$, and let $B$ be the unit polydisc in $\mathbb{C}^3$ centered at the origin.

Consider the open sets:

$$V_1 = \left\{(z_1, z_2, u): \left|z_2 - \frac{1}{64}\right| < \frac{1}{64}, \frac{1}{4} < u < \frac{1}{2}, |z_1| < \frac{1}{16}\right\},$$

$$V_k = 4^{1-k} V_1, k \in \mathbb{N}, \text{ and set } V_k^0 = \{(z_1, z_2, w): (z_1, z_2, u) \in V_k\}.$$

The fractional linear mapping $f$, taking the point $(z_1, z_2, w)$ to

$$((z_1 + w)/(1 - 2iz_2), z_2/(1 - 2iz_2), w/(1 - 2iz_2)),$$

is an automorphism of $M$. We have

$$f^m(z_1, z_2, w) = ((z_1 + mw)/(1 - 2imz_2), z_2/(1 - 2imz_2), w/(1 - 2imz_2)).$$

Let us denote by $V_k^m$ the image $f^m(V_k^0)$, $m \in \mathbb{Z}$. Then, for $m \neq 0$, the inequality

$$|z_1| > \frac{4|m| - 1}{|m| + 4k + 1},$$

is satisfied on $V_k^m$, and consequently, for $m \neq 0$ and $n \geq k$, we have $V_k^m \cap V_k^0 \neq 0$. This means that the sets $V_k^m$ are pairwise disjoint. Moreover, for each $k$, we can find a natural number $N_k$ such that for $|m| > N_k$, $V_k^m$ lies outside of $B$.

We move the hypersurface $M$ onto the set $V_1$ so that the Levi form remains non-degenerate and we change it on the intersection with the sets $V_1^m$, $|m| \leq N_1$, so that the mapping $f$ is also a local automorphism on the hypersurface $M_1$ so obtained.

By $M_2$ we denote a hypersurface agreeing with $M_1$ outside of

$$\bigcup_{m=-N_k}^{N_k} V_2^m,$$

and, in the interior of this same set, differing from a quadric, and such that $f$ is a local automorphism of $M_2$. Continuing this process and choosing the deformation sufficiently small at each step, we obtain, in the limit, a smooth hypersurface $\tilde{M}$ in $B$, with non-degenerate indefinite Levi form, which is not equivalent to a hyperquadric in any neighbourhood of the origin and for which the mapping $f$ is a local automorphism.

The restriction of the differential of $f$ at the origin to the hyperplane $(w = 0)$ is the identity mapping; at the same time, $f$ itself is not the identity.

**10.5. Linearization of Local Automorphisms.** Local automorphisms of a real analytic strictly pseudoconvex hypersurface not locally equivalent to the sphere can be simultaneously linearized via a holomorphic system of coordinates (see 9.3). Generally speaking, this is not the case for smooth hypersurfaces, because the domain of definition of an automorphism can be arbitrarly small. If we restrict our attention to automorphisms defined in a large neighbourhood, then their linearization is possible.

Suppose we are given some family of local automorphisms of a hypersurface $M$ which fix some chain $\gamma$ passing through a point $x$. By assertion 2 of the preceding section, it is not hard to show that if $M$ is not $CR$-diffeomorphic to the sphere in a neighbourhood of $x$, then all points of $\gamma$ are stationary with respect to the automorphisms under consideration. From the existence of such a chain follows:

1. There exists a system of coordinates with respect to which the initial data for each automorphism $f$ of the given family is $U$, $\lambda=1$, $a=0$, $r=0$.
2. For each neighbourhood $V$ of $x$ lying within the common domain of definition of all automorphisms of the family, there exists a neighbourhood $V_0 \subset V$, which is mapped into itself by each of these automorphisms.

The group of automorphisms of the neighbourhood $V_0$ which fix $x$ and have as initial data the unitary matrix $U$ and the parameters $\lambda=0$, $a=0$, $r=0$ is compact and consequently can be linearized. Hence, in order to linearize the local automorphisms of a hypersurface $M$ it is sufficient to find a chain which is invariant for all of these.

**Lemma.** *If a hypersurface $M$ is not CR-diffeomorphic to the sphere in a neighbourhood of a point $x$, then all the local automorphisms of $M$ have a common invariant chain* [24].

First of all, it happens that each local automorphism of the hypersurface has an invariant chain. The proof of this fact proceeds similarly to the proof of 1–3 in the preceding section. Taking into account the earlier remarks, this means that each local automorphism generates a compact subgroup (the closure of the iterations) of the group of all local automorphisms.

Let us denote the group of local automorphisms of $M$ by $\Phi$. Theorem 10.4 shows that there is an embedding of $\Phi$ in the group of unitary matrices. Let us denote by $\Phi^0$ the connected component of the identity in the group $\Phi$, with respect to the topology induced by the representation. Then $\Phi^0$ is compact, and therefore all automorphisms in $\Phi^0$ have a common fixed chain. The remainder of the proof is based on the fact that the factor group $\Phi/\Phi^0$ is isomorphic to some linear group, all of whose elements are of finite order, and which therefore contains a normal commutative subgroup of finite index. From the above lemma, we obtain the following assertion:

**Theorem.** *Let $M$ be a strictly pseudoconvex hypersurface which in no neighbourhood of the point $x$ is locally $CR$-diffeomorphic to the sphere. Then, for each neighbourhood $V$ of $x$, there exists $V_0 \subset V$ and local coordinates $(u, z = (z_1, \ldots, z_{n-1}))$ on $V_0$ such that each automorphisms of $M$ whose domain of definition contains $V_0$ is a unitary transformation in the variable $z$*[24].

# References*


1. Alexander, H.: Holomorphic mappings from the ball and polydisc. Math. Ann. *209*, 249–256 (1974). Zbl. 272.32006 (Zbl. 281.32019)
2. Beloshapka, V.K.: On the dimension of the group of automorphisms of an analytic hypersurface. Izv. Akad. Nauk SSSR, Ser. Mat. *43*, 243–266 (1979); Engl. transl.: Math. USSR, Izv. *14*, 223–245 (1980). Zbl. 412.58010
3. Beloshapka, V.K.: An example of non-continuable holomorphic transformation of an analytic hypersurface. Mat. Zametki *32*, 121–123 (1982); Engl. transl.: Math. Notes *32*, 540–541 (1983). Zbl. 518.32011
4. Beloshapka, V.K., Vitushkin, A.G.: Estimates of the radius of convergence of power series defining a mapping of analytic hypersurfaces. Izv. Akad. Nauk SSSR, Ser. Mat. *45*, 962–984 (1981); Engl. transl.: Math. USSR, Izv. *19*, 241–259 (1982). Zbl. 492.32021
5. Bochner, S.: Compact groups of differentiable transformations. Ann. Math., II. Ser. *46*, 372–381 (1945)
6. Burns, D., Diederich, K., Shnider, S.: Distinguished curves in pseudoconvex boundaries. Duke Math. J. *44*, 407–431 (1977). Zbl. 382.32011
7. Burns, D., Shnider, S.: Real hypersurfaces in complex manifolds, Several complex variables. Proc. Symp. Pure Math. *30*, Pt. 2, 141–168 (1977). Zbl. 422.32016
8. Burns, D., Shnider, S.: Geometry of hypersurfaces and mapping theorems in $\mathbb{C}^n$. Comment. Math. Helv. *54*, No. 2, 199–217 (1979). Zbl. 444.32012
9. Burns, D., Shnider, S., Wells Jr., R.O.: Deformations of strictly pseudoconvex domains. Invent. Math. *46*, 237–253 (1978). Zbl. 412.32022
10. Cartan, E.: Sur la géométrie pseudoconforme des hypersurfaces de deux variables complexes. Oeuvres complètes, Pt . II, Vol. 2, 1231–1304, Pt. III, Vol. 2, 1217–1238, Gauthier-Villars, Paris 1953 (Pt. II). Zbl. 58,83; 1955 (Pt. III). Zbl. 59,153
11. Chern, S.S., Moser, J.K.: Real hypersurfaces in complex manifolds. Acta Math. *133*, 219–271 (1974). Zbl. 302.32015
12. Ezhov, V.V.: Asymptotic behaviour of a strictly pseudoconvex surface along its chains. Izv. Akad. Nauk SSSR, Ser. Mat. *47*, 856–880 (1983); Engl. transl.: Math. USSR, Izv. *23*, 149–170 (1984). Zbl. 579.32034
13. Ezhov, V.V., Kruzhilin, N.G., Vitushkin, A.G.: Extension of local mappings of pseudoconvex surfaces. Dokl. Akad. Nauk SSSR *270*, 271–274 (1983); Engl. transl.: Sov. Math., Dokl. *27*, 580–583 (1983). Zbl. 558.32003
14. Ezhov, V.V., Kruzhilin, N.G., Vitushkin, A.G.: Extension of holomorphic maps along real-analytic hypersurfaces. Tr. Mat. Inst. Steklova *167*, 60–95 (1985); Engl. transl.: Proc. Steklov Inst. Math. *167*, 63–102 (1986). Zbl. 575.32011
15. Faran, J.J.: Lewy's curves and chains on real hypersurfaces. Trans. Am. Math. Soc. *265*, 97–109 (1981). Zbl. 477.32021
16. Fefferman, C.L.: The Bergman kernel and biholomorphic mappings of pseudoconvex domains. Invent. Math. *26*, 1–65 (1974). Zbl. 289.32012
17. Fefferman, C.L.: Monge-Ampère equations, the Bergman kernel, and geometry of pseudoconvex domains. Ann. Math., II. Ser. *103*, 395–416 (1976). Zbl. 322.32012
18. Fefferman, C.L.: Parabolic invariant theory in complex analysis. Adv. Math. *31*, 131–262 (1971). Zbl. 444.32013
19. Griffiths, P.A.: Two theorems on extensions of holomorphic mappings. Invent. Math. *14*, 27–62 (1971). Zbl. 223.32016


* For the convenience of the reader, references to reviews in Zentralblatt für Mathematik (Zbl.), compiled using the MATH database, have, as far as possible, been included in this bibliography.

20. Ivashkovich, S.M.: Extension of locally biholomorphic mappings of domains to a complex projective space. Izv. Akad. Nauk SSSR, Ser. Mat. *47*, 197–206 (1983); Engl. transl.: Math. USSR, Izv. *22*, 181–189 (1984). Zbl. 523.32009
21. Ivashkovich, S.M., Vitushkin, A.G.: Extension of holomorphic mappings of a real-analytic hypersurface to a complex projective space. Dokl. Akad. Nauk SSSR *267*, 779–780 (1982); Engl. transl.: Sov. Math., Dokl. *26*, 682–683 (1982). Zbl. 578.32024
22. Khenkin, G.M., Tumanov, A.E.: Local characterization of holomorphic automorphisms of Siegel domains. Funkts. Anal. Prilozh. *17*, No. 4, 49–61 (1983); Engl. transl.: Funct. Anal. Appl. *17*, 285–294 (1983). Zbl. 572.32018
23. Kruzhilin, N.G.: Estimation of the variation of the normal parameter of a chain on a strictly pseudoconvex surface. Izv. Akad. Nauk SSSR, Ser. Mat. *147*, 1091–1113 (1983); Engl. transl.: Math. USSR, Izv. *23*, 367–389 (1984). Zbl. 579.32033
24. Kruzhilin, N.G.: Local automorphisms and mappings of smooth strictly pseudoconvex hypersurfaces. Izv. Akad. Nauk SSSR, Ser. Mat. *49*, No. 3, 566–591 (1985); Engl. transl.: Math. USSR, Izv. *26*, 531–552 (1986). Zbl. 597.32018
25. Kruzhilin, N.G., Loboda, A.V.: Linearization of local automorphisms of pseudoconvex surfaces. Dokl. Akad. Nauk SSSR *271*, 280–282 (1983); Engl. transl.: Sov. Math., Dokl. *28*, 70–72 (1983). Zbl. 582.32040
26. Loboda, A.V.: On local automorphisms of real-analytic hypersurfaces. Izv. Akad. Nauk SSSR, Ser. Mat. *45*, 620–645 (1981); Engl. transl.: Math. USSR, Izv. *18*, 537–559 (1982). Zbl. 473.32016
27. Lewy, H.: On the boundary behaviour of holomorphic mappings. Acad. Naz. Lincei *35*, 1–8 (1977)
28. Moser, J.K., Webster, S.M.: Normal forms for real surfaces in $\mathbb{C}^2$ near complex tangents, and hyperbolic surface transformations. Acta Math. *150*, 255–296 (1983). Zbl. 519.32015
29. Pinchuk, S.I.: Holomorphic mappings of real-analytic hypersurfaces. Mat. Sb., Nov. Ser. *105*, 574–593 (1978); Engl. transl.: Math. USSR, Sb. *34*, 503–519 (1978). Zbl. 389.32008
30. Pinchuk, S.I.: A boundary uniqueness theorem for holomorphic functions of several complex variables. Mat. Zametki *15*, 205–212 (1974); Engl. transl.: Math. Notes *15*, 116–120 (1974). Zbl. 285.32002
31. Poincaré, H.: Les fonctions analytiques de deux variables et la représentation conforme. Rend. Circ. Mat. Palermo *23*, 185–220 (1907)
32. Rosay, J.-P.: Sur une caractérization de la boule parmi les domaines de $\mathbb{C}^n$ par son groupe d'automorphismes. Ann. Inst. Fourier *29*, 91–97 (1979). Zbl. 402.32001 (Zbl. 414.32001)
33. Segre, B.: Questioni geometriche legate colla teoria delle funzioni di due variabili complesse. Rend. Sem. Math. Roma, II. Ser. *7*, No. 2, 59–107 (1982). Zbl. 5, 109
34. Shiffman, B.: Extension of holomorphic maps into hermitian manifolds. Math. Ann. *194*, 249–258 (1971). Zbl. 213,360
35. Tanaka, N.J.: On the pseudo-conformal geometry of hypersurfaces of the space of *n* complex variables. J. Math. Soc. Japan *14*, 397–429 (1962). Zbl. 113,63
36. Tanaka, N.J.: On generalized graded Lie algebras and geometric structures. I. J. Math. Soc. Japan *19*, 215–254 (1967). Zbl. 165,560
37. Vitushkin, A.G.: Holomorphic extension of mappings of compact hypersurfaces. Izv. Akad. Nauk SSSR, Ser. Mat. *46*, 28–35 (1982); Engl. transl.: Math. USSR, Izv. *20*, 27–33 (1983). Zbl. 571.32011
38. Vitushkin, A.G.: Global normalization of a real-analytic surface along a chain. Dokl. Akad. Nauk SSSR *269*, 15–18 (1983); Engl. transl.: Sov. Math., Dokl. *27*, 270–273 (1983). Zbl. 543.32003
39. Vitushkin, A.G.: Analysis of power series defining automorphisms of hypersurfaces in connection with the problem of extending maps. International Conf. on analytic methods in the theory of numbers and analysis Moscow, 14–19 September 1981, Tr. Mat. Inst. Steklova *163*, 37–41 (1984); Engl. transl.: Proc. Steklov Inst. Math. *163*, 47–51 (1985). Zbl. 575.32010
40. Vitushkin, A.G.: Real-analytic hypersurfaces in complex manifolds. Usp. Mat. Nauk *40*, No. 2, 3–31 (1985); Engl. transl.: Russ. Math. Surv. *40*, No. 2, 1–35 (1985). Zbl. 588.32025

41. Webster, S.M.: Kähler metrics associated to a real hypersurface. Comment. Math. Helv. *52*, 235–250 (1977). Zbl. 354.53050
42. Webster, S.M.: Pseudo-hermitian structures on a real hypersurface. J. Differ. Geom. *13*, 25–41 (1978). Zbl. 379.53016 (Zbl. 394.53023)
43. Webster, S.M.: On the Moser normal form at a non-umbilic point. Math. Ann. *233*, 97–102 (1978). Zbl. 358.32013
44. Wells Jr., R.O.: On the local holomorphic hull of a real submanifold in several complex variables. Commun. Pure Appl. Math. *19*, 145–165 (1966). Zbl. 142,339
45. Wells Jr., R.O.: Function theory on differentiable submanifolds. Contributions to analysis. Academic Press, 407–441, New York 1974. Zbl. 293.32001
46. Wells Jr., R.O.: The Cauchy-Riemann equations and differential geometry. Bull. Amer. Math. Soc. *6*, 187–199 (1982).
47. Wong, B.: Characterization of the unit ball in $\mathbb{C}^n$ by its automorphism group. Invent. Math. 41, 253–257 (1977).

# V. General Theory of Multidimensional Residues

P. Dolbeault

## Contents

## §0. Introduction

**0.1. 1-Dimensional Formula of Residues.** Let $X$ be a Riemann surface and let $\omega$ be a meromorphic differential form of degree 1 on $X$; in the neighborhood of a point where $z$ is a local coordinate, we have $\omega = f(z)dz$, where $f$ is a meromorphic function. Let $Y = \{a_l\}_{l \in I}$ be the set of poles of $\omega$, and $\operatorname{res} a_j(\omega)$ the Cauchy residue of $\omega$ at $a_j$. Then, for $J$ a finite subset of $I$, if for any $j \in J$, $\gamma_j$ is a positively oriented circle whose center is $a_j$ and such that the closed disk of which $\gamma_j$ is the boundary does not meet $Y$ in any point different from $a_j$ and if $(n_j)_{j \in J}$ is a family of elements of $\mathbb{Z}$, $\mathbb{R}$ or $\mathbb{C}$, we have the *residue formula*.

$$\int_{\sum_{j \in J} n_j \gamma_j} \omega = \sum_{j \in J} 2\pi i \; n_j \operatorname{res}_{a_j}(\omega). \tag{0.1}$$

**0.2. 1-Dimensional Residue Theorem.** Moreover if $X$ is compact and connected, the following *residue theorem* is known: For any discrete (hence finite) set $Y = (a_j)_{j \in J}$ of points $X$, and for any meromorphic 1-form $\omega$ whose set of poles is $Y$,

$$\sum_{j \in J} \operatorname{res}_{a_j}(\omega) = 0. \tag{0.2}$$

Conversely, for any such subset $Y$ and any subset $(\alpha_j)_{j \in J}$ of complex numbers such that $\sum_{j \in J} \alpha_j = 0$, there exists a meromorphic 1-form $\omega$ on $X$ having simple poles exactly on $Y$ and such that, for any $j \in J$, $\alpha_j = \operatorname{res}_{a_j}(\omega)$.

H. Poincaré was the first to give a convenient generalization of the notion of residues for closed meromorphic differential forms in several complex variables (1887) [54]. Afterwards E. Picard [53] (1901), De Rham [58] [59] (1932, 1936), and A. Weil [70] (1947) obtained results on meromorphic forms of degree 1 or 2.

**0.3. Theory of Leray–Norguet.** Starting from Poincaré's work, Leray (1959) [45], then Norguet (1959) [49, 50] generalized the situation of 0.1 to the case of a complex analytic manifold $X$ of any finite dimension, where $Y$ is a complex submanifold of $X$ of codimension 1 and $\omega$ is a closed differential form of degree $p$ on $X$, of class $C^\infty$ outside $Y$ and having singularities of the following type on $Y$: every point $x \in Y$ has a neighborhood $U$ in $X$ over which a complex local coordinate function $s$ is defined such that $Y \cap U = \{y \in U;\ s(y) = 0\}$ and $\omega|U$ is equal to $\alpha/s^k$ where $\alpha$ is $C^\infty$ on $U$ and $k \in \mathbb{N}$. More generally, a *semi-meromorphic differential form* on $X$ has a local expression $\alpha/f$ where $\alpha$ is a $C^\infty$ differential form and where $f$ is a holomorphic function.

(a) The topological situation studied by Leray and generalized by Norguet is the following: let $X$ be a *topological space* and $Y$ be a *closed* subset of $X$, we have the following exact cohomology sequence with complex coefficients and compact supports

$$\ldots \to H_c^p(X) \to H_c^p(Y) \xrightarrow{\delta^*} H_c^{p+1}(X \setminus Y) \to H_c^{p+1}(X) \to \ldots \tag{0.3.1}$$

If $Y$ and $X \setminus Y$ are orientable *topological manifolds* of dimension $m$ and $n$ respectively, then the duality isomorphism of Poincaré defines, from $\delta^*$ the homomorphism

$$\delta \colon H^c_{m-p}(Y) \to H^c_{n-p-1}(X \setminus Y).$$

Hence, from the universal coefficient theorem, we have the *residue homomorphism*

$$r = {}^t\delta \colon H^{n-p-1}(X \setminus Y) \to H^{m-p}(Y)$$

and the residue formula

$$\langle \delta h, c \rangle = \langle h, rc \rangle \tag{0.3.2}$$

in which $h \in H^c_{m-p}(Y)$ and $c \in H^{n-p-1}(X \setminus Y)$.

Going back to formula (0.1), it is easy to see that the homology class of $\sum_{j\in J} n_j\gamma_j$ is the image by $\delta: H_0^c(Y) \to H_1^c(X\setminus Y)$ of the homology class $h = \sum_j n_j a_j$ of $Y$ and $a_j \to 2\pi i \operatorname{res}_{a_j}(\omega)$ is the cohomology class image by $r: H^1(X\setminus Y) \to H^0(Y)$ of the 1-cohomology class $c$ of $\omega|X\setminus Y$. So when we consider only homology and cohomology classes, formula (0.1) is a particular case of formula (0.3.2).

(b) Suppose now that $X$ is a complex analytic manifold of complex dimension $n$ and $Y$ a complex analytic submanifold of $X$ of complex dimension $(n-1)$. Given a closed semi-meromorphic differential form $\omega$ on $X$, $C^\infty$ except on $Y$, we say that $Y$ is a *polar set of* $\omega$; if, moreover in the notation of the beginning of 0.3, $\omega = \frac{\alpha}{s}$, we say that $Y$ is a polar set with multiplicity 1. Then, locally, $\omega = (ds/s) \wedge \psi + d\theta$ where $\psi$ and $\theta$ are $C^\infty$; $\psi|Y$ has a global definition and is called the *residue form* of $\omega$ and its cohomology class in $Y$ is the image by $r$, of the cohomology class defined by $\omega|X\setminus Y$ on $X\setminus Y$.

(c) **Theorem of Leray.** For $n = 1$ and $X$ compact connected, consider the exact sequence

$$H^1(X\setminus Y) \xrightarrow{r} H^0(Y) \xrightarrow{j'} H^2(X). \tag{0.3.3}$$

By Poincaré duality $j'$ is transformed into $j''$: $H_0(Y) \to H_0(X)$; the cohomology class of $H^0(Y)$ $c$: $a_j \to 2\pi \operatorname{res}_{a_j}$ is transformed into $c'' = \sum_j (2\pi i \operatorname{res}_{a_j}) a_j \in H_0(Y)$; its image by $j''$ has the same expression in $H_0(X)$; $j''c'' = 0$ if and only if $\sum_j \operatorname{res}_{a_j} = 0$. The residue theorem 0.2 results from the exactness of (0.3.3).

More generally, the exactness of the cohomology sequence

$$\ldots \to H^{2n-p-1}(X\setminus Y) \xrightarrow{r} H^{2n-p-2}(Y) \xrightarrow{j'} H^{2n-p}(X) \to \ldots$$

where $j'$ is induced by the inclusion $Y \to X$ means, in particular, that an element $c \in H^{2n-p-2}(Y)$ belongs to $\operatorname{Im} r$ if and only if $j'c = 0$. This is, trivially, a cohomological residue theorem. The converse of the residue theorem is a particular case of the following:

**Theorem.** *Let $X$ be a complex analytic manifold and $Y$ a submanifold of $X$ of codimension 1, then every class of cohomology of $X\setminus Y$ contains the restriction to $X\setminus Y$, of a closed semi-meromorphic differential form having $Y$ as a polar set with multiplicity* 1.

(d) **Composed residues.** Let $Y_1, \ldots, Y_9$ be complex submanifolds of codimension 1, in general position, and let $y = Y_1 \cap \ldots \cap Y_q$. Then by composition of the residue homomorphism, for $p \geq q$,

$$H^p(X \backslash Y_1 \cup Y_2 \cup \ldots \cup Y_q) \to H^{p-1}(Y_1 \backslash Y_2 \cup \ldots \cup Y_q) \to$$
$$\to H^{p-2}(Y_1 \cap Y_2 \backslash Y_3 \cup \ldots \cup Y_q) \to \ldots \to H^{p-q}(y),$$

we obtain the composed residue homomorphism.

**0.4. Grothendieck Residue Symbol** ([29], cf. also [24], [67]). The Poincaré residue in $n$ variables can be considered as a generalization of the residue in one variable, the singularity being of codimension 1. Another possible generalization is to consider a singularity of dimension zero. This leads to the *point residue* or *Grothendieck residue symbol*: let $f_1, \ldots, f_n$ be holomorphic functions in a neighborhood of 0 in $\mathbb{C}^n$ such that 0 is their only common zero; let $\omega$ be an $n$-meromorphic differential form $\omega = f_1^{-1} \ldots f_n^{-1} g \, dz_1 \wedge \ldots \wedge dz_n$ where $g$ is holomorphic in a neighborhood of 0, and $\Gamma = \{|f_1| = \ldots = |f_n| = \delta\}$ for $\delta$ small enough, then the point residue of $\omega$ at 0 is $\operatorname{res}_{\{0\}} \omega = \left(\frac{1}{2\pi i}\right)^n \int_\Gamma \omega$. This leads to duality theorems

**0.5. Generalizations and Applications.** Since 1968, generalizations of definition (a) of the residue homomorphism have been used for spaces more general than complex manifolds; (b) semi-meromorphic differential forms with polar set having arbitrary singularities and generalizations of such forms have been used to give interpretations of the residue homomorphism; the theorem of Leray (c) and composed residues (d) have been generalized; finally residue formulae of increasing generality have been obtained.

In fact, a small part of this program was realized before Leray's work (1951–1957) [41, 42, 66, 14]. But for a more complete realization the following newer techniques were used: Borel–Moore homology ([8], chapter 5); local cohomology [26, 27]; Hironaka's local resolution of singularities of complex analytic spaces [34, 2]; and properties of semi-analytic and sub-analytic sets (Łojasiewicz [47], Herrera [30], Hironaka [35]).

The Grothendieck residue symbol, introduced in algebraic geometry, has a meaning in analytic geometry and was studied later by Griffiths [24, 25]. It was used by Bott and others [6, 7, 4, 3] for the singularities of holomorphic and meromorphic vector fields [6, 7, 3]. The relation between the Poincaré residue and Grothendieck residue is made explicit by the residual currents introduced by Coleff–Herrera [11]. These currents appear as an interpretation of the composed residue homomorphism (see d), see also [68] and they have applications in a duality theorem [65].

An extensive bibliography on the theory of residues before 1977 can be found in [1], see also the previous survey [20].

## §1. Residue Homomorphism

Homology and cohomology groups are taken with coefficients in $\mathbb{C}$.

**1.1. Homological Residue** [15, 33]. Let $X$ be a *locally compact*, paracompact space of finite dimension and $Y$ be a *closed* subset of $X$. In the Borel–Moore homology (homology of locally compact spaces) [8], we have the following exact homology sequence

$$\ldots \to H_{q+1}(X) \to H_{q+1}(X\setminus Y) \overset{\delta_*}{\to} H_q(Y) \to H_q(X) \to \ldots \tag{1.1}$$

It is transposed from the exact cohomology sequence (0.3.1). When $X\setminus Y$ and $Y$ are orientable topological manifolds of dimension $n$ and $m$ respectively, we have the following diagram

$$\begin{array}{ccc} H_c^p(Y) & \overset{\delta_*}{\to} & H_c^{p+1}(X\setminus Y) \\ | & (Q) & | \\ H_p(Y) & \overset{\delta_*}{\leftarrow} & H_{p+1}(X\setminus Y) \\ & (Q') & \\ P_Y \downarrow & & P_{X\setminus Y}\downarrow \\ H^{m-p}(Y) & \overset{r}{\leftarrow} & H^{n-p-1}(X\setminus Y) \end{array}$$

where the vertical lines mean the pairings between cohomology with compact supports and homology with closed supports, $\delta_* = {}^t\delta^*$, and $P_Y$, $P_{X\setminus Y}$ are the duality isomorphisms of Poincaré. Moreover, the square $(Q')$ is anticommutative and $(Q)$ gives a residue formula. Note that $\delta_*$ is the connecting homomorphism of (1.1); $\delta_*$ is defined under much more general hypotheses than $r$, so it gives a good generalization of $r$ and will be called the *homological residue homomorphism*.

**1.2. Cohomological Residue.** For any topological space $X$ and any closed subset $Y$ of $X$, we have the exact sequence of local cohomology

$$\ldots \to H^p(X) \to H^p(X\setminus Y) \overset{\rho}{\to} H_Y^{p+1}(Y) \to H^{p+1}(X) \to \ldots \tag{1.2}$$

where $H_Y^{p+1}(X)$ is the $(p+1)$-th group of cohomology of $X$ with support in $Y$ [27]. When $X\setminus Y$ and $Y$ are topological manifolds of dimension $n$ and $m$ respectively, the following diagram is commutative

$$\begin{array}{ccc} H^p(X\setminus Y) & \overset{\rho}{\to} & H_Y^{p+1}(X) \\ & \searrow r \quad \approx \nearrow & \\ & H^{p-(n-m)+1}(Y) & \end{array}$$

Thus, the homomorphism $\rho$ is a generalization of $r$ and will be called the *cohomological residue homomorphism* [52].

**1.3. Relation Between the Exact Sequences** (1.1) **and** (1.2). Suppose that $X$ is locally compact, of homological dimension $n$, then $Y$ and $X\backslash Y$ are locally compact. Let $(X)$ be the *fundamental class* of $X$; then applying the cap product by $(X)$ to (1.2) we get the commutative diagram

$$\begin{array}{ccccc} \text{(C)} \ldots \to H^p(X) \to & H^p(X\backslash Y) & \overset{\rho}{\to} & H_Y^{p+1}(X) \to \ldots \\ \downarrow \cap (X) & \downarrow \cap (X\backslash Y) & & \downarrow \cap (Y) \\ \text{(D)} \ldots \to H_{n-p}(X) \to & H_{n-p}(X\backslash Y) & \overset{\delta_*}{\to} & H_{n-p-1}(Y, \mathbb{C}) \to \ldots \end{array}$$

When $X$ is a manifold, $X\backslash Y$ is also a manifold, then $\cap(X)$ and $\cap(X\backslash Y)$ are inverse to Poincaré isomorphisms, and by the five homomorphism lemma, $\cap(Y)$ is also an isomorphism.

## §2. Principal Value; Residue Current

In this section, we suppose that $X$ is a reduced complex analytic space (complex space) and $Y$ a closed complex analytic subvariety of $X$, of pure codimension 1. Note that holomorphic functions, $C^\infty$ (or smooth) differential forms, and currents can be defined on $X$ ([5], [33]).

For the sake of simplicity we shall assume $X$ to be a complex manifold when giving sketches of proofs.

**2.1. Case** $n = 1$. First consider the case where $X$ is a coordinate domain $U$ of a Riemann surface and let $\omega$ be a meromorphic 1-form on $U$ having only one pole $P$ in $U$. Choose the coordinate $z$ on $U$ such that $z(P) = 0$.

Consider the current $\underline{\omega}$ on $U\backslash\{P\}$ such that, for every $\psi \in \mathscr{D}^\cdot(U\backslash\{P\})$

$$\underline{\omega}(\psi) = \int_U \omega \wedge \psi$$

Then for every $\psi \in \mathscr{D}^\cdot(U)$, $Vp(\omega)$ defined by

$$Vp(\omega)(\psi) = \lim_{\varepsilon \to 0} \int_{|z| \geq \varepsilon} \omega \wedge \psi$$

is a current on $U$ whose restriction to $U\backslash\{P\}$ is $\underline{\omega}$; $Vp(\omega)$ is independent of the choice of the coordinate $z$ and of the representation of the meromorphic form $\omega$.

It is called the *Cauchy principal value* of $\omega$. Moreover,

$$dVp(\omega) = d''Vp(\omega) = 2\pi i \operatorname{res}_P(\omega)\, \delta_P + dB$$

where $\delta_P$ is the Dirac measure at $P$ and $B$ a current of type (0, 1) and support $\{P\}$. If $P$ is a simple pole of $\omega$, then $B = 0$. The current $dVP(\omega)$ will be called *the residue current of* $\omega$.

We shall generalize the construction of the principal value and of the residue current to the case of semi-meromorphic differential forms on a complex space $X$.

**2.2. Semi-Holomorphic Differential Operators.** On a complex manifold $X$, a *differential operator* $D$ on the space of currents $\mathscr{D}'_{\cdot}(X)$ is said to be *semi-holomorphic* if, for every $x \in X$, there exists a chart $(U, z_1, \ldots, z_n)$ at $x$ such that on $U$, for every current $T$, the coefficients of $DT$ are linear combinations, with $C^\infty$ coefficients of the partial derivatives

$$\frac{\partial^I}{\partial z^I}, \ (I \in \mathbb{N}^n), \text{ of the coefficients of } T \text{ [66].}$$

Let $Y$ be a complex analytic subvariety of $X$ of codimension 1, and let $\mathscr{S}^{\cdot}_Y$ be the vector space of semi-meromorphic forms whose polar set is contained in $Y$. Then $D$ operates also on $\mathscr{S}^{\cdot}_Y$ and more generally on the space $\mathscr{S}^{\cdot}(X)$ of the semi-meromorphic differential forms on $X$. The set $\Delta$ of the semi-holomorphic differential operators is a ring and $\mathscr{D}'_{\cdot}(X)$, $\mathscr{S}^{\cdot}_Y$ and $\mathscr{S}^{\cdot}(X)$ are $\Delta$-modules.

**2.3. The Case of Normal Crossings.** A complex analytic subvariety $Y$ of $X$ of codimension 1 has *normal crossings* if, for every point $x \in Y$, there exists a chart $(z_1, \ldots, z_n)$ on an open neighborhood $U$ of $x$ such that

$$U \cap Y = \{z \in U;\ z_1(z) \ldots z_q(z) = 0;\ q \leq n\} \tag{2.3.1}$$

*A semi-meromorphic differential form* $\omega$ defined on such a neighborhood $U$ of $x$ in $X$ is called *elementary* if it has a polar set contained in such a subvariety $U \cap Y$; then $\omega = \alpha/z^s$ where $s = (s_1, \ldots, s_q, 0, \ldots, 0) \in \mathbb{N}^n$.

**2.3.1. Theorem.** *Let $(U, z_1, \ldots, z_n)$ be a chart of a complex analytic manifold $X$ of dimension $n$ and $Y$ be a complex subvariety of $U$ defined by* (2.3.1), *then*

(a) *There exists a unique $\mathbb{C}$-linear map $T: \mathscr{S}^{\cdot}_Y \to \mathscr{D}'_{\cdot}(U)$ such that:*

(i) *if $\omega \in \mathscr{S}^{\cdot}_Y$ has locally integrable coefficients, then $T(\omega) = \underline{\omega}$, the current defined by $\omega$;*

(ii) *$T$ is $\Delta$-linear;*

(b) *$T$ is a equal to the Cauchy principal value* (2.1) *with respect to the coordinate functions* [16].

**2.3.2.** We shall give an expression of the current $T$ having an easy generalization. With the notations of 2.2 and 2.3, in the open set $U$, let $g = z^l g_0$ be any function such that $l_1 \ldots l_q \neq 0$ and $g_0$ is holomorphic without zeros in $U$.

Then for every $\varphi \in \mathscr{D}^{\cdot}(U)$, the formula

$$Vp(\omega)(\varphi) = \lim_{\delta \to 0} \int_{|g| \geq \delta} \omega \wedge \varphi$$

defines the current $T(\omega) = Vp(\omega)$ of 2.3.1. Note that $|g| \geq \delta$ is a semi-analytic set. This definition of $Vp(\omega)$ is due to Herrera-Lieberman [33]. One establishes the existence of $Vp(\omega)$ for a particular representation of $\omega$ and a particular $g$ and one verifies that $Vp(\omega)$ satisfies the conditions (i) and (ii) of 2.3.1; for (ii), it suffices to show: $-Vp(\omega)\left(\frac{\partial \psi}{\partial z_i}\right) = Vp\left(\frac{\partial \omega}{\partial z_i}\right)(\psi)$, using the Stokes formula for semi-analytic set. Thus, we obtain the independence of $Vp(\omega)$ with respect to $g$ and to the representation of $\omega$. Moreover, from 2.3.1 (ii), $Vp$ is $\Delta$-linear; the idea of this proof is due to G. Robin [17, 18].

**2.3.3.** When $Y$ is any subvariety of codimension 1 of $X$ with normal crossings, and $\omega \in \mathscr{S}^{\cdot}_Y$, we build up $Vp(\omega)$ locally and glue the local currents by a partition of unity. This makes sense due to the invariance of the definition of $Vp$.

**2.4. General Case.** Let $X$ be a complex space and $Y$ a complex subvariety of codimension 1 containing the singular locus of $X$, globally defined by an equation $g = 0$ where $g$ is holomorphic on $X$. Consider the semi-analytic set

$$X(>\delta) = \{x \in X;\ |g(x)| > \delta\}$$

and the integration current $I[X(>\delta)]$ over $X$ defined by the semi-analytic chain $[X(>\delta), e(>\delta)]$ where $e(>\delta)$ is the fundamental class of $X(>\delta)$. Let $f$ be a holomorphic function on $X$ such that $f = 0$ implies $g = 0$ and let $\omega = \alpha/f \in \mathscr{S}^{\cdot}_Y$ where $\alpha$ is $C^\infty$ on $X$. We set, for every $\varphi \in \mathscr{D}^{\cdot}(X)$

$$Vp(\omega)(\varphi) = \lim_{\delta \to 0} I[X(>\delta)](\omega \wedge \varphi) \tag{2.4.1}$$

After shrinking $X$ if necessary, there exists a proper morphism $\Pi: X' \to X$ such that $X'$ is a manifold and $Y' = \Pi^{-1} Y$ a subvariety of codimension 1 of $X'$ with normal crossings and $\Pi | X' \setminus Y'$ is biholomorphic (Hironaka [34]). Then

$$Vp(\omega)(\varphi) = \lim_{\delta \to 0} I[X'(>\delta)](\Pi^*(\omega \wedge \varphi)) = Vp(\Pi^*\omega)(\Pi^*\varphi) \tag{2.4.2}$$

where $X'(>\delta) = \{x' \in X';\ |\Pi^* g(x')| > \delta\}$. For $\Pi$ fixed, the last member of (2.4.2) is independent of the representation of $\omega$ and of $g$ from 2.3.2; so (2.4.1) makes sense and, from its expression, it is independent of $\Pi$. The use of a partition of unity allows one to define $Vp(\omega)$ for every semi-meromorphic form $\omega$ on $X$.

**2.4.1 Theorem.** *When $X$ is a complex manifold, the map*

$$Vp: \mathscr{S}^{\cdot}_Y \to \mathscr{D}'_{\cdot}(X)$$

*is $\Delta$-linear, $\Delta$ being the ring of semi-holomorphic differential operators.* [18, 19]

**2.5. Residue Current.** We consider the following *residue operator*

$$\text{Res} = dVp - Vpd$$

In the notation of 2.4, it has the following local expression

$$\text{Res}(\omega)(\psi) = \lim_{\delta \to 0} I[X(=\delta)](\omega \wedge \psi) \tag{2.5}$$

where $I[X(=\delta)]$ is the integration current on the semi-analytic set $\{x \in X; |g(x)| = \delta\}$ with the orientation opposite to that of $\partial[X(>\delta), e(>\delta)]$. From the $\Delta$-linearity of $Vp$ (when $X$ is a manifold), we get

$$\text{Res} = d'' Vp - Vpd''$$

Moreover, *the operator* Res: $\mathscr{S}_Y^{\cdot} \to \mathscr{D}'_{\cdot}(X)$ *is* $\Delta$*-linear,* and from (2.5), if $\omega \in \mathscr{S}_Y^{\cdot}$, Supp Res $(\omega) \subset Y$.

## §3. Residual Currents

Coleff and Herrera [11] gave a generalization of the construction of the operators $Vp$ and Res of n.2; it is related to the composed residue homomorphism (0.3, d) and allows us to give an interpretation of this homomorphism. Preliminary results have been obtained by Herrera (1972) [31, 32].

**3.1. Semi-Analytic Chains** [5]. Let $M$ be a locally closed semi-analytic set; its boundary is defined as $bM = \bar{M} \setminus M$; a *q-prechain* $(M, c)$ is a pair where $M$ is as above, dim $M \leq q$ and $c \in H_q(M; \mathbb{Z})$. The coefficient ring $\mathbb{Z}$ may be replaced by $\mathbb{R}$ or $\mathbb{C}$. The boundary of the prechain is $\partial(M, c) = (bM, \partial_{M,bM}(c))$ where $\partial_{M,bM}$ is the homomorphism $H_q(M) \to H_{q-1}(bM)$. The sum of 2 prechains $(M_1, c_1)$, $(M_2, c_2)$ is $(M, c'_1 + c'_2)$ where $M = M_1 \cup M_2 - bM_2 \cup bM_2$ and $c'_j$ is the image of $c_j (j = 1, 2)$. Let $\sim$ be the equivalence relation: $(M_1, c_1) \sim (M_2, c_2)$ means $(M_1, c_1) + (M_2, -c_2) = (M, 0)$; the class of the $q$-prechain $(M, c)$ is denoted $[M, c]$ and is called a *semi-analytic q-chain.* The set of the $q$-chains is a $\mathbb{Z}$-(resp. $\mathbb{R}$-, $\mathbb{C}$-) module denoted $\sigma_q(X)$.

**3.2. Definitions.** Let $X$ be a complex manifold of complex dimension $n$. Let

$$\mathscr{F} = \{Y_1, \ldots, Y_{p+1}\} \ (p \geq 0)$$

be a family of hypersurfaces of $X$. We set $\cup \mathscr{F} = Y_1 \cup \ldots \cup Y_{p+1}$; $\cap \mathscr{F} = Y_1 \cap \ldots \cap Y_{p+1}$; $\dim_{\mathbb{C}}(\cap \mathscr{F}) \geq n - p - 1$. When $\dim_{\mathbb{C}}(\cap \mathscr{F}) = n - p - 1$, $\cap \mathscr{F}$ is a *complete intersection.*

Let $\mathbb{R}_>$ be the set of strictly positive real numbers. Let $\underline{\delta}$ be a continuous map:

$$]0, 1[ \to \mathbb{R}_>^{p+1}; \qquad p \in \mathbb{N}^*$$

$$\delta \to (\delta_1(\delta), \ldots, \delta_{p+1}(\delta)).$$

The map $\underline{\delta}$ is called *admissible* if $\lim_{\delta\to 0} \delta_{p+1}(\delta) = 0$ and for every $j \in [1, \ldots, p]$ and every $q \in \mathbb{R}_>$,

$$\lim_{\delta\to 0} (\delta_j/\delta_{j+1}^q)(\delta) = 0$$

**3.3. Tubes.** Let $X$ small enough so that every $Y_j (j = 1, \ldots, p+1)$ is the zero set of a holomorphic function $g_j$ on $X$. Then

$$g = (g_1, \ldots, g_{p+1})\colon X \to \mathbb{C}^{p+1} \text{ is a holomorphic map}$$

and $|g| = (|g_1|, \ldots, |g_{p+1}|)\colon X \setminus \cup \mathscr{F} \to \mathbb{R}_>^{p+1}$ is a $C^\omega$ (real analytic) map. Consider the sets

$$|T_{\underline{\delta}}^{p+1}(g)| = \{x \in X;\ |g_j| = \delta_j;\ j \in [1, \ldots, p+1]\}$$

$$|D_{\underline{\delta}}^{p+1}(g)| = \{x \in X;\ |g_j| = \delta_j;\ j \in [1, \ldots, p];\ |g_{p+1}| > \delta_{p+1}\}$$

For every $\underline{\delta} \in \mathbb{R}_>^{p+1}$ such that

$$\dim |T_{\underline{\delta}}^{p+1}(g)| \le 2n - p - 1 \tag{3.3.1}$$

$$\dim |D_{\underline{\delta}}^{p+1}(g)| \le 2n - p, \tag{3.3.2}$$

we consider the semi-analytic chains, with $\underline{\delta}' = (\delta_1, \ldots, \delta_p)$,

$$T_{\underline{\delta}}^{p+1}(g) = (-1)^{\frac{p(p+1)}{2}} |g|^{-1} [\underline{\delta}] \in \sigma_{2n-p-1}(X)$$

$$D_{\underline{\delta}}^{p+1}(g) = (-1)^{\frac{p(p+1)}{2}} (|g|^{-1} [\underline{\delta}'] \times [x_{p+1} > \delta_{p+1}]) \in \sigma_{2n-p}(X),$$

where $[\underline{\delta}]$, $[\underline{\delta}']$, $[x_{p+1} > \delta_{p+1}]$ denote the points $\underline{\delta}$, $\underline{\delta}'$ and the open interval $]\delta_{p+1}, +\infty[$ with their canonical orientation and $|g|^{-1}$ the inverse of the map $|g|$ defined on semi-analytic chains. The chains $T_{\underline{\delta}}^{p+1}(g)$ and $D_{\underline{\delta}}^p(g)$ are *tubes around* $\mathscr{F}$. When $\cap \mathscr{F}$ is a complete intersection, then conditions (3.3.1) and (3.3.2) are satisfied.

**3.4. Essential Intersections of the Family $\mathscr{F}$.** Let $\mathscr{F} = (Y_1, \ldots, Y_{p+1})$; we define $V_e(\mathscr{F})$ by induction on $j \in [0, 1, \ldots, p]$. Let $Y'_0 = X$; suppose that $Y'_{j-1} (j \ge 1)$ is defined. Then let $Y'_j$ be the union of the irreductible components of $Y'_{j-1} \cap Y_j$ not contained in $Y_{j+1}$. We call *essential intersections* of $\mathscr{F}$ the two complex subvarieties $\tilde{V}_e(\mathscr{F}) = Y'_p$; $V_e(\mathscr{F}) = Y'_p \cap Y_{p+1}$; then $\dim_{\mathbb{C}} \tilde{V}_e(\mathscr{F}) = n - p$ and $\dim_{\mathbb{C}} V_e(\mathscr{F}) = n - p - 1$.

We shall use the following notation $\mathscr{F}(j) = \{Y_1, \ldots, \hat{Y}_j, \ldots, Y_{p+1}\}$. Then, if $\cap \mathscr{F}$ is a complete intersection,

$$\tilde{V}_e(\mathscr{F}) = \cap \mathscr{F}(p+1);\ V_e(\mathscr{F}) = \cap \mathscr{F}.$$

**3.5. Main Results.** We shall use the notation of n.3 and the following ones: $\mathscr{E}_X^q(* \cup \mathscr{F})$ is the sheaf of semi-meromorphic $q$-forms whose polar set is contained in $\cup \mathscr{F}$. For every semi-analytic chain $\mu$, $I[\mu]$ is the integration current on $\mu$ [30]; it is a generalization of the integration current defined by a complex variety [44].

**Theorem.** *Let $W \Subset X$ be an open set small enough for $Y_j \cap W$ be defined by a global equation $g_j = 0$, then: the tubes $T_{\underline{\delta}}^{p+1}(g)$ and $D_{\underline{\delta}}^{p+1}(g)$ are defined for $\delta$ small enough. Moreover:*

(1) $\forall\, \tilde{\alpha} \in \Gamma_c(W, \mathscr{E}_X^{2n-p}(* \cup \mathscr{F}))$, $R^pP^{p+1}(\tilde{\alpha}) = \lim_{\delta \to 0} I[D_{\underline{\delta}}^{p+1}(g)]\,(\tilde{\alpha})$ *exists,*

(2) $\forall\, \tilde{\beta} \in \Gamma_c(W, \mathscr{E}_X^{2n-p-1}(* \cup \mathscr{F}))$, $R^{p+1}(\tilde{\beta}) = \lim_{\delta \to 0} I[T_{\underline{\delta}}^{p+1}(g)]\,(\tilde{\beta})$ *exists,*

(3) *the limits* (1) *and* (2) *do not depend on the choice of the admissible map $\underline{\delta}$ and of the holomorphic map $g$,*

(4) *the limits* (1) *and* (2) *are zero if $\tilde{\alpha}$ and $\tilde{\beta}$ do not contain terms of type $(n, n-p)$ and $(n, n-p-1)$ respectively; and if $\tilde{\nu} \in \Gamma_c(W, \mathscr{E}_X^{2n-p-1}(* \cup \mathscr{F}))$, then $R^pP^{p+1}(d''\tilde{\nu}) = (-1)^{p+1} R^{p+1}(\tilde{\nu})$.*

(5) *For $\tilde{\lambda} \in \Gamma(W, \mathscr{E}_X^{q-p}(* \cup \mathscr{F}))$, the following linear forms $R^pP^{p+1}[\tilde{\lambda}]$ and $R^{p+1}[\tilde{\lambda}]$*

$$\mathscr{D}^{2n-q}(W) \to \mathbb{C}$$

$$\alpha \mapsto R^pP^{p+1}[\tilde{\lambda}]\,(\alpha) = R^pP^{p+1}(\tilde{\lambda} \wedge \alpha)$$

$$\mathscr{D}^{2n-q-1}(W) \to \mathbb{C}$$

$$\beta \mapsto R^{p+1}(\tilde{\lambda} \wedge \beta) = R^{p+1}[\tilde{\lambda}]\,(\beta)$$

*are currents;*

(6) *the supports of the currents $R^pP^{p+1}[\tilde{\lambda}]$ and $R^{p+1}[\tilde{\lambda}]$ are contained respectively in the complex subvarieties $\tilde{V}_e(\mathscr{F})$ and $V_e(\mathscr{F})$;*

(7) *for every open set $W \Subset X$, and every admissible map $\underline{\delta}$ into $\mathbb{R}^p$, there exists $\delta_{p+1}^0 > 0$ such that, for every $\delta_{p+1} \in ]0, \delta_{p+1}^0[$, $\lim_{\delta \to 0} I[D_{\underline{\delta}';\delta_{p+1}}^{p+1}(g)]\,(\tilde{\alpha})$ exists and*

$$R^p P^{p+1}(\tilde{\alpha}) = \lim_{\delta_{p+1} \to 0} \lim_{\delta \to 0} I[D_{\underline{\delta}', \delta_{p+1}}^{p+1}(g)]\,(\tilde{\alpha}).$$

**3.5.2. Consequences.** The local definitions of $R^p P^{p+1}$ and $R^{p+1}$ respectively, glue together on $X$ (from (3)) and define the currents $R\,P_{\mathscr{F}}\,[\tilde{\lambda}]$ and $R_{\mathscr{F}}\,[\tilde{\lambda}]$ called respectively *p-residue-vp* of $\tilde{\lambda}$ and $(p+1)$*-residue* of $\tilde{\lambda}$.

When $p = 0$, we obtain again $Vp$ and Res of n.2.

**3.6. Proof of Main Results.** To prove 3.5, one shows first, thanks to the properties of the admissible map $\underline{\delta}$, that for $\delta$ small enough, the tubes $T_{\underline{\delta}}^{p+1}(g)$ and $D_{\underline{\delta}}^{p}(g)$ have dimension $(2n-p-1)$ and $(2n-p)$ respectively and, consequently are semi-analytic chains. Then, as in section 2, using Hironaka's morphism, we come back to the normal crossing case. The proof, in this last case rests on the following result:

**3.6.1.** Let $A = [k_1, \ldots, k_p] \subset [1, \ldots, n]$; $z(A) = (z_{j_1}, \ldots, z_{j_{n-p}})$ with $J = [j_1, \ldots, j_{n-p}] = [1, \ldots, n] A$; $z_A = \prod_{k \in A} z_k$; $g_k = h_k z^{\alpha_k}$; $k = 1, \ldots, p+1$, where $h_k$ is an invertible holomorphic function in the unit ball of $\mathbb{C}^n$. We set: $D^{\alpha}_{\underline{\delta}, \delta_{p+1}}(h) = D^{p+1}_{\underline{\delta}, \delta_{p+1}}(g)$; for $h = (1, \ldots, 1)$, we omit $h$; $d\omega_M = dz_{l_1} \wedge d\bar{z}_{l_1} \wedge \ldots \wedge dz_{l_s} \wedge d\bar{z}_{l_s}$; $[l_1, \ldots, l_s] = M \subset [1, \ldots, n]$.

**3.6.2. Main lemma.** *Let $f = f(z(A), \bar{z}(A)) \in \mathscr{E}(\mathbb{C}^n)$, then the iterated limit*

$$\lim_{\delta_{p+1} \to 0} \lim_{\delta \to 0} \int_{D^{\alpha}_{\underline{\delta}', \delta_{p+1}}} z(A)^{-\gamma} z_A^{-1} f dz_A \wedge d\omega_M \tag{3.6.2}$$

*exists, is independent of the chosen admissible map $\underline{\delta}$ and depends continuously on $f$ for the usual topology of $\mathscr{E}(\mathbb{C}^n)$. When the limit is not zero, then* (3.6.2) *is equal to*

$$\pm (2\pi i)^p \lim_{\delta \to 0} \int_{B(A)^{\gamma}_{\delta}} z(A)^{-\gamma} f d\omega_M$$

where $B(A)^{\gamma}_{\delta} = \{z(A); \|z(A)\| < 1; \|z(A)\|^{\gamma} > \delta\}$.

### 3.7. Properties of Residual Currents

**3.7.1.** The definition of the operators $R^p P^{p+1}$ and $R^{p+1}$ is local, so they define morphisms of sheaves of semi-meromorphic differential forms into sheaves of currents whose support is the essential intersection of the family $\mathscr{F}$.

**3.7.2.** From Stokes formula, we get $bRP + (-1)^{p+1} RPd = R$.

**3.7.3.** When $\cap \mathscr{F}$ is a complete intersection,

(1) $R$ depends in an alternating fashion on the order of the sequence $\mathscr{F}$.

(2) Invariance of tubes: Let $\mathscr{F}' = \{Y'_1, \ldots, Y'_{p+1}\}$ such that $Y_k \subset Y'_k$, where $\cap \mathscr{F}'$ is a complete intersection, then the residue operators relative to $\mathscr{F}$ and $\mathscr{F}'$ are equal.

(3) Regularity property: let $\tilde{\lambda} \in \Gamma(W, \mathscr{E}_X^{\cdot -p}(* \cup \mathscr{F}(j))$ for $j \in [1, \ldots, p+1]$, then $R[\tilde{\lambda}] = 0$.

Properties (2) and (3) have been extended with more general hypotheses concerning the essential intersections in [9].

### 3.8. Extension to Analytic Spaces and Analytic Cycles

**3.8.1.** The definitions and results of sections 2 and 3 are valid when $X$ is any paracompact reduced complex analytic space and when the families $\mathscr{F}$ of complex analytic hypersurfaces $Y_j$ satisfy the following condition: for every $j = 1, \ldots, p+1$, $Y_j$ is locally principal, i.e. every $z \in Y_j$ has a neighborhood $U_z$ on which there is a holomorphic function $g_j$ such that $Y_j \cap U_z = \{x \in U_z; g_j(x) = 0\}$.

More generally, we consider a *complex analytic n-cycle* i.e. a pair $\gamma = (X, c)$ where $X$ is a paracompact reduced complex space of pure dimension $n$ and $c \in H_{2n}(X, \mathbb{Z})$. Let $(X_k)_{k \in K}$ be the family of irreducible components of $X$ and let $\gamma_k = (X_k, s_k)$ where $s_k$ denotes the fundamental class of $X_k$, then $\gamma = \sum_{k \in K} (n_k \gamma_k)$, where $n_k \in \mathbb{Z}$.

Then, the tubes of section 3.3 have to be replaced by $T^{p+1}_{\gamma;\underline{\delta}} = \sum_{k \in K} n_k T^{p+1}_{\gamma_k,\underline{\delta}}$; $D^{p+1}_{\gamma;\underline{\delta}} = \sum_{k \in K} n_k D^{p+1}_{\gamma_k;\underline{\delta}}$ where $T^{p+1}_{\gamma_k;\underline{\delta}}$, $D^{p+1}_{\gamma_k;\underline{\delta}}$ are relative to $X_k$.

**3.8.2. Point Residue.** Let $\Omega^n_X(* \cup \mathscr{F})$ be the sheaf of meromorphic $n$-forms having their poles in $\cup \mathscr{F}$. When $p + 1 = n$, then $V_e(\mathscr{F})$ is a discrete set of $X$; let $\tilde{\lambda} \in \Gamma(X, \Omega^n_X(* \cup \mathscr{F}))$, for $y \in V_e(\mathscr{F})$, $W$ an open neighborhood of $y$ in $X$ disjoint from $V_e(\mathscr{F}) \setminus \{y\}$, $\alpha \in \mathscr{D}^0(W)$ such that $\alpha \equiv 1$ in the neighborhood of $y$; let $R^n_\gamma$ be the operator $R^n$ relative to $\gamma$. We define the *point residue* of $\tilde{\lambda}$ as

$$\operatorname{res}_{\gamma;\mathscr{F};y}(\tilde{\lambda}) = R^n_\gamma[\tilde{\lambda}]\,(\alpha).$$

If $\cap \mathscr{F}$ is a complete intersection in $y$, if $X$ is smooth and $\tilde{\lambda}|W = \lambda\, g_1^{-1} \dots g_n^{-1}$; $\lambda \in \Gamma(W, \Omega^n_X)$ it can be shown that $\operatorname{res}_{\gamma;\mathscr{F};y}(\tilde{\lambda})$ is equal to the Grothendieck residue symbol $\operatorname{Res}_y\left[{\lambda \atop g_1 \dots g_n}\right]$.

**3.8.3.** In the notation of 3.8.1., let $\mathscr{F} = \{Y_1, \dots, Y_{p+1}\}$, $Y = \cap \mathscr{F}(p+1)$; $0 < p < n$, such that $\dim \cap \mathscr{F}(p+1) = n - p$; $\dim \cap \mathscr{F} \leq n - p - 1$.

Let $T$ be a reduced complex analytic space of pure dimension $n - p$, oriented by its fundamental class and $\pi: X \to T$ a morphism; let $i: Y \to X$ be the canonical embedding and assume $\dim_{\mathbb{C}} \Pi^{-1}(t) = p$; $\dim_{\mathbb{C}}(\pi \circ i)^{-1}(t) = 0$ $(t \in T)$. Then, for every $t \in \operatorname{Reg} T$(set of regular points of $T$), $\Pi_\gamma^{-1}(t)$ makes sense and is a complex analytic $p$-cycle. Let $\tilde{u} \in \Gamma(X, \mathscr{E}^p_X(* \cup \mathscr{F}))$. Then, for every such $t$, the point residue has a meaning and defines a semi-meromorphic function on $Y$; it is called the *fibered residue* of $\tilde{u}$.

## §4. Differential Forms with Singularities of Any Codimension

The aim of this section is to define locally integrable differential forms generalizing semi-meromorphic differential forms with simple poles.

**4.1. A Parametrix for the de Rham's Complex.** Let $X$ be a smooth real manifold which is denumerable at infinity. Then there exist continuous linear operators on the space of currents $\mathscr{D}'(X)$ such that, for every current $T$,

$$T = dAT + AdT + RT,$$

where $R$ is *regularizing* ($RT$ is $C^\infty$), $A$ is *pseudo-local* (Sing Supp $AT \subset$ Sing Supp $T$); moreover if $T$ is 0-continuous, then $AT$ is locally integrable. The *parametrix* $A$ is different from that of de Rham [61, 62]. In $\mathbb{R}^n$, $A = \delta G$ for currents with compact supports, $G$ being the Green operator and $\delta$ the codifferential in $\mathbb{R}^n$. Then the global construction follows de Rham ([60], §15). In the complex case, the same formula is valid for $d''$ instead of $d$.

**4.2. Definitions.** ([61], cf. also [51]). Let $X$ be a real analytic manifold and $Y$ be a closed oriented subanalytic set of pure codimension $r \geq 1$. Then $Y$ defines an integration current $I_Y$. We call *kernel associated to* $Y$ every locally integrable $(r-1)$-differential form $K$ with singular support contained in $Y$ such that: $I_Y = dK + L$ where $L$ is $C^\infty$ on $X$. From 4.1, $K = A\ I_Y$ is such a kernel.

Let $U = X \backslash Y$ and $\mathscr{S}^p_K(X, *Y)$ be the subspace of $\mathscr{E}^p(X \backslash Y)$ whose elements are the smooth differential forms $\varphi$ on $U$ defined as: $\varphi = K \wedge \psi + \theta | U$ where $\psi$ and $\theta$ are $C^\infty$ on $X$ of respective degrees $p - r + 1$ and $p$; $\varphi$ is said to be $K$-*simple*. Moreover the current $\phi = K \wedge \psi + \theta$ is locally integrable and is the simple extension of $\varphi$ to $X$.

**4.3. Case of Canonical Injection.** Let $i: Y \to X$ be the canonical injection, then $i^*\psi$ is a $C^\infty(p - r + 1)$-form on $Y$ and is independent of the expression of $\varphi$; $\operatorname{res}(\varphi) = i^*\psi$ will be called the residue form of $\varphi$; moreover $d\varphi$ is $K$-simple and $\operatorname{res}(d\varphi) = -d(\operatorname{res} \varphi)$.

**4.4. Complex Case;** $\operatorname{codim}_{\mathbb{C}} Y = 1$. The definitions and results are valid, in particular, when $X$ is a complex analytic manifold and $Y$ a complex subvariety of pure codimension. When $\operatorname{codim}_{\mathbb{C}} Y = 1$, $K$ can be locally defined by $\dfrac{1}{2i\pi}\dfrac{ds}{s}$ where $s = 0$ is a minimal local equation of $Y$. If moreover $Y$ is smooth, we get the results of Leray (0.3.b).

**4.5. Complex Case;** $\operatorname{codim}_{\mathbb{C}} Y \geq 1$. Let $X$ be a complex manifold and $Y$ be a complex submanifold of codimension $k$. For $k = 1$, let $A^{p,q}(X, *Y)$ be the space of locally integrable $(p, q)$-forms $\eta$, $C^\infty$ on $X \backslash Y$, such that $\eta = \dfrac{dt}{t} \wedge \varphi + \psi$ on $U \backslash Y$ where $U$ is a coordinate patch, $U \cap Y = \{t = 0\}$ where $t$ is a coordinate function on $U$ and $[\eta]$ the current defined by $\eta$ on $X$. For $k \geq 1$, let $\sigma: \hat{X} \to X$ be the blow-up of $X$ along $Y$ and $\hat{Y} = \sigma^{-1}(Y)$. Consider

$$A^{p,q}(X, *Y) = \{\sigma_*[\eta];\ \eta \in A^{p,q}(\hat{X}, *\hat{Y})\}$$

Then, given $Y$, there exists $\eta \in A^{k,k-1}(X, *Y)$ and $\xi \in A^{k,k}(X)$ such that $[Y] = d''[\eta] + [\xi]$; $\eta$ is a kernel associated to $Y$. Using the diagonal of $X \times X$, it is possible to construct a parametrix on $X$ and a kernel associated to any subvariety $Y$ of $X$ (J. King [39]).

Let $S$ be the singular set of $Y$ and $N_{\cdot}^{\mathrm{loc}}(Y_\infty, *S)$ be the complex of normal currents on $X$, with supports in $Y$, such that their restrictions to $X\backslash S$, are the direct images of smooth forms on $X\backslash S$. Then for $\omega \in A^{\cdot}(X, *Y)$, Res $\omega$ is defined and belongs to $N_{\cdot}^{\mathrm{loc}}(Y_\infty, *S)$ (Raby [55]).

## §5. Interpretation of the Residue Homomorphisms

The aim of this paragraph is to give an interpretation of the diagram of section 1.3 in terms of differential forms having singularities on $Y$ and of currents. This will be done under various hypotheses on $X$ and $Y$.

**5.1. Complex Case; codim$_{\mathbb{C}}$ $Y = 1$[33]. $X$ is a complex space.**

**5.1.1.** On $X$ let $\mathscr{E}_X^{\cdot}$(resp $\mathscr{E}_X^{\cdot}(*Y)$) be the sheaf complex of $C^\infty$ differential forms (resp. of semi-meromorphic differential forms with polar set $Y$). Let ${}_s\mathcal{Q}_X$ be the quotient sheaf defined by the following exact sequence $0 \to \mathscr{E}_X^{\cdot} \xrightarrow{i} \mathscr{E}_X^{\cdot}(*Y) \to {}_s\mathcal{Q}_X^{\cdot} \to 0$ where $i$ denotes the inclusion.

**5.1.2.** Let $\mathscr{D}'_{\cdot X}$ be the sheaf complex of currents on $X$ and $\mathscr{D}'_{\cdot Y^\infty}$ be the subsheaf of currents with supports in $Y$. The following short exact sequence defines $\mathscr{D}'_{\cdot X/Y^\infty}$

$$0 \to \mathscr{D}'_{\cdot Y^\infty} \to \mathscr{D}'_{\cdot X} \xrightarrow{j} \mathscr{D}'_{\cdot X/Y^\infty} \to 0$$

Principal values and Res define sheaf homomorphisms

$$Vp: \mathscr{E}_X^p(*Y) \to \mathscr{D}'_{2n-p,X}$$

$$\mathrm{Res}: \mathscr{E}_X^p(*Y) \to \mathscr{D}'_{2n-p-1,Y^\infty}$$

In fact Res($\omega$) depends only on the class of $\omega$ modulo $\mathscr{E}_X^p$, and hence defines a homomorphism

$$\mathrm{Res}: {}_s\mathcal{Q}_X^p \to \mathscr{D}'_{2n-p-1,Y^\infty}$$

$$\text{Set } Vp' = j \circ Vp: \mathscr{E}_X^{\cdot}(*Y) \to \mathscr{D}'_{\cdot X/Y^\infty}.$$

From 2.5, we get $d\,\mathrm{Res} = -\mathrm{Res}\,d$, hence a skew complex homomorphism

$$\mathrm{Res}: {}_s\mathcal{Q}_X^{\cdot} \to \mathscr{D}'_{\cdot Y^\infty}$$

we have the following cohomology diagram

$$\begin{array}{llll}
\text{(A)} \ldots \to H^p\Gamma(X, \mathscr{E}_X^{\cdot}) & \to H^p\Gamma(X, \mathscr{E}_X^{\cdot}(*Y)) & \to H^p\Gamma(X, {}_s\mathcal{Q}_X^{\cdot}) & \\
(*) \qquad \bar{V}\downarrow & \overline{Vp'}\downarrow & \overline{\mathrm{Res}}\downarrow(*) & \\
 & & \to H^{p+1}\Gamma(X, \mathscr{E}_X^{\cdot}) \to \ldots & \\
 & & \bar{V}\downarrow &
\end{array}$$

$$\text{(B)} \ldots \to H_{2n-p}\Gamma(X, \mathscr{D}'_{\cdot X}) \to H_{2n-p}\Gamma(X, \mathscr{D}'_{\cdot X/Y^\infty}) \to H_{2n-p-1}\Gamma(X, \mathscr{D}'_{\cdot Y^\infty})$$
$$\to H_{2n-p-1}\Gamma(X, \mathscr{D}'_{\cdot X}) \to .$$

where $\bar{V}_p'$, $\overline{\text{Res}}$ are induced by $V_p'$, Res resp. and $\bar{V}$ is defined by integration over $X$. The squares are commutative, except $(*)$ which is anticommutative.

**5.1.3.** Finally, we have the diagram

$$\begin{array}{ccc} (A) & \longrightarrow & (C) \\ \downarrow & & \downarrow \\ (B) & \longrightarrow & (D)\ , \end{array} \tag{5.1.3}$$

where horizontal arrows come from morphisms of de Rham's theorem in cohomology [5] and homology [33]. The squares occurring in the solid diagram are commutative except one out of three in (A) → (B) and (A) → (C), which is anticommutative. Moreover (B) → (D) is surjective.

This diagram gives an interpretation of the residue homomorphism for homology or cohomology classes in $X \setminus Y$ which can be defined by restriction to $X \setminus Y$ of semi-meromorphic differential forms with poles on $Y$. Moreover, Herrera–Lieberman [33] have a similar result in which (A) is replaced by the corresponding exact hypercohomology sequence for meromorphic differential forms with polar set on $Y$.

**5.1.4.** When $X$ and $Y$ are manifolds, then the morphisms of diagram (5.1.3) are all isomorphisms and if $\text{res}\,(\omega)$ denotes the Leray residue class of $\omega$, then $\overline{\text{Res}}\,(\omega) = 2\pi i\, I_Y \cap \text{res}\,(\omega)$. When $X \setminus Y$ is a manifold, $\overline{Vp'}$ splits canonically. When $X$ is a manifold, $\bar{V}$ is an isomorphism and $\overline{\text{Res}}$ and $\overline{Vp'}$ split canonically.

**5.2. Complex Case**; $\text{codim}_{\mathbb{C}}\ Y \geqslant 2$[11,31,32]. In the situations of section 3, for a family $\mathscr{F} = \{Y_1, \ldots, Y_{p+1}\}$ of complex hypersurfaces of $X$ such that $\cap \mathscr{F} = Y$ is a complete intersection, we have $\text{codim}_{\mathbb{C}} = p + 1$. Then using residual currents, it is possible to give an interpretation of the residue homomorphisms as in 5.1.

**5.3. Real Analytic Case: *Y* Subanalytic of Any Codimension** [61]. Let $X$ be a real analytic manifold and $Y$ be a closed oriented subanalytic set of pure codimension $r \geqslant 1$ without boundary. Then, in the situation of section 4.2., we consider the sheaf $\mathscr{S}_X^{\cdot}(*Y)$ of $K$-simple differential forms. Then we have the analogue of the exact sequences (A), (B), (D) where (A) is replaced by

$$\text{(A}'\text{)} \qquad \ldots \to H^p(\Gamma(X, \mathscr{E}_X^{\cdot})) \to H^p\Gamma(X, \mathscr{S}_X^{\cdot}(*Y)) \xrightarrow{\text{res}} H^{p-r+1}(\Gamma(X, \mathscr{E}_Y^{\cdot})) \to \ldots$$

and res is the map defined by $\varphi \mapsto \text{res}\,\varphi$.

In (B) and (D), $2n$ has to be replaced by the real dimension $n$ of $X$ and $\overline{\text{Res}}$ is equal to res followed by the homomorphism defined by $(\psi \to I_Y \wedge \psi)$.

**5.4. Complex Case Again.** When $X$ is a complex manifold and $Y$ a complex subvariety of pure codimension $k$, then the Residue defined by King and Raby (section 4.5) leads also to an interpretation of the residue homomorphisms ([39], [55]).

**5.5. Relation Between Homology of Currents and Borel–Moore Homology.** The exact sequences (B) and (D) and the morphism (B) → (D) are defined for any $C^\infty$ manifold $X$ of dimension $n$ and any closed set $Y$ of $X$.

$$\begin{array}{lllll}
\text{(B)} & \ldots \to H_{n-p}\Gamma(X, \mathscr{D}'_{\cdot X}) \to & H_{n-p}\Gamma(X, \mathscr{D}'_{\cdot X/Y^\infty}) \to & H_{n-p-1}\Gamma(X, \mathscr{D}'_{\cdot Y^\infty}) & \\
& \quad\downarrow & \quad\downarrow & \quad\downarrow & \\
\text{(D)} & \ldots \to H_{n-p}(X) & \to H_{n-p}(X \setminus Y) & \to H_{n-p-1}(Y) & \\
& & & \to H_{n-p-1}\Gamma(X, \mathscr{D}'_{\cdot X}) \to \ldots & \\
& & & \quad\downarrow & \\
& & & \to H_{n-p-1}(X) \to \ldots &
\end{array}$$

**Theorem.** *When $X$ is a real analytic manifold and $Y$ a closed subanalytic set of any codimension the morphism* (B) → (D) *is an isomorphism* [63, 62].

This theorem is local and easy to prove when $Y$ is a submanifold. When $Y$ is $C^\omega$ with normal crossings, it is proved using the Mayer–Vietoris exact sequence. When $Y$ is any real analytic set, then using a morphism of Hironaka $\pi: \tilde{X} \to Y$, such that $\tilde{Y} = \pi^{-1}(Y)$ has normal crossings, it can be proved that $\pi_*$: $\mathscr{D}'_{\cdot}(\tilde{X}) \to \mathscr{D}'_{\cdot}(X)$ is surjective. The main step is that $\pi^*$: $\mathscr{E}(X) \to \mathscr{E}(\tilde{X})$ is an injective morphism of Fréchet spaces. For the case of semi-analytic sets, the Mayer–Vietoris sequence is used again.

There exists an independent and more algebraic proof in the more general case of subanalytic sets, using the triangulation theorem of Hironaka [36], in [13].

## §6. Residue Theorem; Theorem of Leray

**6.1. Residue Theorem.** In the notation of 4.2 and 5.3, we have the following diagram

$$\begin{array}{ccccc}
H^p\Gamma(X, \mathscr{S}^{\cdot}_X(*Y)) & \xrightarrow{\text{res}} & H^{p-r+1}\Gamma(X, \mathscr{E}^{\cdot}_Y) & \longrightarrow & H^{p+1}\Gamma(X, \mathscr{E}^{\cdot}_X) \\
\downarrow & & \text{Res}\downarrow & & \downarrow \\
H_{n-p}\Gamma(X, \mathscr{D}'_{\cdot X/Y^\infty}) & \longrightarrow & H_{n-p-1}\Gamma(X, \mathscr{D}'_{\cdot Y^\infty}) & \longrightarrow & H_{n-p-1}\Gamma(X, \mathscr{D}'_{\cdot X}) \\
\approx\downarrow & & \approx\downarrow & & \approx\downarrow \\
H_{n-p}(X \setminus Y) & \xrightarrow{\delta_*} & H_{n-p-1}(Y) & \xrightarrow{i_*} & H_{n-p-1}(X)
\end{array}$$

Let $\omega$ be a closed $K$-simple form of degree $p$, $[\omega]$ be its cohomology class and $\operatorname{Res}[\omega]$ be the image of $[\omega]$ in $H_{n-p-1}\Gamma(X, \mathcal{D}'_{\cdot X})$ (or in $H_{n-p-1}(X)$). Then, from the exactness of the lines, $\operatorname{Res}[\omega] = 0$. This is valid, in particular, in the complex case where $Y$ has codimension 1 and $\mathcal{S}^{\cdot}_X(*Y) = \mathcal{E}^{\cdot}_X(*Y)$. Thus we get

**6.1.1. Theorem.** *The image, in $H_{n-p-1}(X)$ of the residue of a closed K-simple p-form is zero.*

**6.1.2.** The same theorem is true for the residue defined in 4.5 and 5.4.

**6.2. Griffiths' Residue Theorem.** *Let $X$ be a compact complex manifold, with $\dim_{\mathbb{C}} X = n$, and $\mathcal{F} = (Y_1, \ldots, Y_n)$ be a family of complex analytic hypersurfaces. Then, in the notation of n.3, let $\operatorname{res}_{\mathcal{F}, Y_0}(\omega)$ be the point residue at $y_0 \in V_e(\mathcal{F})$ of an n-meromorphic form $\omega \in \Gamma(X, \Omega^n_X(*\cup\mathcal{F}))$. Then*

$$\sum_{y_0 \in V_e(\mathcal{F})} (\operatorname{res}_{\mathcal{F}, y_0}(\omega; y_0 \in V_e(\mathcal{F})) = 0. \tag{6.2}$$

The first member of (6.2) is $R^n[\omega]$ (1); the result is a consequence of the identity 3.7.2. for $p = n$ and of $d\omega = 0$, $d1 = 0$.

This theorem is due to Griffiths [24] when $\cap\mathcal{F}$ is a complete intersection and has been extended by Coleff–Herrera to the general case [11].

**6.3. Converse of the Residue Theorem; Theorem of Leray** [61, 63, 64]

**6.3.1.** *Under the hypothesis of* 6.1, *$Y$ being a closed submanifold, let $\omega$ be a closed differential form on $Y$ whose image in $H_{n-p-1}\Gamma(X, \mathcal{D}'_{\cdot X})$ is zero. Then $\omega$ is the residue form of a closed p-form $\varphi \in \Gamma(X, \mathcal{S}^p_X(*Y))$. When $Y$ is a closed subvariety of $X$, there is a canonical homomorphism $e_Y: H^s(Y, \mathbb{C}) \to H^s\Gamma(Y, \mathcal{E}^{\cdot}_Y)$. Let $\beta \in H^{p-r+1}(Y, \mathbb{C})$ such that $i_*([Y] \cap \beta) = 0$. Then any closed $\omega \in \Gamma(Y, \mathcal{E}^{p-r+1}_Y)$ belonging to the class $e_Y\beta$ is the residue form of a closed K-simple form.*

**6.3.2 (Generalization of the theorem of Leray).** *Under the hypothesis of* 6.1 *and if $Y$ is a closed subvariety, the following conditions are equivalent* (*for $p + q = n$*):

(*i*) *$\alpha \in H_q(X \setminus Y; \mathbb{C})$ contains a closed K-simple p-form on $X \setminus Y$;*
(*ii*) *$\delta_* \alpha = [Y] \cap \beta$ where $\beta \in H^{p-r+1}(Y, \mathbb{C})$.*

When $Y$ is a submanifold, condition (ii) is always satisfied.

**6.3.3.** Let $X$ be a complex manifold, and $Y$ be a complex subvariety of $X$ whose singular set is $S$. Let $\sigma': X' \to X \setminus S$ be the blow-up of $X \setminus S$ with center $Y \setminus S$ and $\sigma$ be its restriction to $X' \setminus Y'$, with $Y' = \sigma^{-1}(Y \setminus S)$. *Then every class of $H^p(X \setminus Y; \mathbb{C})$ contains a smooth closed form $\varphi$ on $X \setminus Y$* such that: $\varphi$ is the restriction to $X \setminus Y$ of a locally integrable form of $X$ such that $d\varphi$ is 0-continuous and $\sigma^*\varphi$ is the restriction to $X' \setminus Y'$ of a semi-meromorphic form on $X'$ with simple poles along $Y'$, (cf. 4.5). This is a theorem of Leray more satisfactory than 6.3.2. in the complex case [55].

A more precise result is obtained when $X$ is a complex surface and $Y$ a complex subvariety of codimension one [56].

See also [57].

**6.4. Converse of Griffiths' Residue Theorem.** Let $X$ be a compact complex manifold with $\dim_{\mathbb{C}} X = n$ and let $\mathscr{F} = (Y_1, \ldots, Y_n)$ be a family of complex hypersurfaces such that

(i) $\cap \mathscr{F}$ is a complete intersection;
(ii) for $j = 1, \ldots, n$, the holomorphic line bundle $[Y_j]$ defined by the divisor $Y_j$ is positive.

Let $\cap \mathscr{F} = (P_\nu)_{\nu=1,\ldots,q}$ and $(c_\nu)$ be a family of complex numbers such that $\sum_{\nu=1}^{q} c_\nu = 0$. Then there exists a meromorphic $n$-form $\varphi$ on $M$ whose polar divisor is $\sum_{j=1}^{n} Y_j$ and whose point residue in $P_\nu$ is $c_\nu$, $\nu = 1, \ldots, q$. The main tool of the proof is the Kodaira vanishing theorem [24].

## §7. Residue Formulae

**7.1. The Case of $Y$ be Oriented $C^\omega$ Subvariety of $X$.** Let $X$ be a $C^\omega$ manifold, $Y$ an oriented $C^\omega$ subvariety of $X$ of pure dimension $q$, and $K$ a kernel associated to $Y$ (4.2).

Let $\mathscr{X}$ be a $(p+1)$-subanalytic chain with faithful representative $(Z, \zeta)$ with compact support $\bar{Z}$ in $X$ which properly intersects $Y$ (i.e. such that $\dim Y \cap Z \leqslant \dim Y + \dim Z - \dim X$) and such that $bZ \cap Y = \varnothing$. Then for any K-simple form $\varphi$ in $X$ (4.2) whose differential $d\varphi$ is $C^\infty$ in $X$, the following formula is valid

$$\int_{b\mathscr{X}} \varphi - \int_{\mathscr{X}} d\varphi = \int_{\mathscr{X}\cdot\mathscr{Y}} \operatorname{res} \varphi \tag{7.1}$$

where $\mathscr{X} \cdot \mathscr{Y}$ is the subanalytic intersection of the chains $\mathscr{X}$ and $\mathscr{Y}$, the chain $\mathscr{Y}$ having $(Y, \eta)$ as a faithful representative and $\eta$ being the fundamental class of $Y$.

This formula is a generalization of the classical formula (0.1); it is due to Poly [62]. Particular cases of (7.1) are due to Griffiths [23], King [38] and Weishu Shih [71].

Formula (7.1) is proved by methods of algebraic topology and uses triangulation of subanalytic sets [36], the representation of homology by classes of subanalytic chains [5], [21] and a theory of intersection of subanalytic chains ([62], ch. 4).

### 7.2. Kronecker Index of Two Currents

**7.2.1.** Let $X$ be an oriented $C^\infty$ manifold of dimension $n$ and $\Pi: S^*X \to X$ be the spherical cotangent bundle of $X$. Every current $T \in \mathscr{D}'^{p}(X)$ has the following expression over the domain $U$ of a chart on $X$, with local coordinates $(x_1, \ldots, x_n)$:

$$T|U = \sum_{|I|=p} T_I \, dx_I$$

where $T_I$ is a distribution on $X$.

The subset $WF(T)$ of $S^*X$ such that

$$WF(T) \cap \Pi^{-1}(U) = WF(T|U) = \bigcup_{|I|=p} WF(T_I),$$

where $WF(T_I)$ is the wave front of the distribution $T_I$ (Hörmander [37]), is called the *wave front of* $T$. Denote by $\check{W}F(T)$ the antipodal of $WF(T)$; let $S, T \in \mathscr{D}'^{\bullet}(X)$ such that

$$WF(T) \cap \check{W}F(S) = \varnothing \tag{7.2.1}$$

then the wedge product $T \wedge S$ is defined.

**7.2.2.** Let $(R_\varepsilon)_{\varepsilon > 0}$ be a *regularizing family* of operators on the currents on $X$ [60], [43]. According to de Rham ([60], §20), we say that the *Kronecker index* $\mathscr{K}(T, S)$ *of two currents* $T, S$ on $X$ of respective degree $t, s$, with $t + s = n$, is defined if for any regularizing family $R_\varepsilon$ (resp. $R'_{\varepsilon'}$), $\langle R_\varepsilon T \wedge R'_{\varepsilon'} S, 1 \rangle$ has a limit independent of $R_\varepsilon$, $R'_{\varepsilon'}$ when $\varepsilon, \varepsilon'$ tend to zero.

**7.2.3. Theorem** [43]. *Let $T$ and $S$ be two currents as in 7.2.2. such that* $\operatorname{Supp} T \cap \operatorname{Supp} S$ *is compact. If* $WF(T) \cap \check{W}F(S) = \varnothing$, *then* $\mathscr{K}(T, S)$ *exists and* $\mathscr{K}(T, S) = \langle T \wedge S, 1 \rangle$.

### 7.3. The Case of $Y$ be a Closed $q$-Subanalytic Chain of $X$.

Let $X$, $\mathscr{Z}$ be as in 7.1, $\mathscr{Y}$ be a closed $q$-subanalytic chain, $K$ be a *kernel associated* to $\mathscr{Y}$ (an obvious extension of 4.2), and $\varphi = K \wedge \psi + \theta$ be a $K$-simple form of degree $p$. Then (7.1) is a particular case of the formula

$$\mathscr{K}([b\mathscr{Z}], K \wedge \psi + \theta) = \mathscr{K}([\mathscr{Z}], (L \wedge \psi + (-1)^{n-q-1} K \wedge d\psi + d\theta)) + \mathscr{K}([\mathscr{Z}), [\mathscr{Y}] \wedge \psi),$$

where $[\mathscr{U}]$ denotes the integration current defined by the subanalytic chain $\mathscr{U}$ $[\mathscr{Y}] = dK + L$ where $L$ is $C^\infty$ and when the Kronecker indices under consideration have a meaning ([43], 5).

## §8. A New Definition of Residual Currents ([77], [78])

**8.1.** The present section contains new results obtained since the original Russian edition of this work. We use the notations of section 3. All definitions and results are local. Let $f/g = (f_1/g_1, \ldots, f_{p+q}/g_{p+q})$ be an ordered set of $(p+q)$ globally defined meromorphic functions on a connected complex manifold $X$; $f_j, g_j \in \mathcal{O}(X), j = 1, \ldots, p+q$. Let $\mathcal{F} = \{Y_1, \ldots, Y_{p+q}\}$ be an ordered family of hypersurfaces of $X$, where $Y_j$ has the global equation $g_j = 0$ in $X$. The aim of the new definition is:

1. to replace integration on semi-analytic chains by integration on $X$ after multiplication by smooth functions $\chi_j$ vanishing on shrinking neighborhoods of $Y_j, j = 1, \ldots, p+q$;
2. to replace the use of admissible maps (3.2.) by a mean value on parameters defining the $\chi_j$;
3. using several successive principal values instead of one at most in section 3.

As a result a rather satisfactory exterior product on residual currents will be obtained.

**8.2. Definitions and Tools.** Let $\chi: \mathbb{R} \to [0, 1]$ be a $C^\infty$ function such that $\chi(x) = 0$ for $x \le c_1 < c_2$; $\chi(x) = 1$ for $c_2 \le x$.

Let $\sum_m = \left\{ s \in \mathbb{R}^m; s_j > 0; \sum_{j=1}^{m} s_j = 1 \right\}$ be the $(m-1)$-canonical euclidean simplex in $\mathbb{R}^m$ with Lebesgue measure $S_m$ of total mass 1. For $\varepsilon > 0$, we set $\chi_j = \chi(|g_j|/\varepsilon^{s_j})$. Then, there exists a finite number of non zero vectors $n_1, \ldots, n_N \in \mathbb{Q}^{p+q}$ such that, if $s \in \sum_{p+q}$ is not orthogonal to any of them, then

$$R^p P^q [f/g](\varepsilon^s) = (f_1/g_1) \ldots (f_{p+q}/g_{p+q}) d''\chi_1 \wedge \ldots \wedge d''\chi_p \chi_{p+1} \cdots \chi_{p+q} \quad (8.1)$$

has, in the sense of currents, a limit $R^p P^q [f/g](s)$, when $\varepsilon \to 0$, which is locally constant on $\sum_{p+q} \setminus \bigcup_{l=1}^{N} \{(n_l, s) = 0\}$; $R^p P^q [f/g]$ is the mean value

$$\int_{\Sigma_{p+q}} R^p P^q [f/g](s) \, dS_{p+q}(s).$$

In the expression for $\chi_j$, the function $g_j$ may be replaced by the product of $g_j$ by a nowhere vanishing holomorphic function on $X$, without changing the result. Notations:

$$R^p P^q [f/g] = d''[f_1/g_1] \wedge \ldots \wedge d''[f_p/g_p][f_{p+1}/g_{p+1}] \cdots [f_{p+q}/g_{p+q}] \quad (8.1)$$

The technique is parallel to that of section 3 but more simple and transparent; it also depends on Hironaka's morphisms.

**8.3. Results.** Let $\omega$ be a holomorphic differential form of degree $r$ on $X$.

**8.3.1.** $R^pP^q[f/g] \wedge \omega \in \mathscr{D}'^{(r,p)}(X)$; $\operatorname{spt} R^pP^q(f/g] \subset \cap \mathscr{F}'$, with $\mathscr{F}' = (g_1, \ldots, g_p)$.

**8.3.2.** From (8.1), $d''$ acts on $R^pP^q[f/g]$ as on smooth differential forms.

**8.3.3.** With the multiplication $[f/g] \cdot [f'/g'] = [ff'/gg']$, the principal value currents form a field and $f/g \mapsto [f/g]$ is a field isomorphism.

Furthermore, $R^pP^q[f/g] = d''[f_1/g_1] \wedge \ldots \wedge d''[f_p/g_p][f_{p+1} \cdots f_{p+q}/g_{p+1} \cdots g_{p+q}]$

**8.3.4.** In case $q = 0$ or $1$, and $\cap \mathscr{F}$, (resp. $\cap \mathscr{F}(p+1)$) is a complete intersection, the new residual currents coincide with the ones of Coleff–Herrera (section 3).

One of the main properties is: $R^pP^q[f/g]$ is alternating with respect to the residue factors.

**8.4. Application.** In the complete intersection case, the following properties are equivalent:

(i) $h \in \mathcal{O}(X)$ belongs locally to the ideal generated by $g_1, \ldots, g_p$;
(ii) $hd''[1/g_1] \wedge \ldots \wedge d''[1/g_p] = 0$, [76], [74].

If $X$ is a strictly pseudoconvex domain in $\mathbb{C}^n$ whose defining function belongs to $C^\infty(\bar{X})$ and if the $g_j \in \mathcal{O}(\bar{X})$, there is a formula of the type

$$h(z) = F(z) + hR^p[1/g](\psi(.\,, z)),$$

where $F$ belongs to the ideal and $\psi$ is a certain test form [76].

A similar result, but with growth conditions, holds when $X = \mathbb{C}^n$ and the $g_j$ are polynomials [76].

**8.5. Relation with Meromorphic Extension of $|f|^\lambda$, for $f$ Real Analytic** ([72], [73]). If $f$ is holomorphic, then $\lim_{\varepsilon \to 0} |f|^\varepsilon/f = vp(1/f) = [1/f]$ [75].

More generally, with the notation of 8.1., 8.2., M. Passare [79] proves:

$$R^pP^q[1/g] = \lim_{\varepsilon \to 0} (d''|g_1|^\varepsilon/g_1) \wedge \ldots \wedge (d''|g_p|^\varepsilon/g_p)$$
$$\times (|g_{p+1}|^\varepsilon/g_{p+1}) \cdots (|g_{p+q}|^\varepsilon/g_{p+q}).$$

The steps of the proof are as usual: normal crossing case and general case using morphisms of Hironaka.

Previously, A. Yger used this second definition of residual currents to get applications as in 8.4. [80].

**8.6. Another Application of Residual Currents** [74]. Let $X$ be a complex manifold and $Y$ a complex subspace of pure codimension $p > 0$, which is locally a complete intersection.

A current $R \in \Gamma_Y(X, \mathscr{D}'^{r,p})$ is called *locally residual*, if for every sufficiently small open set $U$ of $X$, $R|U = R^p_{\mathscr{F}}[\tilde{\omega}]$ where $\tilde{\omega} \in \Gamma(U, \Omega^r(* \cup \mathscr{F}))$ and $\cap \mathscr{F} = Y \cap U$.

It follows that for every $T \in \Gamma_Y(X, \mathscr{D}'^{r,p})$ with $d''T = 0$, we have $T = R + d''S$, where $R$ is locally residual, and $S \in \Gamma_Y(X, \mathscr{D}'^{r,p-1})$. Moreover, $R$ and $d''S$ are uniquely determined. Consider the $p$-cohomology group of the complex $(\Gamma_Y(X, \mathscr{D}'^{r,\cdot}), d'')$ (moderate cohomology); locally residual currents are canonical representatives for classes of this group. Moreover holomorphic $p$-chains are residual currents and there exists a duality theorem for such currents.

## References*


1. Ajzenberg, L., Yuzhakov, A.P.: Integral Representations and Residues in Multidimensional Complex Analysis. Nauka, Novosibirsk 1979 [Russian]; Engl. transl.: Am. Math. Soc., Providence, R.I. 1983. Zbl. 445.32002
2. Atiyah, M.F., Hodge, W.V.D.: Integrals of the second kind on an algebraic variety. Ann. Math., II. Ser. *62*, 56–91 (1955). Zbl. 68,344
3. Baum, P.F., Bott, R.: On the zeros of meromorphic vector fields. Essays on topology and related topics, (Mémoires dédiés à G. de Rham), 29–47, Springer-Verlag 1970. Zbl. 193,522
4. Baum, P.F., Cheeger, J.: Infinitesimal isometries and Pontryagin numbers. Topology *8*, 173–193 (1969). Zbl. 179,288
5. Bloom, T., Herrera, M.: De Rham cohomology of an analytic space. Invent. Math. *7*, 275–296 (1969). Zbl. 175,373
6. Bott, R.: Vector fields and characteristic numbers. Mich. Math. J. *14*, 231–244 (1967). Zbl. 145,438
7. Bott, R.: A residue formula for holomorphic vector fields. J. Differ. Geom. *1*, 311–330 (1967). Zbl. 179,288
8. Bredon, G.E.: Sheaf theory. Mac–Graw Hill series in higher Mathematics. 1967. Zbl. 158,205
9. Bucari, N.D., Coleff, N.R., Paenza, A.A.: Residual currents in the non complete intersection case. Preprint 1983
10. Carrell, J., Lieberman, D.: Holomorphic vector fields and Kähler manifolds. Invent. Math. *21*, 303–309 (1973). Zbl. 253.32017 (Zbl. 258.32013)
11. Coleff, N.R., Herrera, M.: Les courants résiduels associés à une forme méromorphe. Lect. Notes Math. *633*, Springer-Verlag 1978. Zbl. 371.32007
12. Coleff, N.R., Herrera, M., Lieberman, D.: Algebraic cycles as residues of meromorphic forms. Math. Ann. *254*, 73–87 (1980). Zbl. 435.14002 (Zbl. 457.14006)
13. Darchen, J.C.: Homologie des ensembles sous-analytiques. C.R. Acad. Sci., Paris, Sér. A *283*, 99–102 (1976). Zbl. 335.32004
14. Dolbeault, P.: Formes différentielles et cohomologie sur une variété analytique complexe. Ann. Math., II. Ser. *64*, 83–130 (1956). Zbl. 72,406; *65*, 282–330 (1957). Zbl. 89,380
15. Dolbeault, P.: Theory of residues and homology. Ist. Naz. di alta Mat., Symp. Math. *3*, 295–304 (1970). Zbl. 193,523


* For the convenience of the reader, references to reviews in Zentralblatt für Mathematik (Zbl.), compiled using the MATH database, have, as far as possible, been included in this bibliography.

16. Dolbeault, P.: Résidus et courants. C.I.M.E., Sept. 1969 (Questions on algebraic varieties). Zbl. 201,539; Sem. Bucarest, Sept. 1969 (Espaces analytiques), 39–56 (1971). Zbl. 238.32009
17. Dolbeault, P.: In Séminaire P. Lelong (1969, 1970, 1971), Lect. Notes Math. *116*, 152–163 (1969). Zbl. 201, 259; *205*, 56–70. Zbl. 229.32.005; 232–243 (1971). Zbl. 218.32003; *275*, 14–26 (1972). Zbl. 239.32007. Springer-Verlag
18. Dolbeault, P.: Valeurs principales sur un espace analytique. Conv. di Geometria, Milano 1971, Acad. Naz. Lincei, 1973, 139–149
19. Dolbeault, P.: Valeurs principales et opérateurs différentiels semi-holomorphes. Coll. inter. CNRS No. 208, 1972, Fonctions analytiques . . . , Gauthiers-Villars, 1974, 35–50. Zbl. 301.32009
20. Dolbeault, P.: Theory of residues in several variables. Global Analysis and its applications, 1972, Inter. Atomic energy agency Vienna, Vol. II, 1974, 74–96
21. Dolbeault, P., Poly, J.: Differential forms with subanalytic singularities; integral cohomology; residues. Proc. Symp. Pure Math. *30*, part I., A.M.S., Providence, R.I. 1975, 255–261 (1977). Zbl. 354.32016
22. Gómez, F.: A residue formula for characteristic classes. Topology *21*, 101–124 (1982). Zbl. 469.57030
23. Griffiths, P.A.: Some results on algebraic manifolds. Algebr. Geom., Bombay Colloq., 93–191, Oxford 1969. Zbl. 206.498
24. Griffiths, P.A.: Variations on a theorem of Abel. Invent. Math. *35*, 321–390 (1976). Zbl. 339.14003
25. Griffiths, P.A., Harris, J.: Principles of algebraic geometry. Wiley, New York 1978. Zbl. 408.14001
26. Grothendieck, A.: Cohomologie locale des faisceaux cohérents et théorèmes de Lefschetz locaux et globaux (SGA 2). Adv. Stud. Pure Math. *2*, 1968. Zbl. 197,492
27. Grothendieck, A.: Local cohomology. Lect. Notes Math. *41*, Springer-Verlag 1967. Zbl. 185,492
28. Grothendieck, A.: On the de Rham cohomologie of algebraic varieties. Publ. Math., Inst. Hautes Etud. Sci. *29*, 95–103 (1966). Zbl. 145,176
29. Hartshorne, R.: Residues and Duality. Lect. Notes Math. *20*, Springer-Verlag 1966. Zbl 212,261
30. Herrera, M.: Integration on a semi-analytic set. Bull. Soc. Math. Fr. *94*, 141–180 (1966). Zbl. 158,206
31. Herrera, M.: Résidus multiples sur les espaces complexes. Journées complexes de Metz (fév. 1972), I.R.M.A., Univ. Louis Pasteur, Strasbourg 1973
32. Herrera, M.: Les courants résidus multiples. Journées de géométrie analytique complexe de Poitiers, Juin 1972, Bull. Soc. Math. Fr., Suppl. Mém. No. 38, 27–30 (1974). Zbl. 293.32007
33. Herrera, M., Lieberman, D.: Residues and principal values on complex spaces. Math. Ann. *194*, 259–294 (1971). Zbl. 224,32012
34. Hironaka, H.: Resolution of singularities of an algebraic variety over a field of characteristic zero. Ann. Math., II. Ser. *79*, 109–326 (1964). Zbl. 122,386
35. Hironaka, H.: Subanalytic sets. Number theory, Algebr. Geom., Commut. Algebra, in honour of Y. Akizuki, Kinokuniya, Tokyo, 453–493 (1973). Zbl. 297.32008
36. Hironaka, H.: Triangulation of algebraic sets. Algebr. Geom., Proc. Symp. Pure Math. *29*, A.M.S., 165–185, Providence, R.I. 1975. Zbl. 332.14001
37. Hörmander, L.: Fourier integral operators. Acta Math. *127*, 79–183 (1971). Zbl. 212,466
38. King, J.R.: A residue formula for complex subvarieties. Proc. Carolina Conf. Holomorphic mappings, Chapel Hill, N.C., 43–56 (1970). Zbl. 224.32009
39. King, J.R.: Global residues and intersections on a complex manifold. Trans. Am. Math. Soc. *192*, 163–199 (1974). Zbl. 301.32005
40. King, J.R.: Log. complexes of currents and functional properties of the Abel–Jacobi map. Duke Math. J. *50*, 1–53 (1983). Zbl. 526.32011
41. Kodaira, K.: The theorem of Riemann-Roch on compact analytic surfaces. Am. J. Math. *73*, 813–875 (1951). Zbl. 54,64
42. Kodaira, K.: The theorem of Riemann-Roch for adjoint systems on 3-dimensional algebraic varieties. Ann. Math., II. Ser. *56*, 298–342 (1952). Zbl. 48,381

43. Laurent–Thiebaut, C.: Produits de courants et formule des résidus. Bull. Soc. Math., II. Ser. *105*, 113–158 (1981). Zbl. 464.58005
44. Lelong, P.: Intégration sur un ensemble analytique. Bull. Soc. Math. Fr. *85*, 239–262 (1957). Zbl. 79,309
45. Leray, J.: Le calcul différentiel et intégral sur une variété analytique complexe (Problème de Cauchy III). Bull. Soc. Math. Fr. *87*, 81–180 (1959). Zbl. 199,412
46. Loeser, F.: Quelques conséquences locales de la théorie de Hodge. preprint, Centre Math. Ecole polytechnique, Déc. 1983
47. Lojasiewicz, S.: Triangulation of semi-analytic sets. Ann. Sc. Norm. Sup. Pisa, Sci. Fis. Mat., III. Ser. *18*, 449–474 (1964). Zbl. 128,171
48. Malgrange, B.: Ideals of differentiable functions. Tata Institute, Bombay $n^0$3, Oxford Univ. Press 1966. Zbl. 177, 179
49. Norguet, F.: Sur la théorie des résidus. C.R. Acad. Sci., Paris *248*, 2057-2059 (1959). Zbl. 133,41
50. Norguet, F.: Dérivées partielles et résidus de formes différentielles sur une variété analytique complexe. Sémin. Anal. P. Lelong 1958–59, No. 10 (1959), 24 pages. Zbl. 197,69
51. Norguet, F.: Sur la cohomologie des variétés analytiques complexes et sur le calcul des résidus. C.R. Acad. Sci., Paris *258*, 403–405 (1964). Zbl. 128,78
52. Norguet, F.: Sémin. P. Lelong (1970), Lect. Notes Math. *205*, 34–55, Springer-Verlag 1971. Zbl. 218.32004
53. Picard, E.: Sur les intégrales des différentielles totales de troisième espèce dans la théorie des surfaces algébriques. Ann. Ec. Norm. Supér. 1901. Zbl. 32,419
54. Poincaré, H.: Sur les résidus des intégrales doubles. Acta Math. *9*, 321–380 (1887). Zbl. 19, 275
55. Raby, G.: Un théorème de J. Leray sur le complémentaire d'un sous-ensemble analytique complexe. C.R. Acad. Sci., Paris, Sér. A *282*, 1233–1236 (1976). Zbl. 353.32013
56. Raby, G.: Formes méromorphes et semi-méromorphes sur une surface analytique complexe. C.R. Acad. Sci., Paris, Sér. A *287*, 125–128 (1978). Zbl. 416.32003
57. Robin, G.: Formes semi-méromorphes et cohomologie du complémentaire d'une hypersurface d'une variété analytique complexe. C.R. Acad. Sci., Paris, Sér. A *272*, 33–35 (1971). Zbl. 207,380; et Sém. P. Lelong (1970), Lect. Notes Math. *205*, 204–215 (1971). Zbl. 222.32005; Springer–Verlag
58. De Rham, G.: Sur la notion d'homologie et les résidus d'intégrales multiples. Ver. Int. Math. Kongress Zürich, 1932, 195
59. De Rham, G.: Relations entre la topologie et la théorie des intégrales multiples. Enseign. Math. *35*, 213–228 (1936). Zbl. 15,85
60. De Rham, G.: Variétés différentiables, formes, courants, formes harmoniques. Hermann, Paris 1955. Zbl. 65,324
61. Poly, J.: Sur un théorème de J. Leray en théorie des résidus. C.R. Acad. Sci., Paris, Sér. A *274*, 171–174 (1972). Zbl. 226.32005; et Formes–résidus (en codimension quelconque), Journées complexes de Metz, fév. 1972, Publ. I.R.M.A. Strasbourg 1973
62. Poly, J.: Formule des résidus et intersection des chaînes sous–analytiques. Thèse Poitiers, 1974
63. Poly, J.: Sur l'homologie des courants à support dans un ensemble semianalytique. Journées de Géom. Anal. compl. de Poitiers, Juin 1972, Bull. Soc. Math. Fr., Suppl. Mém. No. 38, 35–43 (1974). Zbl. 302.32010
64. Poly, J.: Formes et courants résidus. Colloq. Inter. CNRS No. 208 1972, Fonctions analytiques, Gauthiers-Villars, 1974, 204–210. Zbl. 302.32009
65. Ramis, J.P., Ruget, G.: Résidus et dualité. Invent. Math. *26*, 89–131 (1974). Zbl. 304.32007
66. Schwartz, L.: Courant associé à une forme différentielle méromorphe sur une variété analytique complexe. Colloq. internat., Centre Nat. Rech. Sci. *52*, Géométrie différentielle, 185–195, Strasbourg 1953. Zbl. 53,249
67. Severi, F.: Funzioni analitiche e forme differenziali. Atti IV. Congr. Un. Mat. Ital. I, 125–140 (1953). Zbl. 50,307
68. Sorani, G.: Sui residui delle forme differenziali di una varietà analitica complessa. Rend. Mat. Appl., V. Ser. *22*, 1–23 (1963). Zbl. 124,389

69. Mei, Xiang-Ming: Note on the residues of the singularities of a Riemannian foliation. Proc. Am. Math. Soc. *89*, 359–366 (1983). Zbl. 532.57015
70. Weil, A.: Sur la théorie des formes différentielles attachées à une variété analytique complexe. Comment. Math. Helv. *20*, 110–116 (1947). Zbl. 34,358
71. Shih, Weishu: Une remarque sur la formule des résidus. Bull. Am. Math. Soc. *76*, 717–718 (1970). Zbl. 202,548
72. Atiyah, M.F.: Resolution of singularities and division of distributions. Commun. Pure Appl. Math. *23* (1970), 145–150. Zbl. 188,194
73. Bernstein, I.N., Gel'fand, S.I.: Meromorphy of the functions $P^\lambda$. Funkts. Anal. Prilozh, 3, No. 1, 84–85 (1969), Zbl. 208,152 Engl. transl. Funct. Anal. Appl. *3* (1969), 68–69
74. Dickenstein, A. and Sessa C.: Canonical representatives in moderate cohomology. Invent. Math. *80* (1985), 417–434. Zbl. 579.32011
75. El Khadiri, A., Zouakia, F.: Courant valeur principale associé à une forme semi-méromorphe, in El Khadiri, thèse de 3e cycle, Poitiers, 1979
76. Passare, M.: Residues, currents, and their relation to ideals of holomorphic functions, Math. Scand., to appear. Zbl. 633.32005
77. Passare, M.: Produits des courants résiduels et règle de Leibniz, C.R. Acad. Sci. Paris, Sér. I. *301* (1985), 727–730. Zbl. 583.32023
78. Passare, M.: A calculus for meromorphic currents, preprint 1986
79. Passare, M.: Courants méromorphes et égalité de la valeur principale et de la partie finie, Séminaire d'Analyse (P. Lelong, P. Dolbeault, H. Skoda) 1985–1986. Lect. Notes Math. 1295, 157–166 (1987). Zbl. 634.32009
80. Yger, A.: Formules de division et prolongement méromorphe, Séminaire d'Analyse (P. Lelong, P. Dolbeault, H. Skoda) 1985–1986. Lect. Notes Math. 1295, 226–283 (1987). Zbl. 632.32010

# Author Index

# Subject Index